D1472698

GLOBAL ISSUES

MILITARIZATION
OF SPACE

GLOBAL ISSUES

MILITARIZATION OF SPACE

Ann E. Robertson

Foreword by Samuel Black
The Henry L. Stimson Center

Facts On File
An imprint of Infobase Publishing

Facts On File, Inc.
An imprint of Infobase Publishing
132 West 31st Street
New York NY 10001

Library of Congress Cataloging-in-Publication Data
Robertson, Ann E.
    Militarization of space / Ann E. Robertson ; foreword by Samuel Black.
      p. cm. — (Global issues)
    Includes bibliographical references and index.
    ISBN: 978-0-8160-7873-8
    1. Astronautics Military—Juvenile literature. 2. Space warfare—Juvenile literature. 3. Space weapons—Juvenile literature. I. Title.
    UG1520.R63 2010
358.8—dc22 2010015423

Facts On File books are available at special discounts when purchased in bulk quantities for businesses, associations, institutions, or sales promotions. Please call our Special Sales Department in New York at (212) 967-8800 or (800) 322-8755.

You can find Facts On File on the World Wide Web at http://www.factsonfile.com

Excerpts included herewith have been reprinted by permission of the copyright holders; the author has made every effort to contact copyright holders. The publishers will be glad to rectify, in future editions, any errors or omissions brought to their notice.

Text design by Erika K. Arroyo
Composition by Publication Services, Inc.
Cover printed by Art Print, Taylor, Pa.
Book printed and bound by Maple Press, York, Pa.
Date printed: February 2011
Printed in the United States of America

10 9 8 7 6 5 4 3 2 1

This book is printed on acid-free paper.

CONTENTS

PART II: Primary Sources

PART III: Research Tools

Foreword

The beginning of the space age in 1957 marked a fundamental shift in the course of human history. It opened the door to a new level of danger in the cold war, as the rocket technology that allowed access to space also allowed the production of nuclear-tipped long-range missiles. These missiles gave each superpower the ability to hold at risk every inch of the other's territory. But access to space also gave the superpowers a new avenue to gather information about each other. Intelligence provided by reconnaissance satellites gave the United States a peek over the iron curtain and helped U.S. leaders manage some of the uncertainties inherent in a competition with a closed society.

Today, satellites are vital to the economy and national security of the United States. They help rescuers' efforts after natural disasters, transmit communications around the world, help monitor the weather and environment, gather intelligence against threats to the country, and are a force multiplier for the U.S. military.

Having seen the benefits reaped from the utilization of space, other countries are seeking space capabilities of their own. These more recent entrants into the community of space-faring nations are likely to follow in the footsteps of their predecessors in using space to enhance their military capabilities. In doing so, they too will become more dependent on space. This dependency will force these new space-farers to confront the most difficult conundrum associated with militarization of space: Satellites provide services that are central to the conduct of modern military operations, but they are as vulnerable as they are invaluable. Managing this problem is the essence of a field of study known as space security.

Satellites travel in orbits that are bounded by the laws of physics. Their approximate locations can be predicted ahead of time. And while they may be moved, the amount of fuel they carry is limited and must often be used simply to maintain the satellite's intended orbit. Any country with the technological prowess needed to launch a satellite can also launch a weapon that can destroy satellites. So as the demand for satellite services increases and the number of countries that benefit from the use of space increases, so will the number of countries with the capability to destroy satellites.

During the cold war, the United States and Soviet Union took steps to manage the tension between the value and vulnerability of satellites. Several of the arms control treaties signed by the two included clauses that were protective of an important class of satellites—those used to verify the other side's compliance with the terms of the treaty. And neither these satellites, nor any other kind, were ever attacked during a crisis or war.

Part of the reason for this uncommon restraint during a period of acute tension between the United States and Soviet Union was that neither side saw it as in its interest to take a step past the militarization of space. Space has not yet been weaponized. While both sides tested antisatellite (ASAT) weapons during the cold war, these tests were relatively rare. The superpowers tested ASAT weapons approximately 50 times during the cold war. But they tested nuclear weapons more than 1,750 times.

The tacit agreement not to test or use ASAT weapons has survived the end of the cold war. But it is under stress, as there is no guarantee that new space-faring nations will restrain themselves in the same way the United States and Soviet Union did. China successfully tested an ASAT weapon in 2007, the first such test by any country in 22 years. This test created thousands of pieces of large debris, each of which is traveling fast enough to destroy any satellite unlucky enough to be in its path.

The expanding space community thus finds itself at a tipping point. That space is militarized is a fact of life. That it may become weaponized is just one of a number of possible alternatives. Space-faring nations will have to make choices about how to manage the tension between the value and vulnerability of satellites. But we do not yet know what kinds of choices the multitude of newcomers to the space-faring community will make. Those countries with a longer tradition of exploring space will also have to make changes to their policies so that they reflect the changing environment. Again, it is unclear what changes might be made.

Making intelligent choices about the future of humanity's exploration of space will require creative thinking to manage the tensions inherent to space security. But creative thinking must be grounded in an understanding of the past and present of the militarization of space. My hope is that readers will find this book to be a useful foundation for supporting a healthy debate about how best to enhance space security.

—Samuel Black
The Henry L. Stimson Center
Washington, D.C.

List of Maps and Tables

List of Acronyms and Abbreviations

ABM	antiballistic missile
ABMA	Army Ballistic Missile Agency
AHCOPUOS	Ad Hoc Committee on the Peaceful Uses of Outer Space
ALMV	air-launched miniature vehicle
APSCO	Asia Pacific Space Cooperation Organization
ASAT	antisatellite weapon
BMD	ballistic missile defense
Beacon	Bold Experiments to Advance Confidence
CAV	common aero vehicle
CIA	Central Intelligence Agency
CNES	Centre National d'Etudes Spatiales
COPUOS	Committee on the Peaceful Uses of Outer Space
CSG	Centre Spatial Guyanais
DFH	Dong Fang Hong (communication satellites)
DPRK	Democratic People's Republic of Korea
ELDO	European Launcher Development Organisation
EPIRB	emergency position-indicating radio beacon
ESA	European Space Agency
ESRO	European Space Research Organisation
FAA	Federal Aviation Administration
FAS	Federation of American Scientists
FOBS	fractional orbital bombardment system
FY	fiscal year
FY	Feng Yun (meteorological satellites)
GAO	General Accounting Office
GDP	gross domestic product

GEO	geosynchronous or geostationary orbits
GPALS	Global Protection Against Limited Strikes
GPS	global positioning system
HST	*Hubble Space Telescope*
ICBM	intercontinental ballistic missile
IRBM	intermediate-range ballistic missile
ISRO	Indian Space Research Organisation
ISS	*International Space Station*
JPL	Jet Propulsion Laboratory
KARI	Korea Aerospace Research Institute
KCNA	Korean Central News Agency
KSLV	Korea Space Launch Vehicle
LDC	least developed country
LEO	low earth orbit
Liability Convention	Convention on International Liability for Damage Caused by Space Objects
MAD	mutual assured destruction
MDA	Missile Defense Agency
MEO	medium earth orbit
MIRV	multiple independently targetable reentry vehicle
MODS	manned orbital development system
MOL	manned orbital laboratory
Moon Agreement	Agreement Governing the Activities of States on the Moon and Other Celestial Bodies
NASA	National Aeronautics and Space Administration
NORAD	North American Aerospace Defense Command
Outer Space Treaty	Treaty on Principles Governing the Activities of States in the Exploration and Use of Outer Space, including the Moon and Other Celestial Bodies
PLA	People's Liberation Army
Registration Convention	Convention on Registration of Objects Launched into Outer Space
Rescue Treaty	Agreement on the Rescue of Astronauts, the Return of Astronauts, and the Return of Objects Launched into Outer Space
Rumsfeld Commission	Commission to Assess the Ballistic Missile Threat to the United States
SAASS	School of Advanced Air and Space Studies
SALT	Strategic Arms Limitation Talks (or Treaty)
SDI	Strategic Defense Initiative

List of Acronyms and Abbreviations

SDIO	Strategic Defense Initiative Organization
SJ	Shi Jian (scientific research and technological experiment satellites)
SLBM	submarine-launched ballistic missile
SMEC	Strategic Missiles Evaluation Committee
STAR	Significant Technical Achievements and Research
START	Strategic Arms Reduction Treaty
SUPARCO	Pakistan Space and Upper Atmosphere Research Commission
UCS	Union of Concerned Scientists
USIP	United States Institute of Peace
USSPACECOM	United States Space Command
USSTRATCOM	United States Strategic Command
VfR	Society for Space Travel
ZY	Zi Yuan (Earth resource satellites)

PART I

At Issue

1

Introduction

On Friday, October 4, 1957, the Soviet Union successfully placed the first artificial satellite—*Sputnik*—into orbit around Earth. Across the United States and the world, the success brought disbelief, followed by panic. Barely one month later, on November 7, the USSR launched *Sputnik 2*, which had a passenger, Laika the dog. This feat was timed to coincide with the 40th anniversary of the founding of the Soviet Union, and Moscow hailed the space mission as proof of the superiority of communism.

The U.S. public was both shocked and confused about *Sputnik*. Few people fully understood the engineering skills needed to launch, orbit, and recover a satellite. But everyone understood the stark new geopolitical reality: The USSR now had missiles powerful enough to reach U.S. soil. Thanks to its great distance from Europe and Asia, the United States had long felt safe from foreign military attack, a confidence that had made the Japanese attack on Pearl Harbor such a blow in 1941. But now Washington could no longer depend on oceans and air power to defend the country. The ominous *beep . . . beep . . . beep . . .* broadcast by *Sputnik 1* over the next 92 days was easy to pick up on civilian radios and constantly reminded Americans of the new threat overhead.

Sputnik also changed the international balance of power. The end of World War II (1939–45) had led to the beginning of the cold war, as the United States and the Soviet Union began a standoff between different political, economic, and social systems that would last almost five decades. The United States became the protector of Western Europe, deploying nuclear weapons and troops along the border with the Soviet bloc. As those countries rebuilt, they would be protected under the U.S. nuclear umbrella. The United States also stood as the protector of liberal ideas, such as democracy and capitalism, against the spread of communism. Now that the Soviets had missiles powerful enough to reach North America, this arrangement seemed under threat. Perhaps the best summary of the new battlefield came from

3

Senate staffer Charles Brewton. "In previous ages the Romans controlled the world because of their roads, then England controlled the world because of its ships. When humanity moved to the air, the United States was supreme through aviation. *Now the Russians have moved into outer space.*"[1]

The new space age transformed the U.S. government, military strategy, science, and popular culture. *Sputnik* opened a new frontier to conquer, occupy, and deny to rivals, a theme carried into science fiction, which turned it into tales of intergalactic battles and space travel.

Sixty-plus years later, the United States has become highly dependent on military space assets for communication, navigation, and intelligence gathering. The USSR no longer exists, but Russia makes extensive military use of similar technologies, as increasingly does China. Almost all countries use scientific and commercial applications based on space infrastructure and technologies originally designed for military use, from cell phones to Google Earth to pay-per-view movies. But despite the panic engendered by *Sputnik* and science fiction, no weapons have been deployed yet in outer space. Why not?

WHAT IS MILITARIZATION OF SPACE?

The term *militarization of space* refers to military use of space-based assets, especially satellites. During the cold war, the most common uses of satellites were "reassurance, arms control verification, and strategic stability enhancement."[2] Contemporary uses include communications, intelligence gathering, photoreconnaissance, navigation, meteorology, early warning, real-time targeting, remote sensing, and weapons guidance.[3] During the 1991 Gulf War, for example, the U.S. military used satellites to help troops communicate and global positioning system (GPS) data to help them move across featureless deserts. Satellites helped target and maneuver Tomahawk missiles during the U.S. campaign against Serbia in 1999 and Predator drones to attack suspected al-Qaeda affiliates in Afghanistan and Pakistan in Operation Enduring Freedom in 2010.

Military use of space is expected to only increase in the 21st century. U.S. forces used space-based assets to great advantage during the 1991 Gulf War, leading to a change in thinking about how future wars would be fought. The Gulf success seemed to confirm predictions of an ongoing revolution in military affairs, which Andrew Marshall, formerly head of the Pentagon's Office of Net Assessment, defines as ". . . a major change in the nature of warfare brought about by the innovative application of new technologies which, combined with dramatic changes in military doctrine and operational and organizational concepts, fundamentally alters the character and conduct of military operations."[4] Before studying the changes associated with weapons,

Introduction

Michael O'Hanlon of the Brookings Institution cautions, "Despite the tendency of military strategists to rave about defense transformation and a coming revolution in military affairs, many satellite development programs are currently advancing more slowly than once hoped."[5]

Weaponization, in contrast, refers to the deployment of actual weapons into space, such as lasers, space-to-Earth attack weapons, or missile defense system components. Space has not yet been weaponized.

Crossing that threshold, however, could alter the existing global balance of power. The collapse of the USSR in 1991 left only one superpower—the United States. The impact of weaponization largely depends upon who decides to weaponize space. If it is the United States, the move would further increase its hegemony, but it could anger other countries and might cause retaliation. If another state, such as China or Russia, were to weaponize, it might compensate for the current asymmetrical balance of power. Weaponization by any country could also trigger a new arms race, consume a tremendous amount of money, and have dire environmental consequences.

How Satellites Work

The largest category of space-based military assets is satellites. Satellites are objects that orbit around a larger object—typically a planet—and they are held in orbit by the gravitational pull of the larger body. They may be natural, such as moons, or human-made, such as *Sputnik,* a space shuttle, or the *International Space Station (ISS).* Between 1957 and 2003, some 5,450 functional objects were launched into Earth's orbit. The UN estimated in 2003 that 2,800 functional objects were still in outer space and about 25 percent of those were still operational.[6] Seven countries (China, Russia, India, United States, Japan, Iran, and Israel) and the European Space Agency (ESA) currently have the ability to launch satellites into orbit, and many more countries can build satellites for others to launch. As many as 1,100 firms in 53 countries currently use outer space in some capacity.[7]

TYPES OF SATELLITE

Satellites orbit along elliptical routes and are categorized by their purpose, orbital path, and distance from Earth. Most are positioned in low earth orbit (LEO), which is approximately 100–600 miles above Earth, and each orbit takes about 90–120 minutes to complete. Types of satellite include:

Meteorological satellites, which provide long-distance images of weather systems and patterns used in making forecasts.

Communications satellites, which receive telephone calls and data streams and retransmit the information back to Earth via transponders, while **broadcast satellites** relay television signals.

5

Navigational satellites, such as the GPS, Russian GLONASS, and the European Union's Galileo, which use signals to precisely locate objects on Earth.

Rescue satellites, often specialized "pods" that piggyback on other satellites; they receive signals from emergency position-indicating radio beacons (EPIRB), transmitters that broadcast unique signals to help responders locate people in distress.

Scientific satellites record specialized data for research projects, while Earth observation satellites monitor the geophysical features of the planet, such as shifts in the polar ice caps, changes in the rain forests, or crop yields. They can also track refugee flows and pinpoint areas in need of infrastructure such as roads.[8] NASA's *Terra* and *Aqua* satellites helped monitor the 2010 *Deepwater Horizon* oil spill in the Gulf of Mexico and the eruption of Iceland's Eyjafjallajökull volcano.[9]

SATELLITE ORBITAL PATHS

Satellites orbit in one of three regions. Geostationary or geosynchronous orbits (GEO) allow a satellite to remain fixed over a particular location above Earth's equator at all times, at an altitude of about 22,236 miles. Their orbits are synchronized with the rotation of the Earth, so the satellite also completes one cycle in about 24 hours. Asynchronous orbits are divided into LEOs and medium earth orbits (MEOs), which allow satellites to change the time of day that they overpass a particular site on Earth. The most heavily populated asynchronous orbit is LEO, where the space shuttle and space station orbit, while GPS satellites use MEO. Finally, polar orbits have a 90 degree inclination, making them perpendicular to the equatorial plane, and cross the North and South Poles in each cycle. Since Earth rotates horizontally, every part of the planet passes through the polar orbit loop in the course of one day. Photo-reconnaissance satellites tend to use polar orbits because with the correct orbital parameters, solar conditions can be kept constant for each location the satellite images. Some functions require multiple satellites for maximum coverage. The GPS network, for example, needs at least 24 satellites to work, but there are currently six backup satellites in orbit, bringing the total to 30.

GRAVEYARD ORBIT

When satellites fail or simply outlive their usefulness, they can interfere with other satellites. Known as space zombies, these satellites can float around for decades. Controllers can try to move a defunct GEO satellite into a graveyard orbit, several hundred miles above its original orbit and, it is hoped, out of the path of other satellites. In 2006, a Russian communications satellite was moved into a graveyard orbit after being damaged by space debris. Three years later, the Luxembourg-based firm SES had to find a way to bring its

Astra 5A satellite under control, or at least out of the way, when it ceased to function. In May 2010, the Intelsat-owned *Galaxy 15* communications satellite malfunctioned, possibly due to a solar storm, and began to drift. Unlike most damaged satellites, *Galaxy 15* continued to transmit signals as it drifted into the path of another cable television satellite, *AMC 11*. Since both operated on the same frequencies, *Galaxy 15* threatened to disrupt *AMC 11*'s transmissions, but Intelsat was able to help amplify *AMC 11*'s signal and avoid any problem.[10]

In lower orbits, controllers cannot nudge zombies out of the way. Instead, many satellites and spacecraft are intended for eventual cremation. Satellites, Russian cargo ships, the *Mir* space station, and Europe's *Jules Verne* craft were simply left to burn up as they reentered Earth's atmosphere.[11] Stray pieces have survived and have fallen across the Earth. An Oklahoma woman was hit on the shoulder with a piece of a Delta II booster rocket in 1997, while a drinking-water tank from *Gemini V* plunked into Western Australia in the 1960s.[12]

Who Has Satellites?

Since the end of the cold war, the satellite industry has become more complicated, competitive, and lucrative. The earliest satellites were developed by governments at great cost, but over time private firms have entered the satellite market, offering construction, launch, and operational services. Today, satellites may be owned and operated by governments, groups of countries, private firms, or combinations of these groups.

Fifty years after *Sputnik* and the establishment of NASA, other countries have joined Russia and the United States in the space sector. India has launched its own satellites, while South Korea and Pakistan have built satellites that others have launched. China put a man in orbit in late 2008, while Iran launched its own satellite in March 2010. Non-space-faring countries, such as North Korea, or even non-state actors, such as al-Qaeda, could potentially enter the space venue should they gain the technology necessary to interfere with vital satellite operations.

The end of Soviet central planning brought capitalism to the Russian, Ukrainian, and Kazakhstani space sectors, with new private firms offering satellite construction and launch services. Space tourism has begun, as a few private citizens have been willing to spend millions of dollars to ride on a Russian spacecraft and spend a few days at the *ISS*, a multinational endeavor that combined several projects as a cost-saving solution. These included the Russian *Mir 2* and *Freedom*, from a multinational partnership including the United States, the European Space Agency, Japan, and Canada. Russia launched the first module in November 1998, and the first multinational

7

crew arrived in 2000. Originally, the U.S. space shuttle and the Russian Soyuz spacecraft were to share the ferry duties, but the grounding of the shuttle fleet after the *Columbia* disintegrated in 2003 and the retirement of all U.S. shuttles in 2010 will leave Russia as the only ride to and from the ISS. To date, no country currently has announced plans to use or develop a space station for military purposes.

Commercial providers have also entered the market. It is entirely possible that one country purchased a satellite built by a second country and launched by a third. This market diversification blurs ownership and thus liability for accidents. Traditionally, treaties and conventions regarding space-based activities have only been signed by countries, and commercial providers could conceivably claim that they are not bound by those regulations.

Military Use of Space-Based Assets

Space becomes militarized when equipment based in outer space is used for military purposes. Currently, the space technology most widely used by the military is satellites. Satellites may be used to monitor compliance with arms control agreements, track troop movements, gather intelligence, photograph sites of interest, provide early warning of missile launches, and guide weapons. Militaries may develop and deploy their own satellites, join multinational projects, purchase commercially built satellites, use commercial launch firms, or may simply buy relevant data from commercial providers. However, commercial providers may be willing to sell opposing militaries the very same data. Days before the October 2001 invasion of Afghanistan the U.S. government purchased exclusive rights to satellite images of Afghanistan taken by the private firm Space Imaging (now GeoEye) for $2 million per month. Washington feared that the private company might sell photos to the media or to Afghanistan and began to add shutter control provisions, allowing the government to halt the gathering or distribution of satellite imaging, to its contracts with commercial providers.[13]

MONITORING WEAPONS AGREEMENTS

Most international agreements and treaties signed during the cold war included provisions for "national technical means." This odd phrase in the Anti-Ballistic Missile (ABM) and Strategic Arms Limitation (SALT I) treaties refers to a variety of intelligence-gathering opportunities, ranging from reconnaissance satellite imagery to on-site inspections, used to monitor nuclear tests, count missiles, and track troop deployment patterns. Signatories agreed to not interfere in such endeavors.

Space-based assets can also warn of incoming attacks. Information could be relayed from intercepted communications materials, changes in troop movements, or direct observation of a launch. Most interceptor systems try to attack during the boost phase, when it is easier to detect the infrared signature associated with a launch missile.

GATHERING INTELLIGENCE

Governments have changed they way they use satellite data. During the cold war, the United States needed a window into the closed Soviet state. Given the limits of reconnaissance aircraft, President Eisenhower sought to allow unrestricted overflights—open skies—by satellites. After *Sputnik* set a precedent for allowing such overflights, the United States began its CORONA satellite imagery project. In 1962, satellite imagery revealed the presence of Soviet ships and missile pads in Cuba.

In the post–cold war era, satellites have become tactical weapons, providing real-time imagery of battlefield conditions. For example, GPS satellite data are used for blue-force tracking, which allows soldiers to know the location of allied troops to prevent friendly fire incidents.[14]

The CIA uses satellite data to monitor the movements of suspected terrorists. CIA drone aircraft have located terrorists traveling across the Middle East and especially along the mountainous border between Afghanistan and Pakistan.[15] While U.S. forces had four or five reconnaissance aircraft during the Persian Gulf War, today they have hundreds. Observations collected by Predator drones are compared over time to develop "patterns-of-life analysis" to determine whether terrorists are present in a location.[16]

PRECISION-GUIDED WEAPONS

U.S. military forces have used precision-guided cruise missiles and joint direct attack munitions since the 1991 Gulf War. "More than 5,000 were employed in the Afghanistan war of 2001–02, striking as close as five meters from their aimpoints."[17] While most Predator drones had targets in Afghanistan or Pakistan, they are guided from an air force base in Nevada—7,000 miles away.[18]

THREATS AFFECTING SPACE-BASED ASSETS

The increasing U.S. reliance on commercial satellites for communications and reconnaissance brings a great deal of vulnerability. According to Bruce W. MacDonald of the United States Institute of Peace (USIP), "Our space assets are exposed and fragile. They can't run, they can't hide, and today they can't defend themselves. One small object travelling at orbital speeds can destroy them."[19] As satellites are currently the primary space-based asset, antisatellite weapons (ASATs) are the primary threat. ASATs can destroy or disrupt a satellite. However, space-based assets are much harder to target than land-based segments. In its 2001 report, the Commission to Assess the Ballistic Missile Threat to the United States (Rumsfeld Commission) readily admitted, "Attacking or sabotaging the supporting ground facilities has long been considered one of the easiest methods for a U.S. adversary to conduct offensive counter-space operations. Most of these facilities are relatively easy to get in close physical proximity to or access by way of a computer network, making them a prime target."[20]

In theory, satellites might be destroyed by laser or kinetic weapons, but the resulting explosion would create a cloud of debris. Instead, weapons might use electronic transmissions to jam a satellite, blind it with a substance like paint, or take it offline by damaging its solar-energy panels (assuming there is no backup power source). Satellites can even be "hijacked" if the control system is decoded. The Chinese group Falun Gong hacked into the control program of a government satellite and began broadcasting its own antigovernment programming.[21]

During a conflict, a commercial provider could conceivably raise the price for its services or sell them to the highest bidder—which might be a hostile country. Perhaps the biggest threat to space-based assets is the uncertain environment that contains them. "One of the biggest threats we face is what we just don't know," according to MacDonald. "About objects in space, the intentions of those who put them there, and the very strategic landscape of space itself—how it operates, where it poses strategic dangers, and what we need to look out for."[22]

Antisatellite Weapons

The United States, Russia, and China have successfully demonstrated ground-to-space ASAT capabilities. On January 11, 2007, China used a ground-based, medium-range ballistic missile to shoot down one of its own weather satellites. The United States similarly destroyed one of its satellites in 2008. The USSR actually tested a nonnuclear co-orbital ASAT 20 times between October 1968 and June 1982,[23] and in March 2009 Moscow acknowledged that it is developing ASAT capability. In addition, nuclear-tipped missiles that can be launched into LEO could theoretically be used as ASAT, if they were detonated near a satellite.

Other options include deploying small, maneuverable satellites that can crash into other satellites. The U.S. space shuttle could be appropriated for military purposes, such as having astronauts damage opponents' satellites. The U.S. Air Force has researched a common aero vehicle (CAV) that would be deployed into LEO and, when needed, would be maneuvered to de-orbit and release weapons to specific Earth-based targets. Another option is a co-orbital approach—a microsatellite positions itself near the target satellite, enters the same orbit, and then deploys a weapon. However, it takes at least 90 minutes to establish a co-orbit, long enough to give away the ASAT's location.

Alternatively, a stealth weapon could slowly move into position over the course of several weeks, possibly avoiding detection. A bodyguard satellite could deploy near a potential target to protect it from an ASAT attack. Finally, lasers can dazzle or blind satellites, and jamming devices can interfere with communication and surveillance satellites. Weapons designers are still trying to perfect a laser that could provide a stable, potent beam that could reach LEO. The technical demands would be even higher to develop weapons that can target objects in different orbits, because they would have to maneuver across vast distances. "For instance, if a weapon in low earth orbit were to engage military communication satellites in geosynchronous (geostationary) orbits, it would need a range of almost 30,000 km."[24]

Traffic and Debris

Other threats come from traffic and trash. Outer space is much more crowded in the 21st century than it was in 1957. In addition to hundreds of satellites, the *ISS*, and, for the first decade, the U.S. space shuttle, space is littered with thousands of pieces of debris. Yet no space counterpart of the U.S. Federal Aviation Administration (FAA) exists to control the heavens.

Space junk is the detritus left by booster rockets, defunct satellites, dropped tools, and other discarded materials. As of late 2008, there were at least 23,000 objects large enough to be tracked by radar. The United States Strategic Command (USSTRATCOM) currently tracks more than 21,000 objects.[25] Estimates of microscopic debris are terrifying. The Centre National d'Etudes Spatiales (CNES), the French space agency, estimates that there are more than 300,000 pieces of space junk larger than one centimeter in size— and 30 million larger than one millimeter.

Even tiny objects have tremendous damage potential. Objects orbit Earth at tens of thousands of miles an hour, meaning a collision with even a tiny fleck of paint could damage a satellite or spacecraft. According to Michael Krepon of the Henry L. Stimson Center, "Getting hit by a marble-sized piece of debris in low earth orbit has roughly the same impact on Earth as a one-ton safe dropped from a five-story building."[26] NASA has had to replace 60 space shuttle windows due to dings from minute objects.[27] In July 1996, a piece from an old U.S. rocket nicked the stabilizing boom on a French Cerise satellite traveling more than nine miles per second. The satellite's orbit was temporarily disrupted, but later corrected. Another U.S. rocket fragment hit a Chinese rocket in January 2005, creating still more orbiting junk, as did the collision between *Iridium 33* and a discarded Russian military satellite, *Cosmos 2251*, in February 2009.[28]

The largest release of debris to date followed China's 2007 ASAT test, which produced at least 2,500 pieces of detectable debris. "One of those speeding fragments forced NASA to hastily move its $1.3 billion *Terra* Earth-viewing spacecraft out of the way."[29] The significance of the Chinese test was its location; it occurred at a much higher altitude than the 1985 U.S. test, which also produced a lot of debris that took 17 years to clear out. Controllers typically steer dying satellites into LEO, expecting them to burn up on reentry. But the Chinese test did not follow this procedure. "That means the debris will remain in space for tens, thousands, or even millions of years," said Krepon.[30]

One option would be to harvest the debris; another is to explode it into ever tinier particles, although some risk would remain. Other proposals use "space flypaper," sweepers, robot garbage scows, nets, tethers, and attaching small rockets. But, according to Nicholas L. Johnson, NASA's chief debris scientist, "For the near term, no single remediation technique appears to be both technically feasible and economically viable."[31]

Weaponization of Space

In theory, space could also be used to house weapons, to apply force to Earth-based or space-based targets, or to defend against missile attacks. Neither space-based weapons nor space-based missile defenses have been developed or tested thus far.

SPACE-BASED WEAPONS

Space-based weapons could be either satellites deployed far ahead of a conflict, weapons launched in reaction to a specific conflict, or manned or unmanned bases. There are also hybrid weapons, such as land-based and land-targeted bombs and cruise missiles that are guided by satellite systems.

SPACE-BASED MISSILE DEFENSE SYSTEMS

The most commonly proposed defense against incoming weapons—from space or elsewhere—is a missile defense system, which would intercept and destroy the missiles. Both the United States and the Soviet Union have experimented with and deployed ballistic missile defense systems. The 1972 ABM Treaty allowed each side to build two systems—one around a key city, the other around a military installation—to ensure that they had a "'second strike' capability that would deter any state from attempting a crippling nuclear 'first strike.'"[32] The Soviets erected a system around Moscow, while the United States briefly deployed a system around an intercontinental ballistic missile (ICBM) installation in North Dakota. The George W. Bush administration planned to place elements of a new ballistic missile defense (BMD) system in the Czech Republic and Poland, much to Moscow's consternation, but President Barack Obama changed the parameters and schedule of the program in 2009. Instead, Europe will be protected from an attack using the existing ship-based Aegis combat system.

Ballistic missile defense systems can be seen as a security-enhancing strategy, as they could destroy accidentally launched missiles or intercept launches by rogue states or terrorists. However, critics consider BMD systems to be destabilizing, especially if other countries overestimate their effectiveness. "Russian hawks," according to the Union of Concerned Scientists (UCS), "might use the U.S. system to argue against meaningful reductions in Russia's nuclear arsenal . . . China's hawks will be able to make an even stronger case since their country has a much smaller arsenal."[33]

Ronald Reagan's ambitious Strategic Defense Initiative (SDI) conceptualized a series of space-based missile garages. Brilliant Pebbles and Brilliant Eyes were systems of sensors and miniature orbiting weapons platforms. Once a ballistic missile launch was detected, a projectile would be shot downward toward the target, building up sufficient kinetic energy to destroy it on impact. Space mines such as Brilliant Pebbles and Brilliant Eyes could be deployed in advance or on demand. The SDI system included a complex network of laser-equipped satellites that would zap Soviet nuclear-tipped ICBMs in mid-flight, shielding the United States from their destructive capabilities. Space-based weapons could also intercept ballistic missiles launched from Earth to Earth-based targets.

REGULATING SPACE

The current system for regulating space is composed of a series of treaties and bilateral agreements. While some documents specifically ban weapons, many deal with fundamental questions such as What is space? Where does space begin? Who owns space? These simple questions have far-reaching implications for deploying space assets and launching—or defending against—space-based weapons. The *World Book Encyclopedia* says simply: "Space is the near-emptiness in which all objects in the universe move. The planets and the stars are tiny dots compared with the vast expanse of space."[34]

Most space law is based on two sources: (1) the laws of the sea and air, and (2) the laws governing Antarctica.

Laws of the Sea and Air

While the law of the sea formalized customary practices that had emerged over centuries, air laws emerged only in the 20th century, largely in reaction to technological advances. Space law emerged following *Sputnik*. General Howell M. Estes, III, former head of the United States Space Command (USSPACECOM), explains the legal analogy as follows:

> [Military] space operations, like the land, sea, and air operations that evolved before them, will expand [into] the budding new missions already included in the charter of the U.S. Space Command of space control and force application as they become more and more critical to our national security interests.[35]

There are two schools of thought on ownership of space.[36] The *res communis* view sees space as a common zone, to be shared by all humankind. The contrary view, *res nullius*, argues that no person or state currently owns space; ownership will be claimed when a party can effectively control it. Advocates of *res communis* put space in the same category as air, oceans, and sunshine—everyone can enjoy the benefits at the same time without exhausting a seemingly endless resource and everyone has the right of innocent passage. The *res nullius* doctrine considers space (or oceans, animals, or fish) to be up for grabs; the first person or state that can effectively control it and extract benefits from it (such as oil, furs, or dinner) wins control. Advocates of *res communis* would want to keep space a sanctuary free of weapons.

Each view has problems. The ideas of *res communis* lead to the dilemma that the University of California ecologist and microbiologist Garrett Hardin labeled the tragedy of the commons. Specifically, as society expands and populations grow, more and more demands are made on common spaces, which eventually will be claimed and exploited by someone.[37] For example, a town may have beautiful parks for all citizens to enjoy. But if the town becomes overpopulated, a new industry may need to locate a factory there, or a disaster may cause people to seize the common space to satisfy their own needs. Problems affecting the common area, such as pollution and trash, may not be solved as everyone shirks responsibility for cleaning it up. *Res nullius* depends on an international consensus to not claim space, but one day a country or non-state actor may decide to stake a claim of their own.

Before jurists can determine what activities are and are not permitted in outer space, they must establish the basic parameters of space as a realm of law-governed activities. Just as the laws of the sea and air, space law must address four issues: delimitation, sovereignty, registration and liability, and innocent passage.[38] As with civil and criminal law, there often is a differentiation between what you can do—and what you can get away with.

Introduction

DELIMITATION

Delimitation is the issue of where atmosphere ends and space begins. There is no obvious physical barrier, and the distinction is further blurred by craft that can operate in both the atmosphere and in orbit, such as the space shuttle. The military analyst Carol Laymance explains the multiple approaches:

> If trying to define where space begins for biological reasons, one might choose nine miles above the Earth since above this point a pressure suit is required. If concerned with propulsion, 28 miles is important, since this is the limit of air-breathing engines. For administrative purposes, one might find it important that U.S. astronaut wings may be earned above 50 miles. An aeronautical engineer might define space as starting at 62 miles above the Earth's surface since this is where aerodynamic controls become ineffective. Conventional and customary law defines the lower boundary of space as the lowest perigee for orbiting space vehicles, about 93 miles.[39]

This fundamental issue was glossed over in the UN negotiations for the Treaty on Principles Governing the Activities of States in the Exploration and Use of Outer Space, including the Moon and Other Celestial Bodies (the Outer Space Treaty), which entered into force on October 10, 1967, because negotiators feared the entire process would founder otherwise. Two formulations seem to have had the largest following. First, the Karman primary jurisdiction line is the lowest point at which an object can maintain its orbit due to gravitational pull overtaking aerodynamic flight, specifically about 52 miles. The second approach is based on the limits of propulsion systems. Space begins just beyond the maximum altitude for aerodynamic flight. In practice, the standard is 100 kilometers above the Earth's surface—about 62 miles.[40]

SOVEREIGNTY

The second issue that needs to be addressed in the law of space is that of sovereignty, or supreme authority over a defined landmass. According to the 1919 Paris Convention, "Every Power has complete and exclusive sovereignty over the air space above its territory."[41] In practice, the "open skies" principle allowed aircraft to overfly any country during peacetime, provided appropriate advance notice, including a flight plan, has been given. However, no measurable upper limit was established. Sovereignty, however, was applied to air *space,* not the actual air. Following the principle of *res communis,* the air cannot be owned by a single government. With the extreme distances involved in space, it becomes very difficult to determine exactly which country owns the faraway speck of territory as an object

15

orbits Earth. Some scholars have suggested flexible criteria that can adapt to changing technology, setting the boundary at the point where "the scientific programs of any state . . . permits such state to control it." This would mean that overflights by hostile states would be permitted if the overflown country lacks the technology to control its superadjacent space—for example, it meant that the United States did not violate Soviet airspace in the late 1950s and early 1960s by dispatching U-2 planes to fly over the USSR and take photos because the USSR did not have the technology to develop antiaircraft weapons that could defend the high altitude.

Two cases illustrate *res nullius*, in which countries have claimed control over their air space, even if they could not effectively exercise such control. In the Bogotá Declaration of 1976, eight states[42] announced that their sovereignty extended not only across their geographic territory, but also upward to geostationary altitude. Thus, any object flying across their claimed territory or in their claimed airspace would be subject to their national law. This argument flew in the face of the open regime instituted by the Outer Space Treaty. While no other countries have recognized the Bogotá Declaration. Everett C. Dolman, a professor of comparative military studies at the U.S. Air Force's School of Advanced Air and Space Studies (USAFSAASS), sees it as the manifestation of a dangerous shift in how countries perceive space activities. "It remains important," he writes, "because it is representative of a growing desire in the LDCs [least developed countries] to seize a greater share of the common goods."[43] The eight countries themselves appear to backing down, as six have signed the Outer Space Treaty and of that four have ratified it.

The second case comes from the tiny South Pacific Kingdom of Tonga. In 1990, Tonga registered to claim control of 16 geostationary orbital slots. The *New York Times* reported, "The number of these spaces above the earth is sharply limited by international law to prevent interference between satellites, but they can be reserved at no cost by nearly any nation on an essentially first-come, first-served basis."[44] Officially, the International Frequency Registration Board in Geneva, Switzerland, assigns control over these slots, but when Tonga discovered that 16 had not been allocated, it moved to claim them. Tongan officials promptly announced that anyone could lease one for $2 million per year.[45]

Other countries quickly protested Tonga's move. Ultimately, the International Telecommunications Union helped forge several compromises. First, it told Tonga to drop at least 10 of its claims, suggesting they were little more than get-rich-quick schemes. Tonga settled for nine slots, and Indonesia agreed to share one slot with Tongasat, the government's agent responsible for managing satellites.

16

REGISTRATION

Another mechanism used to govern space during the cold war was an international registry of launches and payloads. States must maintain registries of their ships under the Convention on the Law of the Sea, while aircraft must not only be registered but also be de facto citizens of their registered states. Similarly, the Registration Convention of 1976 requires countries to maintain a registry of all objects it launches into space. They must also report that information to the United Nations for inclusion in its registry. Such registries allow for blame and assignment of liability should problems arise. In 1978, the Soviet Cosmos 954 reconnaissance satellite malfunctioned and de-orbited. The satellite was nuclear powered, and it dropped radioactive debris over Canada. Moscow eventually accepted responsibility and paid C$3 million.

INNOCENT PASSAGE

The fourth issue that states need to negotiate is the right of innocent passage, defined as being "not prejudicial to the peace, good order, or security of the . . . state."[46] The 1919 Paris Convention and the 1944 Chicago Convention included "a provision . . . for the right of innocent passage by aircraft not engaged in scheduled international services." The Chicago Convention also prohibited overflights by "pilotless aircraft" and allowed no-fly zones linked to national security. It also allowed states to establish no-aerial-photograph zones. In 1957, *Sputnik* indirectly extended the principle of free passage to include orbiting objects. If Moscow had tried to claim an exemption from this principle it would not be able to use its new technology; if other states objected they would not be able to develop their own versions. Eisenhower seized this precedent to revive his 1955 proposal for an open skies treaty with Moscow.

Laws Governing Antarctica

The second source of legal precedent is the international regime for governing Antarctica. Seven countries have laid claim to separate zones in Antarctica—Argentina, Australia, Chile, France, New Zealand, Norway, and the United Kingdom. However, they have crafted a peaceful, joint-use mechanism through a 12-state[47] governing board established during the 1956–57 International Geophysical Year. More than two dozen states have since agreed to the treaty.

Claims to Antarctica have been based on the principles of sustained occupation or a symbolic claiming act, such as planting a national flag, leaving a plaque, or opening a diplomatic mission. In 1969, the U.S. astronauts Neil Armstrong and Buzz Aldrin, the first and second people to walk on the Moon, carried a flag especially designed to flutter without wind or gravity.

The Soviet lunar probes carried numerous items labeled CCCP (Cyrillic for USSR), and the U.S. and Soviet probes sent to Venus included flags. Similarly, Russia carefully planted its tricolor flag under the North Pole in August 2007.[48] Russian efforts to continuously staff its *Mir* space station between 1987 and 2000 laid the foundation for a claim of sustained occupation.

The body applies to scientific and military outposts, such as space stations, or perhaps to occupation of a specific satellite slot. Thus, in 2003, Tongasat purchased an antiquated satellite from General Dynamics, with the intention of using it in one of its slots.[49]

Treaties

The current international regime for the exploration and use of space comprises five fundamental documents.[50] These are known informally as the Outer Space Treaty (1967); the Rescue Treaty (1968); the Liability Convention (1972); the Registration Convention (1976); and the Moon Agreement (1984). These are supplemented by another four agreements: the Limited Test Ban Treaty (1963); the Anti-Ballistic Missile Treaty (1972); International Telecommunications Convention (1973); and the Convention on the Prohibition of Military or Any Other Hostile Use of Environmental Modification Techniques (1977). Together, the documents create a legal framework formalizing the principles of nonappropriation of outer space by any one country, arms control, free exploration, liability for harm incurred by space activities, notification of activities, and use of the natural resources of outer space.

OUTER SPACE TREATY (1967)

Negotiations for an international regime to govern the use of space began immediately after *Sputnik* launched in October 1957, but the process took a full decade to complete. Initially begun as a bilateral process between the United States and the Soviet Union, on March 15, 1958, Moscow recommended expanding the scope of talks and placing them under the jurisdiction of the United Nations.[51] Washington concurred, and the UN created the Ad Hoc Committee on the Peaceful Uses of Outer Space (AHCOPUOS).

The process then came to a complete standstill, as the United States and the USSR bickered over the number of seats on the committee. Washington proposed nine seats, but the Soviet proposal for 18 members prevailed. One year later, AHCOPUOS was reorganized into a permanent Committee on the Peaceful Uses of Outer Space (COPUOS), but that forum did not convene for two years due to disputes over its composition.

Then on September 25, 1961, U.S. president John F. Kennedy addressed the UN General Assembly and suggested expanding the organization's mandate to include outer space. This move would immediately make international law applicable to outer space. The UN agreed, enshrining the jurisdiction change in UN General Assembly Resolution 1721, "International Cooperation in the Peaceful Uses of Outer Space." Specifically, "International law, including the Charter of the United Nations, applies to outer space and celestial bodies." The resolution also called for a registry of launches and appointed the International Telecommunication Union to allocate radio frequency bands for satellite uses.[52] Next, in 1963, the UN General Assembly adopted a "Declaration of Legal Principles Governing the Activities of States in the Exploration and Use of Outer Space," which became the basis of the landmark Outer Space Treaty.

The Outer Space Treaty builds on the U.S. position that space should be available for all humankind to use for peaceful purposes. Therefore, it bans deployment of weapons of mass destruction in space and prohibits the colonization of the Moon. The weapons ban, however, does not cover conventional weapons.

RESCUE TREATY (1968)

On April 12, 1961, the Soviet cosmonaut Yuri Gagarin became the first man to orbit the Earth. Less than three weeks later, the U.S. astronaut Alan Shepard completed a successful 15-minute suborbital flight. Over the next decade, manned spaceflights became regular missions. Article 5 of the Outer Space Treaty called for mutual assistance to rescue astronauts—"envoys of mankind in outer space"—guaranteeing that they would be given "all possible assistance" should an accident, emergency landing, or other unplanned event occur.

These principles of mutual assistance were expanded and further clarified in the 1968 Agreement on the Rescue of Astronauts, the Return of Astronauts, and the Return of Objects Launched into Outer Space, known as the Rescue Treaty. This document obligates signatories to notify the launching authority and the UN Secretary-General of any spacecraft personnel "experiencing conditions of distress" or having made an emergency or unintended landing on the high seas or the territory of a country other than the launching country. To prevent astronauts from being held as prisoners, Article 4 requires signatories to return astronauts "safely and promptly . . . to representatives of the launching authority."

LIABILITY CONVENTION (1972)

After the Rescue Agreement outlined the appropriate steps to return space personnel to their country of origin, the United Nations adopted the Convention on International Liability for Damage Caused by Space Objects, or Liability Convention, to detail the rights and obligations concerning spacecraft, satellites, missile components, and other nonhuman objects launched into space.

Discussion on this issue began almost as soon as *Sputnik* launched in late 1957. Given the scientific maxim that "what goes up must come down," countries were concerned about potential damage to their territory from man-made objects falling from space. The final agreement concedes that accidents are inevitable, but "A launching State shall be absolutely liable to pay compensation for damage caused by its space object on the surface of the earth or to aircraft flight," according to Article 2. The agreement also establishes a claims commission to adjudicate disputes over damages.

REGISTRATION CONVENTION (1976)

The original Outer Space Treaty called upon states to create national registries that would record pertinent data for all objects launched into outer space. Indeed, the obligations outlined in the liability convention need some type of official record so responsibility can be formally assigned for a launch and, therefore, for any damage caused in conjunction with that launch. The Convention on Registration of Objects Launched into Outer Space, or Registration Convention, of 1976 codifies these requirements. Specifically, Article 2 states that each state is to maintain a registry of launches and to notify the UN Secretary-General of the existence of these registries and individual launches. The UN Secretary-General, in turn, collates these national registries into one open-access registry.

Article 4 specifies the minimum information requirements: (1) name of launching state or states; (2) an appropriate designator of the space object or its registration number; (3) date and territory or location of launch; (4) basic orbital parameters, including nodal period, inclination, apogee, and perigee; and (5) general function of the space object.

MOON AGREEMENT (1984)

Finally, the Agreement Governing the Activities of States on the Moon and Other Celestial Bodies, or Moon Agreement, of 1984 adapts the general coverage of the cosmos included in the 1967 Outer Space Treaty to the specific case of the Moon. The Outer Space Treaty declares that all countries should have equal *access* to the Moon, regardless of their level of economic development. The Moon Agreement changes this principle, stating in Article 4, "The exploration and use of the Moon shall be the province of all mankind and shall be carried out for the *benefit* and in the interests of all countries, irrespective of their degree of economic or scientific development" [emphasis added].[53] Furthermore, section 7 of Article 11 specifies that the Moon's natural resources should be equitably shared among all countries, and "countries which have contributed either directly or indirectly to the exploration of the moon shall be given special consideration."

Introduction

The subtle change from *access* to *benefit* could have tremendous implications concerning who can profit from recovering mineral or other valuable deposits from the Moon, should a claim be registered. While the other four fundamental agreements have been widely ratified (Outer Space: 100 countries; Rescue: 91; Liability: 88; Registration: 53), only 13 countries have ratified the Moon Agreement—Australia, Austria, Belgium, Chile, Lebanon, Kazakhstan, Mexico, Morocco, Netherlands, Pakistan, Peru, Philippines, and Uruguay.[54] None of the countries that have launched Moon probes or made lunar landings (United States, USSR/Russia, India, and China) has signed.

These five UN agreements and treaties are supplemented by a series of arms control agreements.

LIMITED TEST BAN TREATY (1963)

By the 1950s, the nuclear club had three members: the United States, the Soviet Union, and the United Kingdom. Realizing the destructive capacity and associated fallout caused by the testing of these devices in the atmosphere, the public and the international community called for controlling the testing of such weapons. The treaty negotiation process lasted eight years, with verification procedures as the principal sticking point. Specifically, Washington and Moscow disagreed on the number of on-site inspections on their respective territory.

Eventually, the United States, the USSR, and the United Kingdom reached agreement on a very short treaty. Article 1 simply states

Each of the Parties to this Treaty undertakes to prohibit, to prevent, and not to carry out any nuclear weapon test explosion, or any other nuclear explosion, at any place under its jurisdiction or control:
(a) in the atmosphere; beyond its limits, including outer space; or under water, including territorial waters or high seas; or
(b) in any other environment if such explosion causes radioactive debris to be present outside the territorial limits of the State under whose jurisdiction or control such explosion is conducted.

China and France have not signed the treaty, but they have observed its provisions. The treaty does not ban underground testing nor does it address issues related to nuclear-powered spacecraft.

ANTI-BALLISTIC MISSILE TREATY (1972)

The second supplementary treaty is the Anti-Ballistic Missile Treaty of 1972 between the United States and the Soviet Union. These negotiations acknowledged that defending national territory from incoming long-range ballistic missiles could be counterproductive. An ABM system could destroy or disable incoming missiles, theoretically reducing the odds of achieving assured destruction. Thus adversaries would need to stockpile even more weapons to compensate for possibly losing some to ABM defenses, which could lead to an arms race.

According to Article 2, "An ABM system is a system to counter strategic ballistic missiles or their elements in flight trajectory, currently consisting of: (a) ABM interceptor missiles . . . ; (b) ABM launchers, which are launchers constructed and deployed for launching ABM interceptor missiles; and (c) ABM radars . . ."

Article 3 allowed the United States and the USSR to each deploy two ABM systems. One system could have a "deployment area having a radius of one hundred and fifty kilometers and centered on the Party's national capital," while the second would protect missile sites; namely, an "area having a radius of one hundred and fifty kilometers and containing ICBM silo launchers." This provision was reduced to one site each in 1974.

Finally, Article 7 allowed for "national technical means of verification" to monitor compliance. This terminology is diplomat talk for reconnaissance satellites. In addition, the parties to the treaty agreed not to interfere with the other side's verification efforts with "deliberate concealment measures."

The Soviets had already begun installing an ABM system around Moscow in 1966. After China successfully tested a nuclear missile the same year, Washington announced it would begin deploying a "thin" ABM system, but emphasized that it was a response to the threat from China, not the USSR.[55] The Mickelsen Safeguard Complex at Nekoma, North Dakota, was the only operational U.S. ABM site under the terms of the treaty. After the agreement was signed, and though it was allowed to construct a second system around the capital, Washington realized that in a democratic society, any plan to protect the capital with an enormously expensive system, while leaving the rest of the population vulnerable to nuclear annihilation, was a political non-starter.[56] Plans for a second ABM system around Washington, D.C., were quickly shelved.

After Reagan's SDI was announced in 1983, the ABM system became a barrier to deploying new systems. Initially, Washington interpreted the treaty as meaning the United States could test new ABM-type systems, but not deploy them. Even that proved too restrictive.

On December 13, 2001, the United States announced it was withdrawing from the 1972 Anti-Ballistic Missile Treaty, because the "circumstances affecting U.S. national security have changed fundamentally."[57] Specifically, the administration argued that the cold war had ended and Russia was now more often a partner in international arms control than a problem. Now, following the terrorist attacks of September 11, 2001, the Bush administration believed that the greatest security threat came from terrorists or rogue states in possession of weapons of mass destruction, and it needed the option to create antiballistic missile systems to best protect the homeland. The original treaty allowed each signatory to have one antimissile site; the Russian one, around Moscow, is still in operation.

INTERNATIONAL TELECOMMUNICATIONS CONVENTION (1973)

The International Telecommunications Convention is a long-standing multi-country agreement to standardize technology used for global communications. It periodically is reviewed and revised as needed to incorporate changing technology. The first convention was signed by 20 countries in 1865 and addressed telegraph usage.

CONVENTION ON THE PROHIBITION OF MILITARY OR ANY OTHER HOSTILE USE OF ENVIRONMENTAL MODIFICATION TECHNIQUES (1977)

Signed by 51 countries, the Environmental Modification Convention prohibits the deliberate manipulation of "the dynamics, composition or structure of the Earth, including its biota, lithosphere, hydrosphere and atmosphere, or of outer space." Such actions would induce climate change, earthquakes, typhoons, or other natural disasters. While this treaty has limited applications at present, its main value may be in the prevention of new technologies of this kind being deployed in the future.

CURRENT SITUATION

Science has advanced to a point where weapons and technologies that were once the exclusive domain of science fiction have either become reality or are within sight. While Reagan's SDI was dismissed as a fanciful *Star Wars* scenario when it was introduced in 1983, China and the United States successfully destroyed orbiting objects in 2007 and 2008, respectively, albeit using quite different technology. Although neither Washington nor Beijing has yet deployed weapons in space, both have used Earth-based weapons, such as missiles or lasers, to destroy or incapacitate satellites that can be used for tactical military purposes. Specifically, China shot down a defunct weather satellite in January 2007, while the United States destroyed a broken, out-of-control spy satellite that was about to reenter the atmosphere and, according to NASA, posed a possible risk to humans on the ground, in February 2008. These displays of ASAT power could lead to an ASAT weapon arms race that would threaten the nearly 1,000 working satellites currently in Earth orbit—half of which are owned or operated by the United States. Israel, India, and France may be interested in such technology.[58] However, efforts to develop space-based assets, ASAT capabilities, and, to an even greater extent, to deploy weapons into space face technical, financial, and legal obstacles.

Factors Working against Militarization of Space

TECHNOLOGY

The technical requirements behind space-based weapons are huge. Directed-energy weapons require heavy generators to produce powerful, steady laser beams to direct at targets. Chemical lasers are powered using a combustion process that could produce vibrations that would interfere with targeting. The physicist Richard Garwin calculated that "a laser must focus a 25 megawatt beam into a spot 1 meter in area for 6 to 7 seconds if it is to deliver energy to the target fast enough to burn through the casing" of a typical solid-fuel missile.[59] In addition, the laser system needs a highly reflective mirror—probably 33 feet in diameter—and a steerable mechanism to target the beam correctly. "In addition, this mirror must be pointed accurately enough to prevent deviations of more than a few centimeters while the weapon travels about 50 km in one direction and the target travels about 40 km in the other."[60] Finally, BMD satellites would likely orbit for years before they are ever used, and that cumulative inactivity could compromise the accuracy of terminal guidance procedures. This also presumes that there is an adequate power source to activate the weapon.

Like directed-energy weapons, kinetic-energy weapons must be deployed according to extremely precise formulas and have demanding energy requirements. Satellites are difficult targets as they are constantly in motion, but, unlike missiles, their orbits are at least known in advance. Hit-to-kill technologies form the background of the U.S. missile defense system currently in development. One proposed kinetic weapon, nicknamed "rods from God," is based on the idea of kinetic bombardment. Kinetic bombardment is the act of attacking a planetary surface with a projectile that would inflict damage kinetically, causing destruction from impact at very high (orbital) velocities. These weapons would be "a thin, heavy, metallic rod one to two meters in length."[61] The rods would be projected toward buried or hardened targets at a speed sufficient to penetrate and disable or destroy.

COST

Other limitations on space-based assets are quite pedestrian, such as funding. Programs to develop space weapons have both absolute and relative costs.

Absolute costs are enormous. In February 2008, the General Accounting Office (GAO) reported to Congress that the current missile defense program was costing between $8 billion and $10 billion in two-year blocks. For the 2006–07 block, the Missile Defense Agency (MDA) "fielded fewer assets, increased costs by about $1 billion, and conducted fewer tests." They will divert huge sums of money from other possible government programs. Critics insist that such budgetary decisions are immoral, as these "vast amounts of money . . . could be better used on domestic programs like public education, highway and transportation upgrades and the like."[62]

Malfunctioning space assets are costly in terms of lost capital investment. A small design flaw or error in the deployment procedure can render a multimillion-dollar object useless. It can take years to develop a repair solution and deliver it to outer space. Although not a satellite, the history of the *Hubble Space Telescope* (HST) provides a good cost illustration. Launched in 1990, the $2 billion telescope initially returned blurry images, due to a design flaw. The ultimate solution, installed in December 1993, carried a high price: $629 million for the shuttle repair mission, $251 million for Hubble parts, ground operations, and related activities, and another $378 million for the actual shuttle flight.[63] Satellites have been destroyed during launch or lost during deployment, precluding any repair effort, and a blind military satellite is hardly an asset.

What If Space Is Weaponized?

So far, this collection of agreements, economics, and physics has prevented the weaponization of space. Despite decades of ambitious plans, no known weapons have been deployed in outer space. Reality has not matched hype for financial, political, technical, and scientific reasons. An extensive study by the UCS details these obstacles.[64]

Despite these limits and the international cooperation behind the *ISS*, political realists warn that one day individual countries will compete to conquer and control the regions discovered by new explorers. "Efforts to turn space into an entirely non-militarized 'sanctuary' may be commendable," writes Swansea University professor Michael Sheehan, previous director of the Scottish Centre for International Security and researcher at the International Institute for Strategic Studies in London. "But if they were achieved, they would not be a successful defense of space from the looming threat posed by militarization. Rather, they would represent a dramatic reversal of policy," because space has always been militarized.[65]

Deploying weapons to outer space would have strategic, legal, and environmental repercussions. Most analysts envision crossing this threshold to be negative on all counts.

MILITARIZATION OF SPACE

STRATEGIC CONSEQUENCES

According to the RAND analyst Karl P. Mueller, space weaponization would harm the United States. "As both the dominant world power and the preeminent spacefaring state . . . [the U.S.] has the most to lose from space weaponization." U.S. dominance could "encourage other states to align against an apparent assertion of U.S. hegemony." China has developed its space program using partnership agreements with other countries. It could align with them or with Russia against the United States.

Why would Washington embrace space weaponization? Brookings Institution analyst Michael O'Hanlon, among others, points out that the United States is already the dominant military power in the world, so why spend the time and effort to increase its lead?[66] Mueller suggests that if weaponization is inevitable—a debatable proposition—the United States should take the lead. He also notes the argument that Washington should take control of space now in anticipation of advantageous uses in the future. Some policy makers believe that the United States should step into the space arena precisely *because* it is the sole dominant power. They say that a stable, liberal political and economic order will not emerge without help. "Only a hegemon, it is argued, can establish the system of rules, norms, and standards of behavior necessary to proper order."[67]

Weaponization could be highly destabilizing in terms of geopolitics, and it may offer only marginally better firepower. It is also a step that may backfire. As Major William L. Spacy II of the U.S. Air Force commented, "It does not make sense to goad prospective adversaries to greater efforts by developing weapons that offer us only marginal advantages."[68]

The asymmetric balance of power resulting from deployment of space weapons would create a first-strike incentive for the country without them. The unequal side would want to inflict as much damage as possible before being gravely injured by an attack from space. Space weapons could inadvertently fuel nuclear proliferation, as non-space-faring countries seek to acquire the most powerful weapons they can afford.

Michael Sheehan views these programs as products of "techno-nationalism." "The movement into space," he writes, "brought about a new criterion for determining the gradation of power and the allocation of prestige in the global community." But while Dolman and others see the United States as the sole space power—and want it to maintain hegemony—Sheehan identifies a growing multipolarity on Earth with implications for outer space. He points to the growing international space regime that regulates the *ISS* through cooperation, not force. Instead of unchallenged U.S. hegemony, "the gradual emergence of a growing multipolarity is evident and it is noteworthy that the key actors involved, Europe, China, and India, all have vigorous space programs."

LEGAL CONSEQUENCES

The United States is already the first country to land (and return) men on the Moon, and some analysts argue that it should maintain its lead by being the first to weaponize the moon and other celestial bodies. However, this would violate several international agreements, including the Outer Space Treaty and the Moon Treaty.

A convincing case can be made that space should remain weapons free and that it should be a common zone available for all to share equally. Many people believe that space is a common good to be shared by all humankind. The RAND analyst Mueller describes this line of thinking as "sanctuary idealism"; namely, that weapons are bad and should not be introduced into zones that do not already have them. In addition, existing weapons should be eliminated. This point of view does not specifically relate to *space* weapons, but it does call for keeping pristine, weapons-free zones.[69]

The Outer Space Treaty starts with a declaration that "The exploration and use of outer space, including the moon and other celestial bodies, shall be carried out for the benefit and in the interests of all countries." It then asserts that "Outer space, including the moon and other celestial bodies," is not subject to national appropriation. In theory, exploring a distant territory could easily lead to a claim of ownership, especially if other potential claimants lack the ability to physically claim the land—or celestial body—in question. But to date, this has not been the case with the Moon or Mars.

So far an international gentleman's agreement—not to mention technical and financial hurdles—has kept countries from deploying conventional weapons in outer space. The only applicable international treaty is the Outer Space Treaty, which bans the deployment of weapons of mass destruction in space or colonization for military purposes. Conventional weapons, however, are not prohibited. The international community is now at a critical threshold in space policy, due to new players, new weapons, and new frontiers.

ENVIRONMENTAL CONSEQUENCES

In a worst-case scenario, some alarmists predict that a space war would create enormous fields of debris that could hover in outer space, enter Earth or lunar orbit, or rain down on Earth itself. Similarly, a ground-based BMD defense system could cause collateral damage to its own neighborhood if it made impact with a target too near the installation it is protecting. This is the major consideration behind the desire to intercept during the boost phase of flight—let the debris fall on the launching country's territory. If nuclear weapons are involved—either offensive or defensive—radiation becomes another concern.

COUNTERSTRATEGIES

While it may be impossible to ban the future weaponization of space, there are hopes that the expanding boundaries and definition of national security can at least be managed. The existing international treaties relating to exploring outer space and those regulating nuclear weapons can serve as a model. However, as seen with the ABM Treaty, countries can simply withdraw from the treaty and pursue whatever programs they wish. Space policy experts have offered a menu of options, including treaties, deterrence, and hedging. Relevant international treaties have already been discussed.

Deterrence

The cold war was managed using deterrence. Each side built enough nuclear weapons to obliterate the other. Any attack would not only destroy the target country, it would be suicidal for the aggressor. Testifying before Congress, Bruce MacDonald stressed the need to avoid using space-based assets, pointing out that at least in the United States, "Our space technology seems to shape our policy, rather than our policy shaping technical solutions."[70] Instead, he stressed the need for ways to avoid using space-based weapons. U.S. space policy under George W. Bush assumed that "no new arms-control agreements are needed because there is no space arms race."[71] But if one country were to deploy weapons in space, several others have the technology to follow suit, raising the risk of an arms race.

Hedging

A second strategy is hedging, or preparing for the worst. While not actually deploying weapons in space or antisatellite weapons on Earth, countries could work to develop appropriate technology and advertise that fact to potential aggressors. Writing in 2003, Michael Krepon, with contributing author Christopher Clary, calls for a hedging strategy that focuses on in-lab testing, transparency, notification, and cooperation regarding any flight testing.[72] He also discourages testing in actual "ASAT mode," because it could alarm other countries and release a dangerous debris field. Both of his concerns were validated when China conducted an ASAT test in 2007.

Cooperation

States could also draw upon past cooperative endeavors to create new guidelines. As mentioned earlier, the *ISS* and efforts to monitor space junk provide two possible models. Krepon views cooperative strategies as a way to set

Introduction

a global code of conduct regarding weapons in space. He also believes the United States, as the leading power in today's global system, should formulate and obey the code.

Krepon advocates pursuing space assurance, not space dominance. "A space assurance posture requires the adoption of defensive measures to lessen or compensate for satellite vulnerability as well as a hedging strategy against troubling initiatives undertaken by others."[73] In terms of hedging, states should continue to research and possibly develop more powerful offensive and defensive space weapons. However, he opposes deployment and destructive testing. The goal would be to show an enemy that a strong countermeasure is available should an attack occur. Krepon calls for a variety of ways to minimize or compensate for satellite vulnerability, including hardening to protect against radiation or electromagnetic pulsation and explosives. He also advocates arming satellites with defensive weapons and developing technology to make satellites harder to notice. Similarly, he wants beefed-up security at ground stations as well as redundant satellite systems.

A counterstrategy that incorporates legal obligations, deterrence, and a code of conduct should help peacefully manage the increasing use of space for civilian and military purposes. Taken together, this approach in some ways follows the Russian proverb that Reagan liked to quote during discussions with Soviet leaders: "Trust, but verify."

---

[1] Cited in Walter A. McDougall. *The Heavens and the Earth: A Political History of the Space Age.* New York: Basic Books, 1985, p. 149. Emphasis in original.

[2] Michael E. O'Hanlon. "Preserving U.S. Dominance While Slowing the Weaponization of Space." In *Space Weapons: Are They Needed?*, edited by John M. Logsdon and Gordon Adams. Washington, D.C.: Space Policy Institute, George Washington University, 2003, p. 242.

[3] Matthew Mowthorpe. *The Militarization and Weaponization of Space.* Lanham, Md.: Rowman and Littlefield, 2004, p. 3.

[4] Quoted in Paul D. Berg. "Effects-Based Airpower and Space Power." In *Air and Space Power Journal* (Spring 2006): 17.

[5] Michael E. O'Hanlon. *Neither Star Wars nor Sanctuary: Constraining the Military Use of Space.* Washington, D.C.: Brookings Institution, 2004, pp. 62–63.

[6] Christian Amodeo. "Eyes in the Skies: In the Past Ten Years, the Number of People Making Use of Satellite Technology Has Soared." In *Geographical* (February 2003).

[7] Theresa Hitchens. "Weapons in Space: Silver Bullet or Russian Roulette? The Policy Implications of U.S. Pursuit of Space-Based Weapons." In *Space Weapons: Are They Needed?*, edited by John M. Logsdon and Gordon Adams. Washington, D.C.: Space Policy Institute, George Washington University, 2003, p. 142.

[8] Amodeo, "Eyes in the Skies: In the Past Ten Years, the Number of People Making Use of Satellite Technology Has Soared."

[9] "NASA Satellite Imagery Keeping Eye on the Gulf Oil Spill." Available online. URL: http://www.nasa.gov/topics/earth/features/oil_spill_initial_feature.html. Accessed May 14, 2010. For more on *Terra*, see http://www.nasa.gov/mission_pages/terra/.

[10] Michael Weissenstein. "Drifting Satellite Threatens U.S. Cable Programming." *Washington Post* (12 May 2010). Intelset provided regular updates about *Galaxy 15* on its Web site. Available online. URL: http://www.intelsat.com/resources/galaxy-15/operational-status.asp. Accessed June 14, 2010.

[11] Tariq Malik. "What Happens When Satellites Fall." Space.com (January 23, 2009). Available online. URL: http://www.space.com/businesstechnology/090123-falling-satellites.html. Accessed June 1, 2010; Clara Moskowitz. "Satellite Joins Ranks of Space Zombies." MSNBC (May 4, 2010).

[12] "10 of the Most Memorable Pieces of Space Junk That Fell to Earth." Space.com. Available online. URL: www.space.com/missionlaunches/080225-top10-debris.html. Accessed December 19, 2009.

[13] Duncan Campbell. "U.S. Buys Up All Satellite War Images." *Guardian* (17 October 2001), Hitchens, "Weapons in Space: Silver Bullet or Russian Roulette? The Policy Implications of U.S. Pursuit of Space-Based Weapons"; John E. Hyten. "A Sea of Peace or a Theater of War? Dealing with the Inevitable Conflict in Space." In *Space Weapons: Are They Needed?*, edited by John M. Logsdon and Gordon Adams. Washington, D.C.: Space Policy Institute, George Washington University, 2003, p. 306.

[14] O'Hanlon. *Neither Star Wars nor Sanctuary*, p. 3.

[15] Adam Goldman and Matt Apuzzo. "CIA Tracks Al-Qaida Moving from Iran." AP. Available online. URL: http://abcnews.go.com/Politics/wirestory?id=10632757. Accessed June 29, 2010.

[16] Walter Pincus. "Airborne Intelligence Is Growing Component in Fight Against Insurgents." *Washington Post* (28 April 2009), p. A1.

[17] O'Hanlon, *Neither Star Wars nor Sanctuary*, p. 3.

[18] Sally B. Donnelly. "Long-Distance Warriors." *Time*, December 4, 2005.

[19] Bruce W. MacDonald. Testimony before the Strategic Forces Subcommittee of the Armed Services Committee, U.S. House of Representatives, March 18, 2009.

[20] Quoted in Hitchens, "Weapons in Space: Silver Bullet or Russian Roulette? The Policy Implications of U.S. Pursuit of Space-Based Weapons," p. 133.

[21] Joseph Kahn. "China Says Sect Is Broadcasting from Taiwan." *New York Times* (25 September 2002), p. A5.

[22] MacDonald. Testimony before the Strategic Forces Subcommittee of the Armed Services Committee, U.S. House of Representatives, March 18, 2009.

[23] "Co-Orbital ASAT." Federation of American Scientists Space Guide. Available online. URL: www.fas.org/spp/guide/Russia/military/asat/coorb.htm. Accessed May 4, 2010.

[24] William Spacy II. "Assessing the Military Utility of Space-Based Weapons." In *Space Weapons: Are They Needed?*, edited by John M. Logsdon and Gordon Adams. Washington, D.C.: Space Policy Institute, George Washington University, 2003, p. 166.

[25] USSTRATCOM. "Space Control and Space Surveillance Fact Sheet."

[26] Michael Krepon. "Bush's ASAT Test." Stimson Center Commentary (February 21, 2008).

[27] Tony Reichhardt. "Satellite Smashers: Space-Faring Nations: Clean up Low Earth Orbit or You're Grounded." In *Air and Space*, March 1, 2008.

[28] "Russian and U.S. Satellites Collide." BBC News (February 12, 2009). Available online. URL: http://news.bbc.co.uk/2/hi/science/nature/7885051.stm. Accessed June 1, 2010; William J. Broad. "Debris Spews into Space after Satellites Collide." *New York Times* (12 February 2009).

[29] Reichhardt, "Satellite Smashers: Space-Faring Nations: Clean up Low Earth Orbit or You're Grounded."

[30] William J. Broad. "Orbiting Junk, Once a Nuisance, Is Now a Threat." *New York Times* (6 February 2007).

[31] Broad, "Orbiting Junk, Once a Nuisance, Is Now a Threat."; Reichhardt, "Satellite Smashers: Space-Faring Nations: Clean up Low Earth Orbit or You're Grounded."

[32] Everett C. Dolman. "Space Power and U.S. Hegemony: Maintaining a Liberal World Order in the 21st Century." In *Space Weapons: Are They Needed?*, edited by John M. Logsdon and Gordon Adams. Washington, D.C.: Space Policy Institute, George Washington University, 2003, p. 93.

[33] David Wright and Lisbeth Gronlund. "Technical Flaws in the Obama Missile Defense Plan." In *Bulletin of the Atomic Scientists*, September 23, 2009. Available online. URL:http://www.thebulletin.org/web-edition/op-eds/technical-flaws-the-obama-missile-defense-plan. Accessed June 1, 2010.

[34] James E. Oberg. *World Book at NASA*. Available online. URL: http://www.nasa.gov/worldbook/space_exploration_worldbook_prt.htm. Accessed April 20, 2010.

[35] Karl P. Mueller. "Totem and Taboo: Depolarizing the Space Weaponization Debate." In *Space Weapons: Are They Needed?*, edited by John M. Logsdon and Gordon Adams. Washington, D.C.: Space Policy Institute, George Washington University, 2003, p. 28.

[36] Everett C. Dolman. *Astropolitik: Classical Geopolitics in the Space Age*. New York: Routledge, 2001, pp. 97–105.

[37] Everett C. Dolman. "Space Power and U.S. Hegemony: Maintaining a Liberal World Order in the 21st Century." In *Space Weapons: Are They Needed?*, edited by John M. Logsdon and Gordon Adams. Washington, D.C.: Space Policy Institute, George Washington University, 2003, pp. 69–78.

[38] Dolman, *Astropolitik: Classical Geopolitics in the Space Age*, chapter 5.

[39] Cited in Kevin Pollpeter. "The Chinese Vision of Space Military Operations." In *China's Revolution in Doctrinal Affairs: Emerging Trends in the Operational Art of the Chinese People's Liberation Army*, edited by James C. Mulvenon and David Finkelstein. Alexandria, Va.: CNA Corporation, 2005, p. 331.

[40] Peter N. Spotts. "Does Space Need Air Traffic Control?" In *Christian Science Monitor*, March 14, 2008.

[41] "Convention Relating to the Regulation of Aerial Navigation." Signed in Paris, October 13, 1919. Available online. URL: http://www.aviation.go.th/airtrans/airlaw/1914.html-1. Accessed April 20, 2010.

[42] Brazil, Colombia, Congo, Ecuador, Indonesia, Kenya, Uganda, and Zaire.

[43] Dolman, *Astropolitik: Classical Geopolitics in the Space Age,* p. 134.

[44] Edmund Andrews. "Tiny Tonga Seeks Satellite Empire in Space." *New York Times* (28 August 1990).

[45] Andrews, "Tiny Tonga Seeks Satellite Empire in Space."

[46] Dolman, *Astropolitik: Classical Geopolitics in the Space Age,* p. 119.

[47] The other five states are Belgium, Japan, Russia, South Africa, and the United States. These five neither recognize the claims made by the seven states nor have they made their own claims.

[48] Pavel Baev. "Russian Flag Stakes Energy Claim at North Pole." In *Eurasia Daily Monitor,* August 6, 2007.

[49] Edmund L. Andrews. "Tonga's Plan for Satellites Set back by Global Agency." *New York Times* (1 December 1990); "Tonga, Indonesians Agree to Share Pacific Ocean Orbital Position." *Satellite News* (8 November 1993), pp. 1–3; "Tonga Maneuvers to Save Slot." *Satellite News* (13 January 2003).

[50] Full texts are available at the UN Office for Outer Space Affairs, United Nations Treaties and Principles on Space Law. Available online. URL: http://www.oosa.unvienna.org/oosa/en/SpaceLaw/treaties.html. Accessed April 20, 2010.

[51] Dolman, *Astropolitik: Classical Geopolitics in the Space Age,* p. 125.

[52] UN Office for Outer Space Affairs, Resolution 1721. Available online. URL: http://www.oosa.unvienna.org/oosa/SpaceLaw/gares/html/gares_16_1721.html. Accessed April 20, 2010.

[53] Emphasis added.

[54] Agreement Governing the Activities of States on the Moon and Other Celestial Bodies (Moon Agreement). Available online. URL: http://nti.org/e_research/official_docs/inventory/pdfs/moon.pdf. Accessed April 20, 2010.

[55] Arms Control Association. Strategic Arms Limitations Talks (SALT I). Available online. URL: www.armscontrol.org/print/2487. Accessed April 20, 2010.

[56] Dolman, *Astropolitik: Classical Geopolitics in the Space Age,* p. 133.

[57] "ABM Treaty Fact Sheet."

[58] Michael Krepon. "After the ASAT Tests." Stimson Center Commentary (March 24, 2008).

[59] Technical information drawn from Spacy. "Assessing the Military Utility of Space-Based Weapons," p. 161.

[60] Technical information drawn from Spacy. "Assessing the Military Utility of Space-Based Weapons," p. 162.

[61] Technical information drawn from Spacy. "Assessing the Military Utility of Space-Based Weapons," p. 202.

[62] Dolman, "Space Power and U.S. Hegemony," p. 94.

[63] "Hubble Repair Mission Due for Launch Today." *Washington Post* (1 December 1993), p. A8.

[64] David Wright, Laura Grego, and Lisbeth Gronlund. *The Physics of Space Security: A Reference Manual.* Cambridge, Mass.: American Academy of Arts and Sciences, 2005. Available online. URL: http://www.amacad.org/publications/rulesSpace.aspx. Accessed May 5, 2010.

Introduction

[65] Michael Sheehan. *The International Politics of Space.* New York: Routledge, 2007, p. 2.

[66] O'Hanlon, "Preserving U.S. Dominance While Slowing the Weaponization of Space," p. 242.

[67] Dolman, "Space Power and U.S. Hegemony," p. 78.

[68] Spacy, "Assessing the Military Utility of Space-Based Weapons," p. 218.

[69] Mueller, "Totem and Taboo: Depolarizing the Space Weaponization Debate," pp. 11–19.

[70] MacDonald, Testimony before the Strategic Forces Subcommittee of the Armed Services Committee.

[71] Marc Kaufman. "Bush Says Shift Is Not a Step Toward Arms; Experts Say It Could Be." *Washington Post* (18 October 2006).

[72] Michael Krepon with Christopher Clary. "Space Assurance or Space Dominance? The Case Against Weaponizing Space." Stimson Center Report, 2003.

[73] Krepon with Clary, "Space Assurance or Space Dominance?" p. 63.

2

Focus on the United States

While the American public was shocked by the news of *Sputnik*, U.S. government officials were more angry than frightened. Not only were they aware of the Soviet satellite program, they had one of their own. But the USSR would go down in history as first in space.

The U.S. approach to space was closely tied to the cold war rivalry between the United States and the Soviet Union. But since the United States had military superiority over the USSR in the 1950s and 1960s, U.S. leaders chose to publicly emphasize the peaceful, scientific nature of space exploration, while military applications were kept from public view. Unlike the centralized, government-directed Soviet space program, capitalism drove the U.S. program, with a government agency coordinating contacts among government agencies, private companies, and educational institutions. Washington continued to observe the sanctuary approach to space until the 1980s, when the Reagan administration launched the Strategic Defense Initiative (SDI), a technologically ambitious program to create a national shield protecting against a Soviet attack. President George W. Bush would later take steps to actually weaponize space. After the collapse of the USSR and the Eastern bloc, Washington faced new threats from terrorists and rogue states such as Iran and North Korea and focused attention on ballistic missile defense projects. At the same time, more countries, international consortia, and private firms have joined the United States in conducting business in outer space.

EARLY ROCKET AND AVIATION EFFORTS

The early days of aviation and rocketry in the United States were dominated by scientists and inventors working on their own. In 1903, the Wright brothers built the first airplane, the *Flyer*, in their bicycle shop using their own funding, while Robert Goddard toiled away in the desert of Roswell, New Mexico, often using grants from the private Guggenheim Foundation. Many

Focus on the United States

Americans built their own backyard rockets to try out the new technology. The amateur American Interplanetary Society was established in the 1930s, and then changed its name to the American Rocket Society to sound more scientific. Although they had designed and built their plane with their own money, the fact that the Wright brothers were Americans meant that the United States entered this new era before any other countries. However, this lead did not last long, as European governments funded aviation and rocket research, while the U.S. government left such programs to the private sector.

Modern military rocket technology emerged in Germany between the two World Wars. Under the terms of the 1918 Treaty of Versailles, Germany was formally blamed for initiating World War I, and Berlin was ordered to pay $31.5 billion (in 1968 dollars) in damages. Germany was also banned from developing artillery with a range of more than 15 kilometers (9.3 miles), but the treaty did not make any reference to rocketry and missiles. German scientists seized this loophole and began to explore these new technologies.

Assigned to lead the new rocket effort, army captain Walter Robert Dornberger approached the private Society for Space Travel (VfR) club and offered them $400 (in 1932 dollars) to build one rocket.[1] At VfR, Dornberger first met Wernher von Braun, a charismatic engineering student who hoped to one day use his rockets for spaceflight.[2] Although von Braun's military rocket failed a critical test in 1932, Dornberger still respected him and hired him to work at the new military rocket facility in Kummersdorf. The initial program at Kummersdorf designed and built the Aggregate series of rocket. The first version, A1, was a failure, but the second series (two A2 rockets dubbed "Max" and "Moritz") reached an altitude of almost 6,500 feet in December 1932. The research team eventually outgrew Kummersdorf.

The team moved to a new facility, Peenemünde, built on an island in the Baltic Sea. Peenemünde was safely distant from populated areas and had enough space to monitor tests as far as 200 miles away. Over the next three years Peenemünde evolved into two separate facilities, one developing the Luftwaffe's Fieseler Fi 103 bomb best known as the V-1 Buzz Bomb, the other continuing the Aggregate program. Von Braun's team churned out increasingly powerful incarnations, and by October 1938 the A5 successfully reached an altitude of eight miles.

Germany's invasion of Austria in early 1938 changed the military's artillery needs. Instead of surface-to-air missiles, the army needed ground-to-ground rockets. The researchers at Peenemünde adapted, modifying the A4 into a new V series. Formally known as the Vergeltungswaffe ("revenge weapon"), the V-2 was a liquid-fueled long-range ballistic missile 45 feet long. It carried a one-ton warhead and could fly at an altitude of 50 miles.[3] The V-2's first successful launch came on October 3, 1942, when it rose 60 miles and reached its programmed target. According to historian Paul Dickson,

> It was, at once, the world's first launch of a ballistic missile, the first rocket ever to go into the fringes of space, and the ancestor of practically every rocket flown in the world today. It was a true spaceship in that it carried both its own fuel and oxygen and could, if needed, work in a vacuum. But it was also a formidable weapon capable of carrying a warhead of 2,200 pounds; the silent V-2 could reach a velocity of 3,500 miles per hour, making it almost impossible to shoot down.[4]

Adolf Hitler was fascinated by a Dornberger and von Braun presentation on rocketry and immediately ordered 25,000 V-1s and 12,000 V-2s.[5] "This is the decisive weapon of the war, and what is more it can be produced with relatively small resources," Hitler wrote Minister of Armaments Albert Speer. "Speer, you must push the A-4 as hard as you can! Whatever labor and materials they need must be supplied instantly."[6]

Smaller, noisier, and less accurate than the V-2, the V-1 still terrorized England and western Europe. The V-2 traveled at supersonic speed; impact occurred *before* the accompanying sonic boom was audible. With no warning, no countermeasures could be taken to destroy the incoming missile. Between September 8, 1944, and March 30, 1945, 1,115 V-2s struck England, and another 1,524 hit targets on the European continent. Aim and accuracy were still problematic—while 1,115 V-2s made it to England, more than 5,000 were launched. Both the V-1 and V-2 were shy of their targets by an average of nine miles.[7] After the Royal Air Force bombed Peenemünde on August 17, 1943, V-2 production moved to an underground facility near Nordhausen. The new location, named Mittelwerk, was completely under military control.[8]

Allied intelligence was aware of the Nazi V-2 program, and Washington was growing increasingly alarmed that Nazi rockets might soon be able to reach U.S. territory. The Americans launched Project Aphrodite, a series of missions in which bare-bones B-17s loaded with 22,000 pounds of explosives were used as actual warheads. Crews were expected to guide the plane near its intended target—V-2 production sites—set the plane on autopilot, then parachute to safety. An accompanying mother ship would remotely guide the final moments. The navy had a similar program, Project Anvil. Both programs had a very high casualty rate. The first Anvil mission killed Joseph Kennedy, Jr., the older brother of future president John F. Kennedy, and nearly killed President Franklin Delano Roosevelt's son Elliott in a chase plane.[9]

As the tide turned against Germany in early 1945 and surrender seemed inevitable, the United States and the USSR began to search for the rocket labs, factories, and technical specifications. Each side coveted the information, but they especially wanted to prevent the other side from finding the prize. The scientists themselves knew they were being hunted, and they decided they would rather be captured by the Americans than the Russians, the British, or the French. The Americans, they thought, "needed them, could afford them, and would rather let them build rockets than put them on trial."[10] Many feared they would be executed by the German SS, in a last-ditch effort to prevent the Allied countries from capturing German technology. In the end, the German scientists wound up surrendering to U.S. troops on May 2, 1945.[11] Granted asylum by the United States, von Braun's team settled first in Texas, then in New Mexico, and ultimately was transferred to the Redstone Arsenal in Huntsville, Alabama, in 1950.

Von Braun and his team immediately directed U.S. agents to Nordhausen, where they had secretly stashed materials spirited out of Mittelwerk. There the agents discovered enough bomb and rocket parts and completed V-2s to fill 360 railroad cars, as well as blueprints and other technical materials. Soviet forces arrived at Peenemünde three days later, where they found manufacturing equipment and captured the remaining production crew. Meanwhile, Washington unveiled its own leap in weapons technology, dropping atomic bombs on first Hiroshima, then Nagasaki, Japan, causing enormous casualties. On August 15, Japan surrendered, and World War II was over.

The war had produced two new weapons: rocket-launched missiles and the atomic bomb. Together, they created a dramatically more powerful arsenal that could reach targets thousands of miles away.

Nuclear Weapons Are Cost Effective under Truman

President Harry S. Truman backed the atomic weapons program begun under President Franklin D. Roosevelt because he thought it was the most expedient way to end the war, save American lives, and deter future conflicts. The countries of western Europe, still trying to recover and rebuild after World War II, gladly sought protection under the U.S. nuclear umbrella. Both Truman and his successor, President Dwight D. Eisenhower, based their defense policy on the concept of massive retaliation—any attack against the United States or its allies could trigger a nuclear response. Washington also deployed fleets of B-29 bombers at military bases in western Europe, from where they could easily hit targets in the USSR. Moscow, however, did not have long-range aircraft that could hit targets in the United States nor allies near the continental United States.

Still, Washington anticipated that Moscow would try to catch up and searched for ways to maintain its superiority. Truman had embraced deterrence, a national security strategy that seeks to discourage first-strike attacks by amassing a huge stockpile that would annihilate any state that did take offensive action. Given the limited quantity of aging B-29s, the United States needed a new delivery system, which meant either building more bomber jets, pursuing long-distance ballistic missiles, or both. Initially, Washington calculated that 400 nuclear warheads would be sufficient to balance the Soviet threat and ordered production of a new long-range bomber, the B-36.

Truman was not enthusiastic about big-budget government programs and wanted a cost-effective security strategy. Even in peacetime, the military consumed an outsized portion of the budget. "The $37 billion budget for FY [fiscal year] 1947 was down 60 percent from the wartime peak, but still four times the size of prewar budgets."[12] Despite the cost, Washington knew the United States would have to develop longer-range missiles to maintain its lead over Moscow, and rockets would soon have enough power to make a foray into space possible. In October 1945, the navy formed a committee to explore rocketry, and the army air corps hired Douglas Aircraft for a similar task. In May 1946, RAND—then Douglas Aircraft's research division—announced that the United States could have rockets ready to launch satellites by 1951, and the satellite would cost about $150 million. But when the secretary of defense announced the results of the reports in late 1948, public outcry against such profligate spending led to cancellation of the project.

After the Soviet Union successfully tested an atomic bomb in August 1949, Truman ordered production of an even more powerful weapon. The atomic bombs dropped on Japan were based on nuclear fission—splitting an atom; the next generation H-bomb (or thermonuclear bomb) would release enormous amounts of energy through the fusion of atomic nuclei. The heat required to combine atomic nuclei is generated from a fission explosion. Successfully tested in 1952, the bomb was more than 800 times more powerful than the bomb that destroyed Hiroshima.[13] Furthermore, these weapons could be much smaller than atomic bombs and could use unmanned delivery systems—rockets instead of pilots.

Several branches of the military began to experiment with the V-2 technology, albeit with limited results. Washington hired a private contractor, Corvair, to develop a 5,000-mile-range intercontinental ballistic missile, known as Project MX-774. The navy launched a V-2 from the USS *Midway* aircraft carrier, while the army launched 64 V-2 rockets and tested a two-stage V-2 rocket, dubbed "Bumper." But this research was largely suspended when Vannevar Bush, a scientist heavily involved in the Manhattan Project, told Congress that nuclear bombs atop rockets would not work. "In my opinion, such a thing is impossible, I don't think anybody in the world knows how to do such a thing . . . and I feel confident it will not be done for a very long time to come."[14]

Von Braun did not agree with Dr. Bush's pessimistic assessment. After exhausting the inventory of V-2s captured from Germany and stored at the army research center at the White Sands Proving Ground in New Mexico, von Braun's team began to work on the first homegrown U.S. ballistic missile rocket, the Hermes Project, the U.S. version of the V-2. In 1950, his team was relocated to the Redstone Arsenal in Huntsville, Alabama, and their project became the Redstone rocket, which had a range of 500 miles. In von Braun's 1954 report, "A Minimum Satellite Vehicle," he argued that U.S. national pride rested on being the first to launch a satellite. He believed he could accomplish the task with existing hardware (attaching a navy satellite atop an army Redstone rocket) with just an additional $100,000 (in 1954 dollars) in funding. Designated "Orbiter," both the project and his request were denied.[15]

Even when the U.S. government began funding aviation and rocketry research, most of it was done by universities and private contractors. The California Institute of Technology's Jet Propulsion Laboratory (JPL), the University of Chicago, Harvard, and the Massachusetts Institute of Technology were among the first schools to embrace such research projects. On the commercial side, contractors included Northrup, Martin-Marietta, and Curtiss-Wright. While the army and navy created their own research units, eventually they began using private contractors as well.[16]

Rockets and Satellites under Eisenhower

In late 1953, the Pentagon created a Strategic Missiles Evaluation Committee (SMEC) to research rocket delivery options. SMEC reported that the Soviets were rapidly developing their intermediate-range ballistic missiles (IRBMs) and that if the United States did not launch its own rocket program soon, it would quickly fall behind in firepower. Analysts believed Moscow would soon have a larger inventory of missiles than the United States, leading to an impending missile gap. Eisenhower initially ignored the recommendations to begin such a costly research and development program, but by September 1955 he conceded the need for a U.S. rocket program. He began by green-lighting four projects: the Atlas rocket, the Titan intercontinental ballistic missile (ICBM), the Thor IRBM, and the Jupiter IRBM. Jupiter came from the Army Ballistic Missile Agency (ABMA), the agency formed to develop the army's IRBM, and was an adaptation of the German V-2 rocket and U.S. Redstone rocket.

The Eisenhower administration had a three-pronged approach to space. First, in line with the postwar budget constraints, Washington saw satellite reconnaissance as a relatively cheap way to open up the otherwise tightly closed Soviet Union. Thus, in July 1955, at the Geneva summit meeting with

France, Great Britain, and the Soviet Union, President Eisenhower proposed an international open skies policy, which would allow aerial surveillance of other countries' military facilities, as a confidence-building measure. Second, Eisenhower wanted to create an international legal regime that allowed "peaceful" reconnaissance efforts, such as for verification of arms control agreements. Third, in line with the sanctuary approach to space, he wanted to explore and hopefully exploit space for scientific purposes. All of these goals also brought the need for increased rocket research and development, because the United States would need enormously powerful rockets to launch satellites or to power intercontinental rockets.[17]

Initially, Washington was unsure about which branch of the military should be responsible for air-based activities, and the potential expansion into space-based activities brought by rocketry only confused the situation more. The newly created U.S. Air Force took the lead. Air Force general Thomas D. White argued that combat in space would prove to be as effective as had the move to air-based combat, adding that the air force should take the lead as space and air are one undifferentiated zone. However, military involvement would clash with Eisenhower's preferred emphasis on scientific exploration.

In addition to Eisenhower's open skies proposal, the United States decided to participate in the International Geophysical Year satellite program. Not only would launching a satellite be a scientific triumph, it would also establish a legal precedent for future overflights. Washington now needed a satellite and a rocket to launch it into orbit. The ongoing rivalry between the army and the navy offered several options. The army had von Braun's Orbiter satellite and Redstone rocket, while the navy had its Project Vanguard. Even though the Vanguard was still at a very early stage of development, it was chosen because the navy included a slate of scientific testing that could be carried out during the mission and the actual manufacturing would be done by a civilian contractor. The Redstone, a military rocket designed by former Nazis, hardly satisfied the image of peaceful space science that the Eisenhower administration wanted to present to the world.[18]

Although his program technically had been cancelled, von Braun continued his work on Orbiter. He also began a new project, the liquid-fuel Jupiter rocket, a joint effort between the ABMA and the U.S. Navy Special Projects Office. The navy dropped out of Jupiter in 1956, in favor of Polaris, an American submarine-launched ballistic missile (SLBM), and Atlas, the first effort by the United States to develop a land-based ICBM. The air force offered the liquid-fueled Thor IRBM and the solid-fueled Minuteman I—a rocket that could be launched within 60 seconds. Thus, the U.S. rocket program was already well advanced when news came of *Sputnik*. Assistant Secretary of Defense Donald Quarles, reporting to Eisenhower on October 8, 1957, admitted, "There was no doubt that the Redstone, had it been used, could have orbited a satellite a year or more ago."[19]

With the October 1957 launch of *Sputnik*, the Soviet Union achieved the ultimate high ground. *Sputnik* opened the way for placing weapons and other military assets (satellites, antisatellite weapons [ASAT]) in long-term orbit. *Sputnik* also carried the principle of innocent passage into outer space, "The launch of *Sputnik-1* and the lack of international objections to its overflight effectively wrote into international law the right of satellite overflight over national territories."[20]

When President Eisenhower addressed the nation about the Soviet satellite, he calmly reassured the people that the United States was no less safe than it had been before *Sputnik*. Eisenhower and his staff tried to appear unconcerned about *Sputnik*, the president even waited five days, until October 9, to hold a press conference on the subject. He downplayed the significance, brushing off suggestions of a space race or even a link between the satellite and an ICBM. *Sputnik*, he said, had proved that "they can hurl an object a considerable distance." When asked about the security implications of *Sputnik*, Eisenhower answered, "As far as the satellite itself is concerned, that does not raise my apprehensions, not one iota. I see nothing at this moment, at this stage of development, that is significant in that development as far as security is concerned."[21]

In Congress, Senator Lyndon B. Johnson, a Democrat from Texas, seized on *Sputnik* as an opportunity to blast the Republican administration—while gaining national exposure ahead of his own possible run for the presidency in 1960. On November 25, Johnson opened a formal inquiry into satellite and missile programs to investigate why the United States had allowed such a threat to national security to develop. He summoned a variety of officials to testify and made sure there was plenty of media coverage.

The White House hoped to recover some of the lost public faith in U.S. science by inviting the media to broadcast a December 4 test of the Vanguard TV-3 project, the first U.S. attempt to launch a satellite into orbit around the Earth. Anticipation built as the media sat through two days of delays. Then finally, on December 6, the rocket launched; it rose about four feet into the air and then exploded, on live television. The press had a field day, coming up with clever names for the rocket, such as Kaputnik, Stayputnik, and Flopnik.[22]

Following *Sputnik* and the Vanguard fiasco, the White House began to work toward creating a single, unified agency for space issues. The National

Advisory Committee for Aeronautics, which dated to 1915, had been increasingly subordinated to military-funded research, so now there was a push to create a new civilian agency responsible for space research. On October 1, 1958, the National Aeronautics and Space Administration (NASA) began work to coordinate the efforts of multiple agencies, private companies, and educational institutions.

Also in 1958 Eisenhower established the panel that would issue the Purcell Report. It contained recommendations for a space doctrine that would emphasize the sanctuary view of outer space. No country should place weapons in space, but the military could passively use space-based assets such as reconnaissance, communication, and weather satellites.[23] The Department of Defense also began research on antiballistic missile (ABM) systems, assigning the task to the army rather than the air force. The U.S. Army's ABM program modified the existing Project Nike air-defense system, creating a nuclear-tipped, ground-launched Nike-Zeus missile, but it quickly became obsolete as the USSR upgraded its arsenal.

Eisenhower launched at least two reconnaissance programs to observe activities inside the Soviet Union. First, U-2 spy planes began overflights in 1955, but their utility was limited by the range of the plane, the pilot, and Soviet interceptors. Still, the U-2 program provided 90 percent of U.S. intelligence about the USSR at the time.[24] Second, Washington secretly began CORONA, a U.S. spy satellite program that lasted from 1959 to 1972. Designed to replace the U-2 spy planes, CORONA was a joint project of the U.S. Air Force and the Central Intelligence Agency (CIA). The program used satellites to photograph areas of interest and would eject capsules of film to be caught and/or retrieved by air force aircraft. The project was highly classified. The CIA director Allen Dulles insisted that all discussion of the program be oral, ensuring that no paper trail existed. CORONA (later called "Discoverer") was not acknowledged by the government until February 1995, when it was declassified.[25] The superiority of satellite reconnaissance became even more obvious to the United States in May 1960, when the Soviets shot down a U-2 spy plane piloted by Francis Gary Powers. Unlike pilots, satellites could not be arrested and imprisoned.

As Eisenhower neared the end of his second term, Senator John F. Kennedy accused the president of allowing a "missile gap" to develop between the United States and the USSR. Only a few years earlier, there had been alarm that the Soviets had more bombers than the United States; now they appeared ahead in missiles. The Soviet leader, Nikita Khrushchev's boast that his factories were churning out missiles "like sausages," hardly eased the tension. Kennedy used the allegation as cause for criticizing security policy and calling for stepped-up missile production. However, in 1961 intelligence sources concluded that the Soviets only had about 90 ICBMs and 200 bombers and none of the bombers could reach the United States. Years later, the CIA revised its estimate, saying that in 1960 the USSR only had three ICBMs.

The Eisenhower administration launched a variety of other defense programs in the late 1950s. The air force had begun a reconnaissance satellite program, Project Feedback, using a satellite (Pied Piper) carrying a television camera. But in order to legitimate the reconnaissance program, the administration modified its earlier contention that peaceful use of outer space meant "nonmilitary"; now "nonaggressive" became the standard. The army launched its first communications satellite (Courier 1B) on October 4, 1960, while the navy focused on satellites for navigation. The army also produced the first U.S. ABM weapon, the nuclear-tipped Nike-Zeus, which was test-launched on December 16, 1959. The air force began to develop the country's first antisatellite program in 1956, codenamed SAINT. Finally, Washington also ordered the creation of a centralized system for collecting data on domestic and foreign satellites. In time, the Project Shepherd spacetrack network would become the North American Aerospace Defense Command (NORAD).

To the Moon under Kennedy

The new Kennedy administration continued to investigate ways to defend against a potential Soviet missile attack. Incoming secretary of defense Robert McNamara reviewed ongoing ballistic missile defense (BMD) programs but was concerned about the $16 billion price tag on the Nike-Zeus system, which would not even protect the entire country. Consequently, he authorized continued ABM research but not deployment. The Zeus missile was tested 14 times between July 1962 and November 1963; of those tests, nine were rated "complete successes" and another three "partial successes." In one test, on December 22, 1962, a Zeus missile came within 200 meters of its target. Due to the limited capacity of Zeus, McNamara later authorized a multilayered program known as Nike-X. The system used two nuclear-tipped missiles, the exoatmospheric Spartan and the endoatmospheric Sprint. Nike-X missile installations were deployed at about 30 sites across the United States to protect from both Soviet and Chinese attacks. The first Sprint missile was tested in 1965 and the first Spartan in 1968.

Kennedy had barely been in office for four months when cosmonaut Yuri Gagarin became the first man to orbit Earth on April 12, 1961. Again, the Soviets had beaten the United States into space, but the Kennedy White House was determined not to be second again. Vice President Lyndon B. Johnson declared, "Failure to master space means being second best in every aspect, in the crucial arena of our Cold War world. In the eyes of the world first in space means first, period; second in space is second in everything."[26] Kennedy concurred, telling NASA administrator James E. Webb, "I'm not that interested in space . . . Everything we do ought to be tied into getting to the moon ahead of the Russians."[27]

Speaking before Congress on May 25, Kennedy issued his challenge to the people of the United States.

> This nation should commit itself to achieving the goal, before the decade is out, of landing a man on the moon and returning him safely to Earth. No single space project will be more important to mankind or more important for the long-range exploration of space; and none will be so difficult or expensive to accomplish.

"I could hardly believe my ears," Robert Gilruth, future director of the Manned Spacecraft Center in Houston, remembered.[28] Kennedy's call for action led to the greatest peacetime mobilization in U.S. history. "The Apollo project cost more than $25 billion. In 1966, spending on the space program reached its height at nearly $6 billion—more than 4 percent of the entire federal budget and more than Washington spent that year on housing and community development combined. At its peak, more than 400,000 people in government and industry labored to push the project forward."[29]

Even with this tremendous new push, Kennedy followed Eisenhower's lead and put space science before space weapons. When the Soviets began to boast of putting nuclear weapons into orbit, however, the White House and Pentagon took another look at ASAT capabilities. In 1962, Secretary of Defense McNamara ordered the army to modify the Nike-Zeus ABM as a possible antisatellite weapon, known as Project 505 or Mudflap. The next year Kennedy approved the air force Program 437, a ground-launched ASAT using the Thor IRBM.[30] Kennedy also went before the United Nations to push for the negotiations that eventually produced the 1967 Outer Space Treaty (see chapter 1), which bans the deployment of military facilities or weapons of mass destruction in space.

Further downplaying the military side of space research, in November 1961 the White House issued an order excluding the press from attending military launches and banning the release of technical information such as orbital characteristics. "No government officials would even admit that many of the programs existed," according to air force publications. "The military programs sank into obscurity, known only to a select few, while NASA's up-and-coming manned program seized and held the spotlight for the next decade."[31]

Focus on the United States

Tasked with Kennedy's lunar mission, NASA began a manned spaceflight program in three phases. Phase one, Mercury, would put manned spacecraft into Earth's orbit. Phase two, Gemini, would concentrate on the maneuverability of the spacecraft. Phase three, Apollo, would take men to the moon and back. Mercury's first mission launched on May 5, 1961, and Gemini on April 8, 1964. Three Apollo astronauts were killed in a launchpad fire on January 27, 1967; the mission was later designated as *Apollo 1.*

In 1962, the air force offered its own program to research man's ability to work in space, the manned orbital development system (MODS). MODS consisted of a space station with a four-person crew, attached to a Gemini or Apollo capsule. Six missions were planned under the code name "Blue Gemini," as air force officers wore blue uniforms. The program would be able to carry scientific equipment into space, such as ground-mapping radar, but air force personnel also wanted to focus on rendezvous and inspection techniques. "But the ultimate goal might be to reach a 65-degree orbit, which the Blue Gemini planning document noted was the orbit used by Soviet Vostok vehicles." Faced with competing visions for Gemini, McNamara opted for NASA's program, but with the air force allowed to include some of its projects on the mission schedule. At the same time, he canceled the air force's X-20 space plane, which was running over budget and behind schedule. Instead he authorized the manned orbiting laboratory program, an updated version of MODS.[32]

Kennedy did not live to see the United States become the first—and so far only—country to put a man on the Moon in July 1969. Even so, Kennedy's vision had a tremendous impact on U.S. society. Technology developed for various stages of the U.S. space program has found widespread civilian and commercial applications, including Teflon, Tang orange drink mix, Velcro, cell phones, and automated teller machines (ATMs). The material developed for spacesuits has been used for stadium roofs and even the roof at the Denver International Airport, while the Dustbuster hand vacuum is a descendant of the cordless drill that took lunar soil samples.[33] Space technology also developed sensors that are now used in global positioning systems (GPS), washing machines, and other products. According to the White House, "More than 1,300 NASA and other U.S. space technologies have contributed to U.S. industry, improving our quality of life and helping save lives." The lengthy list includes

- image processing, used in computerized axial tomography (CAT) scanners and magnetic resonance imaging (MRI) technology in hospitals worldwide, which came from technology developed to computer-enhance pictures of the Moon for the Apollo programs

- kidney dialysis machines, which came from a NASA-developed chemical process, and insulin pumps, which were based on technology used on the Mars *Viking* spacecraft
- programmable heart pacemakers, first developed in the 1970s using NASA satellite electrical systems
- fetal heart monitors, developed from technology originally used to measure airflow over aircraft wings
- surgical probes used to treat brain tumors in children, which resulted from special lighting technology developed for plant growth experiments on space shuttle missions
- infrared handheld cameras used to observe blazing plumes from the shuttle, which have helped firefighters point out hot spots in brush fires[34]

Don't Invite a Soviet Attack, Says Johnson

Kennedy's call to put a man on the Moon was accomplished in July 1969. But after five subsequent lunar landings, the American people and government began to lose interest in the space program. Kennedy's goal had been met, and it seemed time to spend taxpayer money to solve Earth-based issues, such as poverty, civil unrest, and the ongoing war in Vietnam. While President Johnson's time in the White House was dominated by these terrestrial issues, he nevertheless continued U.S. ASAT research as a hedge against the Soviet orbital weapons system supposedly in development. However, he insisted that ASAT programs remain defensive, fearing that an offensive attack against Soviet satellites would escalate into all-out war.

The Soviets had already begun installing an ABM system around Moscow in 1966, and the White House was considering similar protection for Washington, D.C. After China successfully tested a nuclear missile the same year, Washington announced it would begin deploying a "thin" ABM Sentinel system that only protected major cities, but emphasized that it was a response to the threat from China, not the USSR.[35] Johnson met with Soviet premier Alexei Kosygin at the Glassboro Summit Conference in New Jersey from June 23 to June 25, 1967, and tried to convince his counterpart to abandon the Soviet missile defense program because it would inevitably result in a U.S. buildup of nuclear weapons. Kosygin refused, declaring, "Defense is moral; offense is immoral!"

Nuclear Limits and Reusable Spacecraft under Nixon

Elected in 1968, President Richard Nixon did not want to despoil outer space by deploying weapons there, but he did see the value of using satellite surveillance as a relatively cheap confidence-building measure. Prudent spending was a dominant theme in Nixon's space policy.

Nixon established his own advisory space task group on February 13, 1969. The group issued a report that September, recommending a reusable shuttle instead of the current one-use spacecraft and a manned mission to Mars no later than 1981. The overwhelming theme of the report was reconciling costs and benefits, with a belief that the space program could make tremendous scientific discoveries with a much smaller budget. "Much of the negative reaction to manned spaceflight, therefore, will diminish if costs for placing and maintaining man in space are reduced and opportunities for challenging new missions with greater emphasis upon science return are provided." The key would be a new, reusable space shuttle. "The new system will differ radically from all existing booster systems," according to the report, "in that most of this new system will be recovered and used again and again—up to 100 times. The resulting economies may bring operating costs down as low as one-tenth of those present launch vehicles."[36] As for military space programs, the "Department of Defense would only be permitted to embark on new space programs when they could show it to be more cost effective to carry out the task in space."[37]

Early in the Nixon presidency, intelligence revealed an ongoing Soviet effort to develop ASAT weapons. Specifically, Moscow seemed to be developing a co-orbital ASAT. Such weapons enter the same orbit as their target, waiting months or even years before detonating a weapon that would destroy or disable the satellite on impact. The Pentagon concluded that a "U.S. counter would not deter Soviet use of an ASAT because of greater U.S. dependency on its space assets. In such a contest, the U.S. would be hurt by an ASAT more than the Soviets would be."[38] Furthermore, the United States would likely suffer consequences were it to launch an attack using weapons from Program 437. Data from the 1962 Starfish Prime high-altitude nuclear test revealed that the explosion "released a sizable amount of radiation into space. This radiation, trapped by the Earth's magnetic field, created artificial radiation belts 100 to 1,000 times stronger than background levels and damaged a number of satellites."[39] This discovery, plus a dwindling inventory of the necessary Thor rockets, led the Pentagon to put Program 437 on a standby basis; now, with a 30-day period required to activate the weapons for an interception, the program was effectively dead.

Nixon's most important contributions to U.S. defense policy were negotiating limits on ground-based nuclear weapons and preserving the relative balance of power by renouncing antiballistic missile defense systems. On May 26, 1972, Nixon and Soviet general secretary Leonid Brezhnev met in Moscow to sign the landmark Interim Agreement on the Limitation of

Strategic Offensive Arms (known as SALT I) and the Antiballistic Missile (ABM) Treaty. The Interim Agreement capped the number of strategic ballistic missile launchers that either side could include in their stockpile. According to Article V, "For the purpose of providing assurance of compliance with the provisions of this Interim Agreement, each Party shall use national technical means of verification at its disposal in a manner consistent with generally recognized principles of international law." The phrase "national technical means" refers to the use of reconnaissance satellites, a key space-based military asset. The ABM Treaty limited the number of antimissile defensive installations in each country to two: one to protect the capital city and one to protect a site of the country's choice. Each site could have up to 100 interceptors.

The logic behind the ABM Treaty is complicated and connected to the concept of mutual assured destruction (MAD). MAD relies on second-strike capabilities to deter a first-strike nuclear attack. The logic is that if, for example, Moscow knew Washington could launch a deadly retaliatory attack while Soviet missiles were still in flight, Moscow might not launch at all knowing that it would be destroyed as well. "The logic was flawless, but not comforting. It meant that if a missile were launched, accidentally or on purpose, it would hit and destroy its target," Everett Dolman wrote in 2003. "There was (and still is) no means to protect the citizens of this or any other country (except possibly the city of Moscow) from ballistic devastation."[40]

In addition to a system to protect Washington, D.C., Nixon authorized repurposing Sentinel as the Safeguard Project to be used to protect the Mikkelsen ICBM complex at Grand Forks Air Force Base, North Dakota. But after the ABM Treaty was signed, Washington realized that in a democratic society any plan to protect the capital with an enormously expensive system, while leaving the rest of the population vulnerable to nuclear annihilation, was a political nonstarter.[41] Plans for an ABM system around Washington, D.C., were quickly shelved. Two years later, in 1974, the ABM Treaty was amended to allow only one site per signatory.

The Grand Forks ABM site was operational between October 1, 1975, and October 2, 1975. Congress abruptly canceled the project due to its cost—a projected $40 billion, concerns that future treaties would make it unusable, and changes in Soviet weaponry. The USSR had begun using missiles with multiple independently targetable reentry vehicle (MIRV) warheads, making it even easier to overwhelm any U.S. ABM system.

Nixon made a dramatic shift in defense and space policy by authorizing the development of space shuttles to replace rockets as the country's primary spacecraft. NASA developed plans to create a permanent space station and a reusable space shuttle to ferry crews and equipment between the space station and Earth. "NASA's original justification for the space shuttle was that it was an all-purpose, largely reusable vehicle that could launch cargo from commercial satellites to space stations more cheaply than expendable boosters."[42] Congress balked at the price tag and only approved the shuttle program.

NASA pursued two major space projects under Nixon. First, *Skylab* launched from the Kennedy Space Center on May 14, 1973. A descendant of the air force manned orbital laboratory (MOL) program of the 1960s, *Skylab* would be the first manned U.S. space station, but it had a rocky start. One of its protective outer shields and one solar panel came off during the launch, the other solar panel refused to deploy. Astronauts subsequently repaired the damage and made three visits to *Skylab* in 1973 before abandoning the station. *Skylab* eventually fell out of orbit in 1979, largely burning up during reentry. The second Nixon-era project was the 1975 Apollo-Soyuz program. U.S. astronauts and Soviet cosmonauts cooperated to develop a common docking system for manned spacecraft, to facilitate emergency rescues. In July 1975, the U.S. *Apollo 18* and Soviet *Soyuz 9* successfully docked in low earth orbit (LEO), and the two crews spent three days conducting joint scientific and medical tests.

Satellite Vulnerability Recognized under Ford

When Nixon resigned on August 8, 1974, Moscow was concerned that the arms control efforts begun by Nixon and Brezhnev might falter. Four months later, President Gerald Ford met with Brezhnev in Vladivostok, where they agreed on a list of instructions for the next round of Strategic Arms Limitations Talks (SALT II) in Geneva.

In 1975, Ford convened two expert panels to evaluate the U.S. military's use of satellites. Both panels ultimately released reports that noted how dependent the U.S. military had become on satellites, and they warned that an ASAT program would not enhance survivability of U.S. satellites nor would deterrence, given the high level of dependence. On top of this gloomy conclusion, intelligence indicated that the USSR had resumed testing of its co-orbital ASAT in 1976. Thus, three days before his term ended in January 1977, Ford issued National Security Memorandum 345, instructing the Pentagon to develop an operational nonnuclear ASAT. This project reached the testing phase in 1985.

Air-Launched ASAT Weapons under Carter

President Jimmy Carter pursued a two-track national security policy. First, he continued his predecessors' dialogue with Brezhnev. At the same time, Carter ordered the Department of Defense to continue research on a U.S. ASAT system as a potential bargaining chip. He hoped Soviet awareness of the U.S. program would push Moscow toward an agreement. Certainly the United States was aware of the Soviet ASAT program.

On October 4, 1977—exactly 20 years after Sputnik—U.S. secretary of defense Harold Brown held a press conference to announce that the USSR now had an operational antisatellite weapon. "I find that troublesome," he commented. Press accounts at the time described the ASAT as a "hunter-killer" satellite that would "fly close to its quarry and then detonate a non-nuclear explosive to destroy its target."[43] As a result, Brown instituted a Space Defense Program that would focus on developing ASAT technology, hardened satellites, and better space surveillance. The program also continued work on the miniature homing vehicle, a kinetic ASAT launched from a booster rocket or airplane.

The United States awarded a $58.7 million contract for its ASAT, described by the *Washington Post* as a "flying tomato can." *Time* magazine was even more descriptive: "The U.S. plan is to leap the relatively crude Soviet ASAT technology and put into space by the mid-1980s hunter-killer satellites armed with lasers that could vaporize metal in 20 billionths of a second."[44]

Carter issued the first U.S. national space policy on May 11, 1978. Presidential Decision on National Space Policy 37 affirmed that outer space should be used for peaceful purposes and must have the right of free passage. But it also clarified that "peaceful purposes" included allowing for "military and intelligence-related activities in pursuit of national security and other goals." The document also announced that "the United States will pursue activities in space in support of its right of self-defense," acknowledging for the first time that outer space could become a battlefield theater.[45]

Finally, on June 18, 1979, Carter and Soviet general secretary Brezhnev signed the SALT II treaty. In that follow-up to the 1972 Interim Agreement, the two sides agreed to set limits on the number of bombers, missiles, and nuclear weapons each could carry. Each side was permitted to develop one new land-based ICBM. Like all treaties, SALT II needed Senate ratification before becoming law. But when the USSR invaded Afghanistan in late December 1979, Carter asked the Senate to delay consideration of the matter. As a result, the United States never ratified the treaty, but both sides agreed to abide by its provisions.

Strategic Defense Initiative under Reagan

Ronald Reagan ran for the presidency in 1980 on a platform that emphasized Soviet threats to national security. According to him, former secretary of defense McNamara had left the country unable to defend itself against incoming Soviet missiles by rejecting an ABM program. One of his military advisers, Lieutenant General Daniel O. Graham, recommended a space-based defense system to intercept Soviet missiles.[46]

Space would play a large role in Reagan's presidency. Only a few months into his first term, the first U.S. space shuttle lifted off from the Kennedy Space Center on April 12, 1981. Reagan subsequently instructed the National Security Council to review U.S. space policy in August 1981, which led to the formulation of National Security Decision Directive 42, released on July 4, 1982. According to the new policy:

> The Space Shuttle is to be a major factor in the future evolution of United States space programs. It will continue to foster cooperation between the national security and civil efforts to ensure efficient and effective use of national resources. Specifically, routine use of the manned Space Shuttle will provide the opportunity to understand better and evaluate the role of man in space, to increase the utility of space programs, and to expand knowledge of the space environment.[47]

The document also called for additional research toward establishing permanent space facilities. But the principal shift from the Carter space doctrine was the declared intention to use ASAT and ABM technology as a deterrent. "The United States will pursue activities in space in support of its right of self-defense." Furthermore, "The United States will oppose arms control concepts or legal regimes that seek general prohibitions on the military or intelligence use of space."

On March 23, 1983, Ronald Reagan announced SDI, a major shift in U.S. space policy. The goal was deceptively easy: "What if free people could live secure in the knowledge that their security did not rest upon the threat of instant U.S. retaliation to deter a Soviet attack; that we could intercept and destroy strategic ballistic missiles before they reached our own soil or that of our allies?" He called for the creation of a "nuclear shield" around the United States that would prevent any deadly weapon from striking U.S. targets. The program has been alternately described as a brilliant strategic concept, a technological fantasy, the nail in the USSR's coffin, and an elaborate hoax. Ten years later, the Soviet Union no longer existed, few components of the defense package had been developed, much less deployed, and Reagan's successors reoriented focus to ground-based interceptors.

SDI—dubbed "Star Wars" by Senator Edward Kennedy—was an antiballistic missile system made up of a constellation of ground- and space-deployed defensive weapons. A new agency within the Department of Defense, the Strategic Defense Initiative Organization (SDIO), was established in 1984 to oversee the programs, which would focus on destroying weapons during the three separate stages of flight: booster, midcourse, and terminal. Starting with a budget of more than $1 billion, research was to focus on five areas: systems; surveillance, acquisition, tracking, kill assessment; directed energy weapons (lasers); kinetic energy weapons; and supporting technologies such as survivability, logistics, communications, and computers.

Originally, the signature SDI program was the space-based interceptor. These orbiting platforms or "garages" would each hold 10 interceptors. Once a hostile launch was detected, an interceptor rocket would be shot downward toward the target, building up sufficient kinetic energy to destroy on impact. Targets would be detected using the many satellites making up the space-based surveillance and tracking system. The SDIO selected five contractor teams, awarded each contracts of $4.5 million to $7 million to carry out testing and requested a report of the results in one year. To facilitate testing, construction began for a new national test facility at Falcon (now Schriever) Air Force Base in Colorado. In late 1988, the Pentagon announced it would reduce the amount of planned space-based interceptors while increasing the number of planned ground-based interceptors. Current technology was more reliable for ground-based systems and the shift would slash the cost estimates by almost half.[48]

In August 1983, Soviet general secretary Yuri Andropov offered a unilateral moratorium on tests of the Soviet ASAT system. He also said the USSR was "prepared to dismantle the existing antisatellite weapons systems" if the United States would agree on banning "future development of such weapons." The White House dismissed the offer, saying Moscow only sought to preserve its ASAT lead by keeping the United States from developing its own system.[49]

Reagan refused to discuss either ASAT or ABM issues with Moscow, prompting considerable criticism. The White House had ignored the first Soviet overture in August 1981, which had offered a draft treaty to ban the deployment of any kind of weapon in outer space. When the USSR proposed another ASAT treaty—which offered to dismantle the Moscow ABM site—still Washington refused. Reagan explained that such an agreement would prohibit any military use of the space shuttle and any verification would be difficult. Congress did not accept the White House rationale and cut funds for ASAT testing.

Reagan's advisers dodged accusations that SDI violated the ABM treaty by noting that SDI was, for now, limited to lab research only. But as research—and criticism—mounted, the Pentagon explained that there was no violation, as the treaty only applied to systems that existed when the treaty was signed. This was a switch from the 1972 negotiations. "The complication in all this was that the U.S. had tried to ban futuristic technology during the original ABM negotiations, but the Soviets were unwilling to agree to such restrictions."[50]

Arms control negotiations, frozen since the Soviet invasion of Afghanistan, resumed during Reagan's second term and culminated in the Strategic Arms Reduction Treaty (START). Many Reagan fans believe that SDI was the death knell of the Soviet Union, as Moscow realized it could never afford to spend the amount of money necessary to defeat the new U.S. system. However, this view is based on the questionable assumption that Reagan had such a complex, long-range plan. It also overlooks domestic Soviet politics at the time. Three Soviet leaders died between 1981 and 1985, and government decisions—and reactions—slowed to almost inaction.

On May 30, 1985, Reagan issued National Security Directive 172, a fact sheet on SDI. The document formally incorporated criteria from policy adviser Paul Nitze to evaluate the program. The "Nitze criteria" ask whether the system is effective, can survive a direct attack, and is cost effective at the margin, meaning "any deployed defensive system would create a powerful incentive not to respond with additional arms, since those arms would cost more than the additional defensive capability needed to defeat them."

The SDI program scored one major success in 1985, successfully testing the first U.S. ASAT weapon. On September 13, 1985, a U.S. Air Force F-15 fighter jet launched an ASM-135 ASAT that slammed into the U.S. Navy's Soliwind meteorological satellite. The test used an air-launched miniature vehicle (ALMV) atop a missile and mounted beneath the F-15 and the vehicle locked onto its target using on-board infrared sensors. The Pentagon noted that the Soviets had conducted similar tests, but they relied on a fixed-position rocket, meaning "the Soviets would have to wait until the target was in proper position over the launching site, rather than carry the missile aloft in a small fighter." Some U.S. scientists were not impressed by the F-15 test, however. John Pike, an expert from the Federation of American Scientists (FAS), suggested the target was chosen because it was "the only thing big enough and low enough for them to hit."[51]

As early as 1986, Congress cut $1 billion from the White House's $3.7 billion budget request for SDI.[52] In December 1985, Congress attached a provision to the defense budget that banned further ALMV testing on targets in space unless the USSR resumed ASAT tests.[53] Secretary of Defense Frank Carlucci ultimately canceled the F-15 ALMV program in 1988, "because Congress, for the third straight year, had imposed a moratorium on testing the system against objects in space," he said. "I could not justify developing a system if it could not be tested adequately."[54]

Even Reagan backtracked. National Security Study Directive 4-86, issued in October 1986, recognized the slow progress made to date. While the ALMV was "a good first step towards deterring the Soviet Union from using its ASAT capability," according to the document, "support for our program has eroded due to its relatively high cost and limited capability."[55] The ALMV, for example, was initially priced at $500 million in 1983; by 1985 it was up to $5.3 billion.

On January 28, 1986, the U.S. manned spaceflight program was shaken by the explosion of the *Challenger* space shuttle just 73 seconds after liftoff. All seven crewmembers perished. The shuttle fleet was grounded while the disaster was investigated, and flights did not resume until October 1988. The delays meant both civilian and military payloads had to wait.

The Reagan administration announced a new U.S. space policy in January 1988. The program had four basic tenets: 1) deter or defend against enemy attack; 2) not allow hostile forces to deny U.S. access to U.S. space assets; 3) negate hostile space systems; and 4) enhance operations of U.S. and allied military forces. Two panels were created to study related systems. The Fletcher Panel, composed of 50 engineers and academics and led by former NASA director James C. Fletcher, focused on research and technological feasibility, while the Future Security Strategy Study, chaired by Fred S. Hoffman, examined how SDI would affect U.S.-Soviet relations. The two groups reached quite different conclusions. The Fletcher Panel was optimistic that missile defense was feasible, if not in the immediate future. "The scientific community may indeed give the United States the means of rendering the ballistic missile threat impotent and obsolete." However, the report included a major caveat: "It is not technologically credible to provide a ballistic missile defense that is 99.9% leakproof." The study chaired by Hoffman was even more skeptical, concluding, "nearly leak-proof defenses are required to provide a high level of protection for the population." But "nearly leak-proof defenses may take a very long time or may prove to be unattainable in a practical sense."[56]

NASA personnel complained that military demands were driving the space shuttle program, compromising scientific research, and forcing potential customers to turn to the European Space Agency for launch services. From 1986 to 1994, one-third of scheduled shuttle payloads belonged to the Pentagon, and 70 members of the 102 in the astronaut corps were military men, including all of the shuttle pilots. During Reagan's first term, "up to 20% of NASA's budget [was] spent directly on military application[s]," and in 1984 the "military bumped a showcase *Skylab* mission off a shuttle flight after a glitch occurred in its own orbiter."[57] Both scientists and former military personnel connected to SDI accused the Pentagon of seeking style over substance. Standard policy was to identify and publicize flashy tests to convince the public of the program's validity; such events were code-named Beacon (Bold Experiments to Advance Confidence) and STAR (Significant Technical Achievements and Research).

POST–COLD WAR AMERICA, 1990 TO THE PRESENT

The international political system changed following the collapse of the Soviet Union in 1991. Instead of two superpowers, the United States faced a variety of rivals, including Russia, China, and non-state actors such as terrorist groups. U.S. security policy had to change accordingly.

Scaled-back SDI Programs under Bush

In 1988, George H. W. Bush was elected president. During his one-term presidency, the Berlin Wall fell, allied forces drove Iraq out of Kuwait, and the Soviet Union collapsed. He scaled back SDI programs as the cold war ended. At the same time, the Gulf War in 1991 confirmed the value of space-based military assets, especially reconnaissance and communications satellites. "The result is that investment in space systems is taking an increasingly larger share of a shrinking total DoD [Department of Defense] investment budget," according to a Senate report. "In fiscal year 1993, space investment will exceed 15 percent of total investment, a double of the share since fiscal year 1986. For comparison purposes, the space investment budget now exceeds total investment in the Army by 20 percent, whereas in fiscal year 1986 it was less than half."[58] However, most of this investment was for space-based military assets, not weapons.

Bush came to office aware of the pressing need to reduce the U.S. military budget, including SDI funding, to pull the country out of an economic recession. The research projects envisioned under SDI had low start-up costs, but now they were reaching the much more expensive testing stage and there simply was not enough money to go around. The gap between projects and funds was projected at more than $250 billion.[59] Moreover, a 1988 Pentagon review of SDI highlighted the many technical shortcomings of the program. Despite the $12 billion already sunk into the project, the report predicted a "catastrophic failure" if the system were ever activated.[60]

Bush revised national security policy accordingly. Instead of Reagan's lasers zapping incoming missiles, Bush turned to less-exotic technology. The space-based interceptor was abandoned in favor of Brilliant Pebbles, a boost-kill program under development by the Lawrence Livermore National Laboratory in conjunction with the Brilliant Eyes sensor system. Unlike the original garaged interceptors, Brilliant Pebbles would protect space-based assets using 4,000 space-based interceptors. Costing roughly $11 billion for the first 1,000 units, the interceptors would be primarily controlled from the ground, but would have limited ability to communicate with other units and act autonomously. Advocates believed Brilliant Pebbles could be deployed in five years for about $25 billion.[61]

SDI's narrow focus on the USSR was abandoned in January 1991 in favor of a Global Protection Against Limited Strikes (GPALS) model, which would shield the United States from "purposeful strikes by various Third World powers developing ballistic missiles, or accidental or unauthorized launches from the USSR." While SDI sought to protect 3,500 potential U.S. targets, GPALS was designed to intercept 200 warheads coming from rogue states. President Bush announced the shift in his January 29, 1991, State of the Union address: "I have directed that the Strategic Defense Initiative program be refocused on providing protection from limited ballistic missile strikes, whatever their source. Let us pursue an SDI program that can deal with any future threat to the United States, to our forces overseas and to our friends and allies."[62]

At the end of 1991, Bush signed the Missile Defense Act of 1991. This new law required the Pentagon to "aggressively pursue the development of advanced theater missile defense systems, with the objective of down selecting and deploying such systems by the mid-1990s." The Department of Defense was to "develop for deployment by the earliest date allowed by the availability of appropriate technology or by fiscal year 1996 a cost-effective, operationally effective, and ABM Treaty–compliant antiballistic missile system at a single site as the initial step toward deployment of an antiballistic missile system." This system was to be "designed to protect the United States against limited ballistic missile threats, including accidental or unauthorized launches or Third World attacks."[63] The following summer, Secretary of Defense Dick Cheney reported that a theater missile defense system would be operational by 1996.

Smaller Systems under Clinton

President Bill Clinton, elected in November 1992, had run on an anti-SDI platform, and one of his earliest policy plans was to drop the program. Once in office, however, he was persuaded to fund at least the research portion of SDI.

In the post–cold war era, the United States would not likely have to defend against thousands of ICBMs coming from the Soviet Union, but a handful of medium-range missiles coming from rogue countries such as North Korea, Libya, or Iraq. At the same time, instead of focusing on defending the continental United States with long-range missiles, now allies and foreign bases could be protected by shorter range, theater missiles. Consequently, Clinton ordered a major shift in national security policy. Instead of protecting the entire country, he called for smaller, highly mobile systems to protect U.S. regions, allies, and forces deployed abroad from attacks by specific countries. Known as theater missile defense, this approach uses ground-based intercepts.

Clinton also sought to promote innovation and competition within the U.S. space industry. In August 1993, he ordered all White House science-related councils be consolidated into the National Science and Technology Council. The Council standardized the national weather satellite system and moved to quell a looming gap in remote-sensing data. One year later, the White House Council issued a new statement on space transportation, designating the Pentagon to modernize the current launch vehicle fleet and NASA to develop the next generation of reusable spacecraft. The directive also established guidelines for using foreign technology and to consider U.S. commercial options for meeting national needs.

In 1996 President Clinton released a new National Space Policy that assumed that space would become more and more important to the economy, education, and national security. His space policy had five goals:

- Enhance human knowledge of Earth, the solar system, and the universe
- Strengthen and maintain U.S. national security
- Enhance U.S. economic competitiveness as well as scientific and technical capacity
- Encourage private sector, state, and local sources to embrace space technology
- Promote international cooperation to further U.S. domestic, national security, and foreign policy goals

The list suggests a return to Eisenhower's emphasis on science over national security. The Clinton policy also emphasized cooperative endeavors and does not claim any unique right to space assets or deny access by other countries. Finally, the full text of the document affirms the need to use space in accordance with existing treaty obligations.

The Pentagon's Strategic Defense Initiative Organization was subsequently renamed the Ballistic Missile Defense Organization and downgraded to report to the undersecretary of defense for acquisition to reflect the policy shift. Brilliant Pebbles was allowed to die through a series of budget cuts, the same fate as the army's ground-based kinetic ASAT program.

In March 1999, the Senate passed the National Missile Defense Act of 1999, which states, "It is the policy of the United States to deploy as soon as is technologically possible an effective National Missile Defense System capable of defending the territory of the United States against limited ballistic missile attack (whether accidental, unauthorized, or deliberate)." Upon reviewing the legislation, Clinton agreed to study the proposal but said that any system he approves would need to be effective in terms of cost, technology, deployment, and adherence to the ABM Treaty. Ultimately, Clinton passed on missile defense, leaving the issue for his successor.

Near the end of Clinton's second term, members of the House and Senate Armed Services Committees and the secretary of defense in consultation with the director of Central Intelligence appointed a Commission to Assess United States National Security Space Management and Organization to evaluate U.S. space capabilities and recommend future policy. Ford's secretary of defense Donald Rumsfeld was named chair, and the panel is generally known as the Rumsfeld Commission. The panel's report, released January 11, 2001 (nine days before George W. Bush was inaugurated), called for the United States to immediately focus its attention on gaining control over outer space and—for the first time in U.S. history—allowing the president to deploy weapons in space, otherwise a "Space Pearl Harbor" was inevitable.[64]

Militarization under George W. Bush

Given that President George W. Bush appointed Donald Rumsfeld as defense secretary, it was hardly surprising that the Bush administration embraced a more militaristic approach to space weapons, acting on many of the recommendations of the Rumsfeld Commission. According to Theresa Hitchens, director of the UN Institute for Disarmament Research, the second Bush administration took concrete steps "aimed at making the United States the first nation to deploy space-based weapons," but Congress refused to fund such projects.[65] On May 8, 2001, Rumsfeld announced the creation of a Policy Coordinating Committee for Space under the National Security Council. He also discussed making space the fourth arena (land, sea, air, space), with plans for the United States to be the dominant space power. The 2001 Quadrennial Defense Review was released on October 1, 2001, and clearly stated Bush's intention: "A key objective . . . is not only to ensure U.S. ability to exploit space for military purposes, but also as required to deny an adversary's ability to do so."[66]

The terrorist attacks of September 11, 2001, seemed to confirm arguments that the United States now faced hostile threats coming from new sources. Bush and Russian president Vladimir Putin met in fall 2001, and the former cold war enemies agreed to work together to confront new enemies, such as terrorists. The subsequent wars in Iraq and Afghanistan drew heavily upon space-based military assets for communication, real-time imaging, and navigation. Military personnel can use real-time satellite imaging to guide Predator drones to targets halfway around the world.

As Washington prepared to ramp up a missile defense system and possibly deploy space-based missile defenses, President Bush withdrew from the ABM Treaty on June 13, 2002. The Pentagon's Ballistic Missile Defense Organization was reconfigured into the Missile Defense Agency (MDA) in 2002. The MDA promptly requested $1.33 billion to revive the Brilliant Pebbles program.[67] The agency also oversaw development of the space-based laser, a weapon that could conduct ASAT operations, deny access to space by other countries, disrupt information flows with satellites, destroy high-altitude aircraft or weapons, and hit hardened, deeply deployed targets on Earth.

The entire U.S. space program was reevaluated when the space shuttle *Columbia* disintegrated upon reentry on February 1, 2003, following two weeks in orbit. As with the *Challenger* disaster in 1986, both military and civilian payloads had to wait while the incident was investigated. Based on recommendations by the investigation panel, NASA decided to retire the shuttle program and replace it with a new system that would have separate launch options for crew and cargo.[68] The system, Constellation, would use a new Ares-series rocket, and the crew module was designated *Orion*.

In 2004, Bush announced a plan to return to the Moon and use it as a staging area for a Mars mission but provided no additional funding. NASA administrator Michael Griffin told a receptive Congress that for $2 billion in additional funding, NASA could cut the shuttle gap from five years to three, but the White House refused funding.[69] By mid-2008, NASA officials were saying that even if the $2 billion were suddenly available, it now would be too late to close the gap.[70] Griffin had already admitted that the United States may lose this round of the lunar race. "If China wants to put people on the moon," he said, "and if it wishes to do so before the United States, it certainly can. As a matter of technical capability, it absolutely can."[71]

In October 2006, the White House announced a new U.S. national space policy, replacing the 1996 Clinton doctrine.[72] The text indicates a clear shift in U.S. policy toward weaponizing outer space. Specifically, it states that the United States "will preserve its rights, capabilities and freedom of action in space . . . and deny, if necessary, adversaries the use of space capabilities hostile to U.S. national interests." This statement directly contradicts the provisions of the Outer Space Treaty, namely that space should be used for nonaggressive purposes.[73] The new doctrine also reaffirms that intelligence activities should be considered a legitimate, peaceful purpose, and that the United States "will oppose the development of new legal regimes or other restrictions that seek to prohibit or limit U.S. access to space."[74]

The new Bush space policy alarmed the international community, with some countries interpreting it as the first overt step in weaponizing outer space. China, for example, cited hostile-sounding phrases as justification for developing its own space defense. In fact, argues China space policy expert Kevin Pollpeter, "The U.S. Defense Department may unwittingly be producing a security dilemma where its own efforts to protect its systems may be driving others to develop systems to counter U.S. space efforts."[75]

In 2007, the White House announced plans for a U.S. missile defense system that could intercept incoming missiles. System components were to be deployed in several different locations, including Poland, with a key radar site in the Czech Republic. The possibility of having U.S. military assets deployed so near its borders enraged Moscow. Washington repeatedly stressed that the system was intended to intercept ballistic missiles originating in Iran or North Korea, but that explanation did not calm Russia's objections. The Czech and Polish components were only part of a much larger system that called for radars in California, Alaska, Greenland, and Fylingdales, a British air base. Forty interceptor missiles were slated for Alaska, four for California, and 10 for Poland. Plans also called for 130 interceptors to be deployed on ships.

Then president, now prime minister, Vladimir Putin's initial response to the plan was to threaten to redeploy nuclear warheads aimed at Europe. He toned down his position in June 2007, offering to let Washington use a radar site that Moscow rents in Azerbaijan, along Iran's northern border. When Bush dismissed that idea, on July 2, 2007, Putin offered another radar facility in southern Russia and proposed a joint Russia-U.S.-Europe defense arrangement.[76]

While focusing on missile defense, Bush neglected a vital component of the U.S. space program—finding reliable rides to the *International Space Station*. At that time, the space shuttle program was due to be phased out by 2010; but *Orion*, the replacement program, was not scheduled to be ready before 2015. Instead, after spending close to $100 billion to develop the *International Space Station*, the United States would now have to rely on Russia to launch supplies and crew to use it.[77]

The United States found itself in such a vulnerable position for several reasons. First, the loss of *Columbia* in 2003 led to the decision to scrap the Nixon-era program. Second, NASA funding has been insufficient. "What we have here is an agency that has been given a lot to do but has been starved for funds," says Senator Bill Nelson, a Florida Democrat and member of the Science Committee. "I think the gap is largely due to the [Bush] administration's refusal to give NASA the funds it needs. And now we'll be forced to give billions to the Russians because we didn't spend millions before. It's the worst of all worlds."[78] Finally, Bush showed a decided lack of interest in space exploration. In six years as governor of Texas and eight as president, he visited the Johnson Space Center in Houston only once—for the *Columbia* memorial service.

Aside from the blow to national pride associated with asking Russia for help with the U.S. space program, Washington was legally barred from doing business with Moscow due to a law banning technology purchases from countries that deal with "unfriendly countries" with nuclear programs, such as Iran. (Russia is helping build a nuclear reactor in Iran.) Congress has waived that ban, allowing the United States to purchase launch services from Moscow through 2011.[79]

There could be alternatives to Russia. On March 8, 2008, the European Space Agency (ESA) launched the *Jules Verne*, a new automated transfer vehicle (ATV) from French Guiana. The spacecraft went to the *International Space Station*. If successful, the *Jules Verne* could provide NASA with a non-Russian alternative route for reaching the *ISS* while waiting for the next generation of shuttles to come online.[80]

Obama Rethinks Missile Defense, Return to Moon

Upon entering the White House, President Barack Obama began to publicly question the rationale for the missile defense system. Experts were already downplaying potential threats. In May 2009, the EastWest Institute, "a global think-and-do tank that devises innovative solutions to pressing security concerns," issued a joint U.S.-Russian report on the Iranian program entitled "Iran's Nuclear Missile Potential: A Joint Threat Assessment by U.S. and Russian Technical Experts." The panel concluded that, "The missile threat from Iran to Europe is thus not imminent." They predicted that Iran would need six to eight years to produce a nuclear missile that could reach southern Europe and up to a decade for enough power to hit the United States.[81]

In September, President Obama announced a major shift in U.S. plans for missile defense systems in Europe. Instead of fixed locations for interceptors and radars, he offered a flexible "adaptable architecture" consisting of sea-based interceptors and forward-based radar arrays to provide advanced warning of attacks.

Obama empaneled an advisory board headed by Norman Augustine, the former head of Lockheed Martin, to evaluate the current space program. Obama stressed the need for cost-effective, technologically sound programs, saying he would "not divert resources from other national security priorities until we are positive the technology will protect the American public."[82] The panel offered a new strategy—flexible path. "The idea would be to focus on building a heavy-lift rocket that could take astronauts to a variety of interesting destinations."[83] In addition, the panel recommended a partial privatization of space expenses. Instead of developing its own equipment, the United States should contract with commercial providers for launch services and other needs.

The Obama White House issued a new national space policy on June 28, 2010. Building on the flexible path principle, the policy encourages the use of commercial sources whenever possible, including launches of U.S. payloads. The policy also stresses the importance of diplomacy, peaceful uses of space, and making space available to all responsible parties. While the George W. Bush administration refused to consider limiting its own activities in space, the Obama policy is amenable to arms control measures that are equitable, effectively verifiable, and enhance U.S. national security. The declaration also pledges continued support to the ISS, and calls for NASA to set far-reaching exploration milestones, including crewed mission beyond the Moon by 2025 and orbiting Mars by the mid 2030s.

CURRENT AND FUTURE SPACE POLICY

Once the world leader in manned spaceflight and astronomy, the U.S. government has put less and less emphasis and smaller and smaller budgets on the space program since the end of the cold war. At the same time, more players have entered the outer space realm, potentially challenging U.S. dominance.

Technology

One of the most successful demonstrations of U.S. ASAT prowess took place in February 2008 when the United States destroyed one of its satellites threatening to de-orbit. The Pentagon made every attempt to alleviate international concerns about the maneuver, emphasizing it was needed as a safety measure and offering to share information with China and releasing a video of the explosion.[84] However, Russia and China—the latter having been extremely secretive about its own destruction of a dying Chinese satellite in 2007—remained convinced that the satellite was either destroyed to prevent some supersecret technology aboard from possibly falling into the wrong hands or merely as a ruse to justify testing new antisatellite technology. Even some U.S. experts found the decision to destroy the satellite overly vigilant. "It was a surfeit of caution," commented James Lewis of the Center for Strategic and International Studies.[85] Others were more blunt. Michael Krepon of the Stimson Center think tank accused the Bush administration of waiting until the last moment possible to announce the activity and called the environmental explanation a ruse. "Advocates succeeded in carrying out a destructive ASAT test that would have otherwise not been approved by Congress."[86]

The shoot down was also the most public indication that the intercept technology promised by SDI was feasible. "The technical degree of difficulty was significant here," said General James Cartwright of the Joint Chiefs of Staff. "You can imagine that at the point of intercept there were a few cheers that went up."[87] The successful test was even more welcome after a 2004 test had failed.[88] While in 1983 the military described SDI as "trying to hit a bullet with a bullet," by 2008 the head of the MDA, Lt. Gen. Henry Obering, crowed, "Not only can we hit a bullet with a bullet, we can hit a spot on the bullet with a bullet."[89] Yet many experts think the military is again overselling its abilities in tests that do not reflect the real conditions of battle. Roger Hagengruber of the Sandia National Laboratory calls such demonstrations "strap-down chicken tests, where you strap the chicken down, blow it apart with a shotgun and say shotguns kill chickens. But that's quite different from trying to kill a chicken in a dense forest while it's running away from you."[90]

Funding

While increasing the use of space-based military assets, the United States has not correspondingly increased NASA's budgets. If current trends continue, the United States will have to buy a ticket from a commercial launch provider, Russia, or perhaps one day China, to go and service satellites or ASAT components or visit the space station. While Europeans and Asians are enthusiastic about new ventures, possible new space programs have not captured America's attention. In a 2008 study, the Futron technology consulting firm listed the most significant U.S. space weakness as "limited public interest in space activity."[91]

As early as the 1990s, NASA's informal motto was no longer "Higher, faster, stronger," but "faster, better, cheaper."[92] With scant funds for its own programs, NASA even instituted prizes as a way of outsourcing research. In 2008, Griffin announced that NASA would "re-balance the agency's science priorities," deemphasizing exploration of Mars, while instead focusing on the outer solar system. The 2009 NASA budget allocated $343 million for Mars, compared with $620 million in 2007.

Several private companies in the United States have tried, with varying degrees of success, to create commercial launch and cargo services. NASA has expressed interest in using one such company, SpaceX, to fill the gap between the old and new shuttle fleet. After three failed attempts in August 2008, which involved losing three government satellites and the ashes of former astronaut Gordon Cooper and *Star Trek*'s James "Scotty" Doohan, the fourth SpaceX dummy payload finally reached orbit on September 27, 2008. Celebrating the long-awaited success, Elon Musk, an entrepreneur and the CEO of SpaceX, declared, "This really means a lot. There's only a handful of countries on Earth that have done this. It's usually a country thing, not a company thing. We did it."[93] On June 4, 2010, SpaceX successfully launched its own rocket, *Falcon 9*. The company plans to offer cargo transport services to the *ISS* beginning in 2011.[94]

The reality that the United States may have to rely on Russia to ferry crew to the *ISS* for at least five years has been widely condemned. "This is a very serious betrayal of American interests," declared Senator Nelson. "This will be the first time since *Sputnik* when the United States will not have a significant space superiority. I remain dumbfounded that we've allowed this serious threat to our national security to develop." The former NASA administrator Griffin agreed, saying the situation is "unseemly, simply unseemly, for the United States—the world's leading power and leading space power—to be reduced to purchasing services like this. It affects, in my view, how we are seen in the world, and not for the better."[95]

PROS AND CONS OF SPACE WEAPONIZATION FOR THE UNITED STATES

The United States is already the dominant military power on Earth and has more space-based military assets than any other country. What would it gain from placing weapons in space?

- **Protect Its Lead:** By staking a formal claim on space, Washington could prevent other countries—whether hostile or friendly—from seizing the ultimate high ground. This appears to have been the rationale behind George W. Bush's 2006 space policy.

- **Keep Space Safe for All:** Some policy makers suggest that the United States is the ideal guardian for space, because it historically has not demonstrated imperial desires for expansion.

- **Protect Existing Military Assets:** The U.S. military depends on its space-based assets and would be severely handicapped by their loss. Indeed, "Army leaders are keenly worried about the possibility that allowing hostile forces free access to space-based assets could erase the edge U.S. forces now enjoy through exploiting satellite imagery, communications, and precision targeting."[96]

But there also could be negative effects from weaponizing space.

- **Trigger Space Race:** If the United States enters space, other large, technically advanced countries could feel threatened. "China and Russia long have been worried about possible U.S. breakout on space-based weaponry."[97] This is one reason Beijing and Moscow have been pushing for talks to prevent an arms race in outer space.

- **Trigger Nuclear Arms Race:** Countries that cannot afford ASAT or missile defense programs may see a nuclear buildup as their best defense from U.S. weapons and space-deployed assets.

- **Trigger New Alliances:** Non-space countries and even non-state actors could choose to pool resources to balance a space-based United States.

The U.S. space program is in a period of wholesale change. Missions are changing, funding is changing, and both NASA and the Pentagon have rivals in the fields of space exploration and space militarization. To date, the United States and all other countries have not crossed the threshold into space weaponization. However, there is no guarantee that degree of restraint will endure in the future.

[1] Paul Dickson. *Sputnik: The Shock of the Century.* New York: Walker, 2001, pp. 50–53.

[2] Heike Hasenauer. "Rocket Pioneers." The Official Homepage of the United States Army, October 6, 2008. Available online. URL: http://www.army.mil/-news/2008/10/06/13102-rocket-pioneers/. Accessed June 9, 2010.

[3] Walter A. McDougall. *The Heavens and the Earth: A Political History of the Space Age.* New York: Basic Books, 1985, p. 41.

[4] John L. Frisbee. "Valor: Project Aphrodite." *Air Force Magazine* 8 (1997). Available online. URL: http://www.airforce-magazine.com/MagazineArchive/Pages/1997/August%201997/0897valor.aspx. Accessed June 9, 2010.

[5] Matthew Brzezinski. *Red Moon Rising: Sputnik and the Hidden Rivalries That Ignited the Space Age.* New York: Times Books, 2007, pp. 235–237.

[6] "Walter Dornberger." Available online. URL: http://www.spartacus.schoolnet.co.uk/GER-dornberger.htm. Accessed June 9, 2010.

[7] National Museum of the U.S. Air Force. V-2 with Meillerwagen. Available online. URL: http://www.nationalmuseum.af.mil/factsheets/factsheet.asp?id=511. Accessed June 6, 2010.

[8] V2 Rocket.com. Available online. URL: http://inventors.about.com/gi/dynamic/offsite.htm?site=http://www.v2rocket.com/start/deployment-start.html. Accessed June 6, 2010.

[9] Dickson, *Sputnik,* pp. 54–55.

[10] Dickson, *Sputnik,* pp. 56–57; Hasenauer, "Rocket Pioneers."

[11] Dickson, *Sputnik,* pp. 56–58.

[12] McDougall, *The Heavens and the Earth,* p. 95.

[13] Everett C. Dolman. *Astropolitik: Classical Geopolitics in the Space Age.* New York: Frank Cass, 2002.

[14] Richard A. Hand, Bonnie Houchen, and Lou Larson, editors. *Space Handbook: A War Fighter's Guide to Space*. Maxwell Air Force Base, Ala.: December 1993. Available online. URL: http://www.au.af.mil/au/awc/awcgate/au-18/au180003.htm. Accessed June 1, 2010.

[15] Dolman, *Astropolitik*, p. 92.

[16] McDougall, *The Heavens and the Earth*, p. 81.

[17] Matthew Mowthorpe. *The Militarization and Weaponization of Space*. Lanham, Md: Rowman & Littlefield, 2004, p. 13.

[18] McDougall, *The Heavens and the Earth*, pp. 122–123.

[19] Cited in McDougall, *The Heavens and the Earth*, p. 134.

[20] Mowthorpe, *Militarization and Weaponization of Space*, p. 14.

[21] Stephen E. Ambrose. *Eisenhower: Soldier and President*. New York: Simon and Schuster, 1991, p. 451.

[22] "Enoughnik of This." *New York Times* (8 December 1957), p. 36.

[23] Mowthorpe, *Militarization and Weaponization of Space*, p. 15.

[24] Hand, Houchen, and Larson, *Space Handbook*.

[25] "President Orders Declassification of Historic Satellite Imagery." National Reconnaissance Office Press Release (February 24, 1995). Available online. URL: http://www.nro.gov/PressReleases/prs_rel.html. Accessed June 1, 2010.

[26] Cited in William J. Cook, Gareth G. Cook, and Jim Impoco, "When America Went to the Moon." *U.S. News and World Report*, July 3, 1994.

[27] Nigel Hey. *The Star Wars Enigma: Behind the Scenes of the Cold War Race for Missile Defense*. Dulles, Va.: Potomac Books, 2007, p. 100.

[28] Cook, Cook, and Impoco, "When America Went to the Moon."

[29] Cook, Cook, and Impoco, "When America Went to the Moon."

[30] Mowthorpe, *Militarization and Weaponization of Space*, p. 15.

[31] Hand, Houchen, and Larson, *Space Handbook*.

[32] Dwayne A. Day. "The Blue Gemini Blues." *Space Review*, March 20, 2006.

[33] Cook, Cook, and Impoco, "When America Went to the Moon."

[34] White House. "A Renewed Sense of Discovery." Available online. URL: http://www.whitehouse.gov/infocus/space/index.html. Accessed June 6, 2010. Also Mary Hennock, Adam B. Kushner, and Jason Overdorf. "The New Space Race." *Newsweek*, October 20, 2008.

[35] State Department. SALT I background materials. Available online. URL: www.armscontrol.org/print/2487. Accessed April 22, 2010.

[36] NASA. "Report of the Space Task Group, 1969." Available online. URL: http://www.hq.nasa.gov/office/pao/History/taskgrp.html. Accessed May 20, 2010.

[37] Mowthorpe, *Militarization and Weaponization of Space*, p. 16.

[38] Hand, Houchen, and Larson, *Space Handbook*.

[39] Hand, Houchen, and Larson, *Space Handbook.*

[40] Everett C. Dolman. "Space Power and U.S. Hegemony: Maintaining a Liberal World Order in the 21st Century." In *Space Weapons: Are They Needed?* edited by John M. Logsdon and Gordon Adams. Washington, D.C.: Space Policy Institute, George Washington University, 2003, p. 93.

[41] Dolman, *Astropolitik,* p. 133.

[42] Cook, Cook, and Impoco, "When America Went to the Moon."

[43] George C. Wilson. "Brown Says Some U.S. Satellites Are Vulnerable to Soviet Hunters." *Washington Post* (5 October 1977), p. A2.

[44] "Targeting a Hunter-Killer." *Time,* October 17, 1977.

[45] Presidential Decision NSC-37. "National Space Policy." May 11, 1978. Available online. URL: http://www.fas.org/irp/offdocs/pd/index.html. Accessed June 6, 2010.

[46] Frances FitzGerald. *Way Out There In the Blue.* New York: Simon and Schuster, 2000, especially chapter four.

[47] National Security Decision Directive 42. "National Space Policy." July 4, 1982. Available online. URL: http://www.hq.nasa.gov/office/pao/History/nsdd-42.html. Accessed June 6, 2010.

[48] John H. Cushman, Jr. "Pentagon Plans Cut in Weapons in Missile Shield." *New York Times* (7 October 1988).

[49] Dusko Doder. "U.S. Senators Told of Space Moratorium." *Washington Post* (19 August 1983), p. A1.

[50] Hand, Houchen, and Larson, *Space Handbook.*

[51] James Gerstenzang. "ASAT Destroys Target in Space Test of Pentagon's Anti-Satellite Weapon Called a Flawless Success." *Los Angeles Times* (14 September 1985), p. 1.

[52] William J. Broad. "Science Showmanship: A Deep 'Star Wars' Rift." *New York Times* (16 December 1985), p. A1.

[53] Walter Pincus. "Pentagon to Continue ASAT Development." *Washington Post* (25 December 1985), p. A1.

[54] Frank C. Carlucci. "Our Unfinished Business: An Antisatellite Capability." *Los Angeles Times* (1 March 1989), p. 7.

[55] Quoted in Dwayne Day. "Blunt Arrows: The Limited Utility of ASATs." *Space Review,* June 6, 2005.

[56] Reports quoted in FitzGerald, *Way Out There In the Blue,* p. 253.

[57] Mark Zieman. "NASA Clash: Growing Militarization of the Space Program Worries U.S. Scientists." *Wall Street Journal* (15 January 1986), p. 1.

[58] U.S. Senate Committee on Armed Services. (102nd Congress, second session), *National Defense Authorization Act for Fiscal Year 1993 Report,* Report 103-352. Washington, D.C.: GPO, 1992, p. 85.

[59] Leonard Silk. "The New Guns-and-Butter Battle." *New York Times* (22 May 1988).

60 David Johnston. "Congressional Report Casts Doubt on Feasibility of Anti-Missile Plan." *New York Times* (24 April 1988).

61 Donald R. Baucom. "Missile Defense Milestones, 1944–1997." Available online. URL: www.fas.org/spp/starwars/program/milestone.htm. Accessed April 20, 2010.

62 Baucom, "Missile Defense Milestones, 1944–1997."

63 Baucom, "Missile Defense Milestones, 1944–1997."

64 Quoted in Larry Chin. "'Deep Impact' and the Militarization of Space: Official Policy, Not Science Fiction." Available online. URL: via www.globalresearch.ca. Accessed June 6, 2010.

65 Theresa Hitchens. "Weapons in Space: Silver Bullet or Russian Roulette? The Policy Implications of U.S. Pursuit of Space-Based Weapons." In *Space Weapons: Are They Needed?*, edited by John M. Logsdon and Gordon Adams. Washington, D.C.: Space Policy Institute, George Washington University, 2003, p. 116.

66 Quoted in Hitchens, "Weapons in Space: Silver Bullet or Russian Roulette?" p. 118.

67 Hitchens, "Weapons in Space: Silver Bullet or Russian Roulette?" p. 126.

68 Brian Berger. "Space Shuttle Program Begins Last Chapter." MSNBC, August 15, 2005.

69 Marc Kaufman. "U.S. Finds It's Getting Crowded Out There." *Washington Post* (9 July 2008), p. A1.

70 Paul Rincon. "China 'Could Reach Moon by 2020.'" BBC News, July 15, 2008.

71 Quoted in Paul Rincon, "China 'Could Reach Moon by 2020.'"

72 The policy is actually dated August 31, 2006, and some reports use that date instead of the October 6 release date.

73 Laura Grego and David Wright. "Bush Administration National Space Policy." October 13, 2006. Available online. URL: http://www.ucsusa.org/nuclear_weapons_and_global_security/space_weapons/policy_issues/bush-administration-national.html. Accessed June 6, 2010.

74 Grego and Wright, "Bush Administration National Space Policy."

75 Kevin Pollpeter. "The Chinese Vision of Space Military Operations." In *China's Revolution in Doctrinal Affairs: Emerging Trends in the Operational Art of the Chinese People's Liberation Army*, edited by James C. Mulvenon and David Finkelstein. Alexandria, Va.: CNA Corporation, 2005, p. 351.

76 "Q&A: U.S. Missile Defense." BBC News, October 12, 2007. Available online. URL: http://news.bbc.uk/1/hi/world/europe/6720153.stm. Accessed June 6, 2010.

77 Joel Achenbach. "Where Will NASA's Next Giant Step Take Us?" *Washington Post* (27 October 2009).

78 Marc Kaufman. "NASA Wary of Relying on Russia." *Washington Post* (7 March 2008), p. A1.

79 Marc Kaufman. "Two Bills Aim for the Skies." *Washington Post* (27 September 2008), p. A2.

80 Kaufman, "NASA Wary of Relying on Russia"; Traci Watson. "NASA Chief Requests Russian Craft." *USA Today* (19 September 2008), p. A8.

81 Joby Warrick and R. Jeffrey Smith. "U.S.-Russian Team Deems Missile Shield in Europe Ineffective." *Washington Post* (19 May 2009).

[82] Greg Bruno. "National Missile Defense: A Status Report." Council on Foreign Relations Backgrounder, September 17, 2009. Available online. URL: http://www.cfr.org/publication/18792/national_missile_defense.html. Accessed June 6, 2010.

[83] Joel Achenbach. "Where Will NASA's Next Giant Step Take Us?" *Washington Post* (27 October 2009), p. E1.

[84] Marc Kaufman. "Satellite Fuel Tank Thought Destroyed; Pentagon Points to Analysis of Debris." *Washington Post* (26 February 2008).

[85] Peter Grier and Gordon Lubold. "U.S. Missile Shoots Down Satellite—but Why?" *Christian Science Monitor* (22 February 2008), p. 2.

[86] Michael Krepon. "After the ASAT Tests." Stimson Center Commentary (March 24, 2008). Available online. URL: http://www.stimson.org/pub.cfm?ID=586. Accessed June 10, 2010.

[87] Grier and Lubold, "U.S. Missile Shoots Down Satellite."

[88] "Missile Defense Shield Test Fails." BBC News, December 15, 2004.

[89] Bruno, "National Missile Defense: A Status Report."

[90] Quoted in William J. Broad. "Science Showmanship: A Deep 'Star Wars' Rift." *New York Times* (16 December 1985), p. A1.

[91] Kaufman, "U.S. Finds It's Getting Crowded Out There."

[92] J. R. Minkel. "NASA Turns 50 Today." *Scientific American*, July 29, 2008.

[93] "Private Rocket Launch Succeeds on 4th Try." *Washington Post* (29 September 2008), p. A3.

[94] Marc Kaufman. "Launch of SpaceX *Falcon 9* Private Rocket Is a Success." *Washington Post* (5 June 2010), p. A3.

[95] Kaufman, "NASA Wary of Relying on Russia."

[96] Hitchens, "Weapons in Space: Silver Bullet or Russian Roulette?" p. 123.

[97] Hitchens, "Weapons in Space: Silver Bullet or Russian Roulette?" p. 138.

3

Global Perspectives

The international space scene has dramatically changed since the Soviet Union shocked the world by launching the first artificial Earth satellite in October 1957. The United States won the race to the Moon in 1969, and the Soviet Union ceased to exist in December 1991.

At the beginning of the 21st century, there are multiple international players in the space business. In addition to the United States and Russia, China, Japan, India, Iran, and Israel can launch satellites on their own, and many more countries can build their own satellites. The European Space Agency (ESA), an intergovernmental organization of 18 member states, also has satellite launch capability and has tested an unmanned cargo vehicle. More states have the capacity to launch medium- or intermediate-range ballistic missiles, including Israel, Saudi Arabia, India, and Pakistan.

With the end of the cold war, geostrategic rationales for going into space have shifted. According to Marc Kaufman, a reporter for the *Washington Post*, "The global competition today is being driven by national pride, newly earned wealth, a growing cadre of highly educated men and women, and the confidence that achievements in space will bring substantial soft power as well as military benefits. The planet-wide eagerness to join the space-faring club is palpable."[1] The developed world, especially China, is rapidly catching up to the United States thanks to massive investment in space technology. According to the former NASA administrator Michael D. Griffin, "We spent many tens of billions of dollars during the Apollo era to purchase a commanding lead in space over all nations on Earth. . . . We've been living off the fruit of that purchase for 40 years and have not . . . chosen to invest at a level that would preserve that commanding lead." Furthermore, "A Chinese landing on the Moon prior to our own return," he added, "will create a stark perception that the U.S. lags behind not only Russia, but . . . also China, in space."[2]

However, China's professed interest in space exploration may be a smoke screen for developing a national weapons program using space-based assets.

According to Everett C. Dolman, author of *Astropolitik: Classical Geopolitics in the Space Age*, "A state that demonstrates a working space launch/orbital payload capability fully demonstrates the capability of an ICBM [intercontinental ballistic missile]. For this reason, many states have used the guise of developing a space launch capability when their true intention has been to develop an operational ICBM. In this manner, they skirt international sanctions in the transfer of ICBM technology, or at least paint for themselves a portrait of peaceful cohabitation with other states, via the acceptable pursuit of scientific—as opposed to military—knowledge."[3]

USSR/RUSSIA

While *Sputnik* and cosmonaut Yuri Gagarin are perhaps the most famous players in the Soviet space program, Russia has a rich tradition of scientific research in rocketry. Konstantin Tsiolkovsky is considered to be the father of the Soviet/Russian space program, if not the father of modern space science. His work laid the theoretical foundation for liquid-fueled rockets and multistage booster rockets. His 1895 book *Dream of Earth and Sky* was standard reading for 20th-century rocket scientists, and in 1898 he devised a formula for propelling an object into Earth despite the pull of gravity.

But Tsiolkovsky was a loner, toiling away at his home. Decades would pass before the totalitarian reach of the Soviet regime would scoop up the best minds and group them in well-funded laboratories at specially built scientific complexes. In his Pulitzer Prize–winning book *The Heavens and the Earth: A Political History of the Space Age*, Walter A. McDougall argues that the highly centralized structure of the USSR explains why the Soviets reached space first. Central planning allowed Moscow to invest millions, if not billions, of rubles into scientific research, while ignoring consumer needs. Censorship not only kept a tight lid on anyone who might protest the distorted budget, it also covered up disasters in the space program.[4]

Russia remains a formidable space presence today. While the commercial and scientific aspects of its space program may be more visible, that knowledge and equipment could easily be converted for military purposes.

History of Space Science in USSR/Russia

From its establishment in 1917, the Soviet Union pursued aviation and rocketry. Just one year after the Bolshevik Revolution, Moscow established a Central Aerodynamics Institute in 1918, while the Zhukovsky Academy of Aeronautics opened in 1919.

ROCKET SCIENCE UNDER LENIN

In 1920, one of Tsiolkovsky's students, F. A. Tsander, met with the Soviet leader Vladimir Lenin to discuss rocketry. Lenin appeared enthusiastic about the new technology, and four years later Moscow established the Central Bureau for the Study of the Problems of Rockets. The Bureau was charged with bringing together all Soviet scientists working on the topic, compiling information on Western research, gathering and publishing information on spaceflight, and conducting research, especially regarding possible military uses for rockets.[5] In 1934, the USSR became the first country to officially pursue research on spaceflight.

BOMBS AND ROCKETS UNDER STALIN

Following Lenin's death in 1924, Joseph Stalin became the supreme leader of the USSR. He launched a massive campaign to rapidly industrialize the country, and science was drafted into the race. Stalin would have a mixed relationship with the scientific community over the next 30 years. At times he poured money into their laboratories, believing their achievements would confirm the superiority of communism. Other times, he saw independent-minded researchers as a threat to absolute party control, purging and arresting scientists.

Stalin was particularly fascinated by aviation and pushed the development of Soviet airplanes, using scientific successes to justify his ruthless regime. "By 1939 the Soviet Union was turning out 700 to 800 planes per month, more than Japan and roughly equal to the amount produced by Britain, Germany, or the United States."[6] To celebrate communism's achievements in aviation, Stalin designated August 18 as "Aviation Day," when Soviet pilots would give demonstrations and compete in endurance contests. The Soviet media played up the connection between the aviation feats and the national leader, publishing comments by pilots such as, "He is our father. The aviators of the Soviet Union call Soviet aviation 'Stalinist aviation.' He teaches us, nurtures us, warns us about risks like children who are close to his heart." Similarly, Stalin's close associate Lazar Kaganovich declared, "Aviation is the highest expression of our achievements. Our aviation is a child of Stalinist industrialization; flyers are our proud falcons, raised lovingly and with care by Stalin."[7]

Some of the hyperbole may have been a deliberate effort to draw public attention away from the ongoing political purges, which sent over 1 million citizens to prisons or their death. Inevitably, the repression and arrests affected the aviation industry. When German planes upstaged Soviet aircraft during the Spanish civil war in 1936, Stalin responded by ordering the arrest and execution of Marshal Mikhail Tukhachevsky, the head of the air force, and other top-ranking officials in 1937. The Soviet air force was still struggling to recover from these losses in 1941 when Germany invaded and destroyed what was left of the fleet of aircraft based on the Soviet Union's western frontier.

As World War II drew to a close, the Soviets raced against U.S. troops to acquire as much information as possible about the Nazi V-2 rocket. The German engineers themselves knew they were being hunted, and they decided to surrender to America rather than to Russia. Moscow had to be content with one mid-level German designer, Helmut Gröttrup, and some discarded equipment.

Another major international development at the end of World War II came when the United States dropped atomic bombs on the Japanese cities of Hiroshima and Nagasaki. Moscow would not detonate its own nuclear bomb for another four years, and it would take years to rebuild the Soviet air force. Stalin realized that the newly acquired Nazi rocket research might help develop another method to deliver weapons to their long-distance targets. But while the V-2 was sufficient for delivering bombs from Germany to England, the USSR needed a much greater range if its bombs were to reach U.S. soil. Deputy Prime Minister Georgy Malenkov summed up the situation: "The flying bomb is an outmoded weapon now. . . . The point is that the V-2 is good for 400 kilometers [about 250 miles], and no more. And after all, we have no intention of making war on Poland. Our vital need is for machines which can fly across oceans!"[8] Stalin agreed and in 1947 ordered production of a new intermediate range ballistic missile (IRBM) with a maximum range of 1,000 miles. In 1949, the USSR debuted the Pobeda (T-1) rocket, an upgraded, nuclear-tipped version of the V-2 with a range of 550 miles. Its U.S. equivalent, the Redstone, was first tested in 1953.

SATELLITES AND SPACE UNDER KHRUSHCHEV
Nikita Khrushchev succeeded Stalin after he died on March 5, 1953. Just five months later, on August 12, 1953, the USSR successfully tested its own hydrogen bomb. Khrushchev launched a broad reform program that promised to ease censorship, provide more consumer goods, and open new lands to agriculture. To fund these ambitious programs, he downsized the military expenditure for research and development and doubled the size of the conventional forces to maintain readiness at a much lower cost.

Khrushchev was a paradox for the West. At times he seemed to offer a friendlier face to the international community. Although continually touting the superiority of communism over capitalism, he insisted that war between the two systems was no longer inevitable and called for an era of peaceful coexistence. At other times, he was full of bluster, sputtering that the USSR "will bury you!" or bragging that his country was churning out long-range bombers and missiles "like sausages." The reality behind the iron curtain was somewhere in between.

Khrushchev "was eager to boast world firsts in ICBMs, space launchers, artificial satellites, manned spaceflight, Moon probes, spacewalks, and space stations" and soon had an opportunity to exaggerate the size of the Soviet bomber fleet.[9] Moscow continued to develop cost-effective ways of delivering its nuclear bombs to the continental United States. In 1949, Stalin had ordered the Bison bomber, an airplane capable of flying to the United States and back. The bomber debuted in the 1954 May Day parade, but its most infamous flyover was on July 13, 1955. Although the USSR only had nine or 10 Bison at the time, the pilots made three separate passes over the reviewing stand, allowing the audience to believe there were 30 or more in total. That trick gave rise to the alleged missile gap—the belief that the Soviets had more bombers than the United States.[10] Later imagery from U2 spy planes revealed the deception.

As in the United States in the 1950s and early 1960s, Soviet leaders sought to design and deploy powerful rockets that could carry either military or civilian payloads. The highly centralized structure of the Soviet command economy allowed vast amounts of resources to be quickly allocated to the rocket programs, while in the United States NASA had to coordinate among dozens of independent agencies, firms, and universities. The early Soviet rocket program had to find a way to deliver a nuclear warhead to the continental United States, some 4,000 miles away. The engineer Sergei Korolev came up with a possible method of producing enough thrust to launch the two-ton Soviet nuclear bomb. Instead of producing a single, massive rocket, he clustered 20 engines together, creating the R-7 rocket. The rocket was four times as powerful as the V-2 and could carry payloads of up to 1.5 tons.

Like the United States, Moscow agreed to participate in the International Geophysical Year (IGY) that ran from July 1, 1957, to December 31, 1958. The IGY was a prestigious program to bring together scientists from around the world to address common issues. The highlight of plans for 1957–58 was the agreement to launch an artificial satellite into orbit around the Earth.

Although the Soviets spoke confidently of succeeding in this task by late 1957, few Western observers believed them. One popular joke among American scientists asked, "Why couldn't Russia smuggle a suitcase bomb into the United States? Because the Soviets hadn't yet perfected the suitcase."[11] Even when the USSR test-launched its own ICBM on August 3, the reports were dismissed. *Sputnik* got their attention.

Space Exploration

The Soviet Union won the first three battles of the space race, successfully launching the first artificial Earth satellite on October 4, 1957, putting a human in orbit on April 12, 1961, and conducting the first tethered space-walk on March 18, 1965.

But over all, the USSR lost its technological edge in space exploration during the 1960s. Realizing that it would lose the race for the Moon to the United States, the Soviet program instead engaged in a variety of "stunt" missions, including the first dog in space and the first female cosmonaut and a series of endurance missions aboard space stations in Earth's orbit.

Barely one month after *Sputnik*, on November 3, 1957, the USSR launched a spacecraft carrying a two-year-old dog named Laika to analyze how the launch, orbit, and weightlessness would affect a live mammal. But because technology did not yet exist to allow a spacecraft to safely reenter the Earth's atmosphere, Laika was doomed. *Sputnik-2,* her space capsule, was outfitted with food and cooling fans, as well as an automatic lethal injection, programmed to be administered after one week. Laika became a celebrity, depicted on postage stamps and bubblegum cards; she even had her own brand of cigarettes.[12] However, some of the publicity value was undercut when it became known that Laika was doomed. Decades after the florid eulogies about the brave dog being humanely euthanized aboard her spacecraft, the truth came out—she died of heat and stress only a few hours after the launch.[13]

Another three years would pass until an unmanned (but "animaled") Soviet spacecraft safely returned its passengers to Earth; this honor went to the two-dog team of Belka and Strelka in August 1960. Next, the first female in space, Soviet cosmonaut Valentina Tereshkova, began a historic three-day mission on June 16, 1963, during which she orbited the Earth 48 times over the next three days. Many observers believe she was more a passenger and publicity stunt than a trained pilot, citing her background as a textile worker and amateur parachutist. As a result, Tereshkova is frequently omitted from Western cosmonaut rosters. However, with her single mission, lasting nearly 71 hours, she logged more time in space than all of the U.S. Project Mercury astronauts combined. After a lull in the mid-1960s, the Soviet program became a leader in endurance flight technology, a subfield vital to the establishment of space colonies.

The exact parameters of the early Soviet space programs have always been difficult to discover, due to the closed nature of the political system. Launches, including *Sputnik,* were never announced in advance or open to the press in case of failure. While the world knew of the launchpad fire that killed the *Apollo 1* crew in 1967, not until the glasnost policy of the Gorbachev era did news of Soviet disasters become known. The worst was the Nedelin explosion at the Baikonur Cosmodrome in October 1960, when at least 120 people were incinerated in an enormous launchpad fire.

Military Implications

Both sides believed the other was ahead in the ICBM race, and the poor reconnaissance options in the late 1950s encouraged worst-case assumptions. The United States was fortunate that allies allowed Washington to deploy missiles, bombers, and personnel forward near the western edge of the Soviet bloc, but Moscow did not have that advantage. The giant boosters unveiled for the U.S. space program further unnerved Moscow. The USSR's efforts to deploy missiles in Cuba in 1963 failed disastrously when U.S. reconnaissance discovered launch facilities under construction and weapons en route. The international humiliation caused by the Cuban missile crisis played a major role in the ouster of Khrushchev in October 1964. The new general secretary, Leonid Brezhnev, would take a more conservative approach than the brash Khrushchev.

ABMs, SALT, AND SPACE STATIONS UNDER BREZHNEV

Having lost the race to the Moon, in 1969 the Soviet government sought to develop three military space applications. First, space satellite systems could monitor the operations of all branches of the armed forces to improve their effectiveness. Second, space assets could prevent other countries from utilizing space. Third, strategic offensive systems would be needed if space were to become a theater of war.[14] An antiballistic missile (ABM) system could serve the latter two roles by intercepting any incoming airborne weapons.

The USSR had established a missile range at Kasputin Yar, some 75 miles east of Volgograd in 1946 to test V-2 rockets; in 1956 Moscow added a missile test site at Sary Shagan in Kazakhstan. The two facilities were over 1,000 miles apart, so that Sary Shagan could intercept missiles fired from Kasputin Yar.

U.S. intelligence data suggested the early Soviet ABM program was successful. The Soviet Air Defense Forces added an antimissile unit, and U.S. reconnaissance intelligence from the U-2 spy plane program noted a growing radar system along the USSR's external borders. In March 1961, a nuclear-armed ICBM was successfully launched to test the Sary Shagan ABM system. Another ICBM weapon, the fractional orbital bombardment system (FOBS), was also deployed in the 1960s. Unlike regular ICBMs, FOBS nuclear-tipped missiles would briefly enter low earth orbit (LEO) en route to their target, but not complete a full orbit of Earth before descending near its destination. This technicality led Moscow to argue that the Outer Space Treaty's ban on deploying nuclear weapons in space was not applicable.

Next Soviet military engineers began to establish rudimentary ABM systems to protect Leningrad and Moscow. Located on the USSR's western border, launch sites around Leningrad would provide early detection and interception of any attack from the west, while the Moscow system would protect the capital city. The Leningrad system used 30 surface-to-air Griffon[15] missiles, but performed poorly and the system was cancelled and dismantled by 1964.[16] The Griffon interceptor missile, paraded in Red Square in 1964, proved not adequate to the task.[17]

The Moscow ABM project was slightly more promising. Begun in 1962, it used the Galosh, a nuclear-armed exoatmospheric interceptor missile, and a series of early warning and targeting radar sites dubbed the "Hen House," "Dog House," and "Cat House." The Galosh interceptors were deployed in a ring 40–50 miles outside Moscow and were oriented to detect incoming attacks from the west and from China. U.S. intelligence concluded that the Galoshes worked by producing an X-ray effect that scrambled the incoming ICBM's guidance system.[18] Unfortunately, the accompanying radars had spotty coverage, and the Galosh was not especially reliable or accurate.[19] Original plans called for eight ABM complexes and 96 launchers. But interest in the program faded, so that by 1970 only four complexes and 64 launchers were in place.

The Soviets also developed an antisatellite weapon (ASAT) program. The Istrebitel Sputnikov (satellite exterminator) was a nonnuclear weapon designed to collide with its target satellite, making it one of the first weapons systems designed to be deployed in space against a space-based target. There was a small interceptor missile designed to be fired from a MiG-31 aircraft, and reportedly ground-based laser ASATs were established at the Sary Shagan ABM facility in Kazakhstan and the Nurek space tracking center in Tajikistan.[20] To adhere to the Outer Space Treaty, the ASAT missiles used the "hot-metal kill" method; the weapon detonated near the target and sent a cloud of metal shards to damage the satellite. Beginning in 1968, Moscow tested ASAT weapons by launching Cosmos series satellites and then destroying them. U.S. reconnaissance satellites noted interception tests in 1970 and 1971, but the tests appeared to cease during the Strategic Arms Limitation Talks (SALT). The tests resumed in 1976, possibly as a negotiating strategy ahead of a new round of disarmament talks.

By the late 1960s, the escalating nuclear arms race prompted calls for placing limits on the number of weapons each superpower could have. In 1970, representatives from the USSR and the United States began a series of negotiations known as SALT. On May 26, 1972, Brezhnev and Nixon signed the landmark Interim Agreement on the Limitation of Strategic Offensive Arms (known as SALT I), which set limits on future weapons acquisitions. Under the agreement, the United States could have 1,054 ICBM silos and submarine-launched ballistic missile (SLBM) tubes; the USSR would be allowed 1,607 ICBM silos and 740 SLBM tubes. The agreement would be enforced through national technical means—meaning reconnaissance satellites. Brezhnev and Nixon agreed to pursue further limits in a second round of negotiations, to be called SALT II.

The two leaders also signed an Anti-ballistic Missile Treaty, which addressed the problem of defending national territory from incoming ballistic missile systems. Each side was allowed to deploy two ABM systems: one near their capital city and one to protect an ICBM missile site. A revision in 1974 reduced this provision to one site each.

Rather than building an entirely new ABM site around Moscow, Soviet leaders worked to upgrade the existing Galosh system. Testing between 1972 and 1976 sought to improve interceptor accuracy; and later in the decade the aboveground Galosh launchers were dismantled and replaced with less-vulnerable underground silos. Upgrades in the 1980s added short-range Gazelle (SH-08) missiles as well as the Gorgon (SH-04) and Grumble (SA-10). By 1989, the Soviets had the maximum number of launchers allowed under the ABM treaty and had also begun work on a new phased-array radar system dubbed "Pillbox." As of 2004, the Moscow ABM system consisted of 100 interceptors, arranged in silos at eight launch sites, and three control radars. The system is unlikely to be used except in the event of a global nuclear war. While the ABM program of U.S. president George W. Bush planned to use direct interceptors—which smash into their targets—the Russian missiles approach their target and detonate a nuclear weapon. Thus using a nuclear-based ABM system could cause as much damage as the incoming warhead.[21]

The Soviet manned space program focused on the Salyut series of space stations in the 1970s. The unmanned *Salyut-1* long-duration orbiting station was launched from the Baikonur cosmodrome on April 19, 1971, using the heavy-duty Proton launcher. Four days later, the *Soyuz 10* spacecraft blasted off for *Salyut,* carrying a three-man crew. The *Soyuz-10* was unable to successfully dock with *Salyut-1* and returned home the next day. In a second attempt, *Soyuz-11* launched on June 6 and successfully docked, and the three-man crew remained aboard *Salyut-1* for 23 days. Unfortunately the *Soyuz-11* was damaged as it left the space station, and the resulting loss of air pressure killed the crew. *Salyut-1* burned up when it de-orbited on October 10.

Over the next three years, the Soviet team redesigned the Salyut model. During that process, a military space station, known as the *Almaz,* was apparently launched under the designation *Salyut-2* and was intended to be a manned spy satellite. Ultimately, the civilian Salyut and military Almaz designs were merged into a hybrid, and *Salyut-3* launched in June 1974. Because of secrecy surrounding the military missions, it is difficult to know precisely how many Salyut stations were deployed, but experts estimate the number at seven or eight. Several of the Almaz-type stations reportedly were equipped with space cannons.

Multiple Salyut stations spent extended periods orbiting Earth for the next decade. *Salyut-7,* launched on April 19, 1982, had multiple manned and unmanned spacecraft dock at the station, including astronauts from France, India, East Germany, and Poland. The Soviet cosmonaut Svetlana Savitskaya became the first female spacewalker, leaving *Salyut-7* for three hours and 35 minutes on July 25, 1984. The Salyut program made significant technological contributions, such as the use of unmanned space ferries to supply the facility and multiple docking stations, and it studied the physical and psychological consequences of prolonged stays.

The Soyuz program was very reliable in the 1980s, and Soviet scientists surpassed the United States in terms of astronomical research, space medicine, long-duration stays, and number of annual launches. The powerful Proton launchers made Soviet services relatively affordable. Former NASA administrator James Beggs commented, "There's been a habit in this country of thinking of the Soviets as stupid and that they steal all their technology. That's just not so."[22]

The Soviet space shuttle was less successful. Soon after U.S. president Richard Nixon approved the U.S. space shuttle program, the USSR began its own reusable space plane program. The *Buran* space shuttle bore a strong exterior resemblance to the U.S. shuttle, but the program quickly became bogged down in financial and technical issues. Ultimately, only two *Buran* orbiters were built, and only one shuttle ever flew, a 206-minute test mission on November 15, 1988.

Nixon's sudden resignation on August 8, 1974, alarmed Moscow, as Brezhnev had established a good working relationship with the president. He had even visited Nixon's home in California during a tour of the United States in June 1973, when they jointly agreed upon "Basic Principles of Negotiations on the Further Limitation of Strategic Offensive Arms." Nixon had returned to Moscow in June 1974 hoping to move toward SALT II, but his domestic problems overshadowed the discussions. According to Anatoly Dobryin, USSR ambassador to the United States from 1962 to 1986, "Watergate not only undid Nixon's presidency, but also destroyed any chances during the visit of a breakthrough on SALT II."[23]

Taking office in August 1974, President Gerald Ford immediately sent word to Brezhnev that he meant to stay the course in bilateral talks. The new president was preparing for a trip to Japan in November and asked to meet

Brezhnev afterward in Vladivostok, a city in the Russian Far East. After two days of intense discussions, they agreed on a list of instructions for the next round of Strategic Arms Limitations Talks (SALT II) in Geneva that involved setting limits to be achieved by 1995.

Ford's tenure in the White House was short-lived, as he lost the 1976 presidential election to former Georgia governor Jimmy Carter. Moscow was wary of Carter's emphasis on human rights, especially his support for Soviet dissidents. "His policy was based on linking détente to the domestic situation in the Soviet Union. This represented an abrupt departure from the policy followed by preceding administrations," recalled Dobrynin, "thus invariably making his relations with Moscow tense."[24] Carter wanted to replace the Vladivostok framework with a more ambitious project, but Moscow made clear that it would not agree to starting over. Finally, Carter and Brezhnev signed the SALT II Treaty on June 18, 1979, which imposed limits of 2,400 delivery vehicles on each side. Carter asked the Senate to postpone ratification of SALT II when the USSR invaded Afghanistan in December 1979.

ANDROPOV AND CHERNENKO

Moscow was ill-prepared to respond to Ronald Reagan's crusade against the evil empire and his Strategic Defense Initiative (SDI) when it was announced in March 1983. Brezhnev had died in late 1982, and the Soviet leadership was fragile. His successor, Yuri Andropov, died in 1984, and then Andropov's successor, Konstantin Chernenko, died in March 1985 after only 13 months in office. All three men were in very poor health, and decision-making came to almost a complete standstill. Soviet arms control experts stormed out of negotiations about intermediate-range missiles in Europe, cutting off dialogue with the United States. No progress toward resumption occurred until the much younger Mikhail Gorbachev became general secretary in March 1985.

The Soviet leadership was unsure what to make of Reagan. Ambassador Dobryin recalled how Reagan would make belligerent public statements about the Soviet Union but then send friendly overtures through private channels. But Reagan's 1983 Star Wars speech alarmed Moscow because it seemed to move the arms race into space. "Our leadership was convinced that the great technical potential of the United States had scored again and treated Reagan's statement as a real threat," Dobrynin explained.[25] But within a year, Moscow adopted a "less paranoid interpretation to the evil empire, Star Wars, and the like."[26]

SUMMITS, COMMERCE, AND COLLAPSE UNDER GORBACHEV

With a dynamic new leader in Moscow, high-level diplomatic talks between the United States and the USSR resumed in May 1985. But the two sides immediately clashed over SDI. The Soviet negotiator Victor Karpov held firm that he would only discuss reducing medium-range missiles if Reagan would agree to keep SDI confined to research—no testing. He even offered to cut the number of the USSR's strategic missiles by 25 percent if the United States would cancel Star Wars, but to no avail. Reagan and Gorbachev met face-to-face for the first time in November 1985, but neither side would budge.

The two leaders met again in Reykjavik, Iceland, in October 1986, and Gorbachev brought a slate of new proposals with him. Most important, he offered to cut the number of long-range nuclear weapons in half now and eventually eliminate all of them—if the United States would agree to keep SDI testing confined to the laboratory. He also sought to apply the 1972 ABM Treaty guidelines to SDI research. Reagan refused the deal, saying he could not deny himself and "future Presidents . . . the right to develop, test, and deploy a defense against nuclear missiles for the people of the free world. This we could not and will not do." Gorbachev countered that only "a madman" would consent to an agreement that allowed testing in outer space.[27] As Reagan's second term neared an end in 1988, both sides found a compromise. Reagan and Gorbachev signed a treaty on December 8, 1987, to eliminate intermediate-range nuclear weapons and agreed to begin Strategic Arms Reduction Talks (START). While the SALT process placed limits on future weapons acquisitions, START involved reducing existing arsenals. They agreed that SDI research could continue so long as it strictly adhered to the ABM Treaty. Gorbachev explained, "We will not build an SDI, we will not deploy SDI, and we call upon the United States to act likewise."[28]

Under Gorbachev, the Soviet space station program also advanced with the launch of the first *Mir* module on February 19, 1986. *Mir* means both "peace" and "world" in Russian, and the launch took place shortly after Gorbachev established his openness (glasnost) policy. Information was quickly released to the media, although still only after the launch took place. The first crew made a pit stop at the still-orbiting *Salyut-7* station to pick up equipment. *Mir* had six docking stations, allowing easy crew and cargo access, and additional modules were added during the station's life span. The cosmonaut Yuri Romanenko praised the relative spaciousness of the new station. "There is even room that can be used for living room. Atmospheric conditions are better, and all the instruments provide for good fresh air. It's much better than Salyut."[29] *Mir* was occupied almost continuously between 1986 and March 2001, but its last years were marred with mechanical and electrical problems. An unmanned Russian cargo ship docked with *Mir* in January 2001 and nudged it out of orbit, allowing it to disintegrate when it reentered Earth atmosphere.

Current Program

The collapse of the Soviet Union in December 1991 resulted in a situation whereby critical components of the Soviet space program were located in three independent countries: Russia, Ukraine, and, most important, Kazakhstan. Each country created its own national space administration. The Baikonur facility, the USSR's main launch site, was in now-independent Kazakhstan. Ultimately, a deal was reached whereby Russia would lease the facility. The introduction of capitalism brought new opportunities for commercial ventures, at a time when the state budget for space programs plummeted. The Baikonur facility fell into disrepair as skilled workers fled the remote location for other jobs, and remaining workers rioted in 1994 over their poor living conditions. The Russian space budget for that year was only 10 percent of the 1989 level, and the agency scrapped almost half of its national projects.[30]

Mir was still awaiting its final two modules when the USSR broke up, and Russia could not afford to launch them for installation. The cosmonaut Sergei Krikalev even had his stay aboard *Mir* extended by five months because Russia had no money to launch a spacecraft for his ride home.[31] The U.S. president George H. W. Bush reached out to Russian president Boris Yeltsin, and they concluded an agreement to cooperate on completing *Mir* and work toward a new international facility. Russian Energia rockets could send crews and cargo to a space station in fewer missions than the U.S. shuttle, potentially saving as much as $25 billion.[32]

Post-1991 Russia wound up with a sizable inventory of Soviet-made space-launch vehicles and rocket engines. Like most enterprises, parts of the Soviet space program were privatized. The firm Energia became *Mir's* owner. In an effort to keep the station afloat, Energia and a Dutch firm, MirCorp, introduced space tourism, offering rides for paying customers and even sponsoring a television game show with the grand prize of a stay aboard *Mir.* They never raised enough funds, however, to send a space tourist before *Mir* was deactivated.

In 1995, the Russian state-owned Khrunichev Enterprises, Lockheed, and Energia established International Launch Services (ILS) based in McLean, Virginia, to market and manage Russia's commercial launch business, particularly use of the Proton rockets as well as the Angara series rockets. Angara rockets have been stuck in development since 1995 due to funding shortages.[33] In May 2008, the Khrunichev State Research and Production Center, a state-owned firm, bought the majority interest in ILS. At the time of the purchase, ILS had a backlog of 22 rocket orders worth a total of $2 billion. John Logsdon, of the George Washington University Space Policy Institute, speculated that Khrunichev bought ILS so that "any profit from the commercial use of Proton stays in Russia."[34]

In the end, the aging *Mir* became too expensive to operate, and Energia allowed it to drop from orbit. The *Salyut* and *Mir* projects provided extensive information about space station construction, long-term life in space, and regular transport links that would be applicable should a military outpost in space ever be considered. The international cooperation that kept *Mir* running also laid the foundation for the *International Space Station* (ISS). The ISS combined several rival space station projects as a cost-saving solution, bringing together the space programs of Russia, the United States, the ESA, Japan, and Canada. The space shuttle *Endeavour* launched the first module in November 1998, and the first, multinational crew arrived in 2000.

Russian firms have cut deals with a variety of foreign partners, including a multiyear $780 million NASA contract for crew and transport services to the *ISS*.[35] Russia has launch-service agreements with the ESA and Arianespace, and together Russia and the ESA have built a launchpad capable of using Soyuz and Ariane rockets at the Centre Spatial Guyanais (CSG), while China and South Korea have incorporated Russian designs in their space programs.

While Russian firms regularly launch satellites for other countries, Russia's military satellite fleet is nearly depleted. According to the head of Rosaviakosmos, the Russian Space Agency, in June 2001, "68 of the country's 90 orbiting satellites were near or at the end of their service lives. Many of the military's 43 satellites were simply too old to be considered reliable. Russia's civilian space budget of $193 million covered only half of the agency's needs."[36] Also in 2001, all remaining military space assets were consolidated into the Military Space Forces, which would also coordinate commercial activities. Only 68 percent of the budget allocations for 2001–05 actually made it to the military, and existing spacecraft are increasingly being lost in accidents.[37]

Russia's Stance on the Militarization of Space

Soviet leaders and their Russian successors have officially all backed the sanctuary view of outer space, as was evident in their rejection of SDI testing. They agree on the same set of goals regarding space security, namely "equitable, sustainable, and secure access to the use of space, and freedom from space-based military threats."[38] Moscow has long advocated UN and other international efforts to conclude a treaty that enshrines into law the peaceful use of space. It was one of the driving forces behind the 1967 Outer Space Treaty, which banned the deployment of nuclear weapons in space. Along with China, Moscow has been active in the UN Conference on Disarmament, supporting the PAROS (prevention of an arms race in outer space) resolution, and even proposing the creation of a World Space Organization in 1986. "Russia's position on this issue has not changed for decades," Defense

Minister Sergei Ivanov announced in 2005. "We are categorically against militarizing outer space."[39]

Russia did not welcome the missile defense program advocated by George W. Bush, seeing it as a bid to weaponize outer space. Russian president Vladimir Putin harshly criticized Bush's related decision to withdraw from the 1972 ABM Treaty, accusing Washington of "seeking to untie their hands in order to take weapons to outer space, including nuclear weapons."[40] Although Bush insisted the missiles were not aimed at Russia, but at rogue states such as Iran, the planned deployment was too close for comfort. In addition, Moscow feared that the United States would later build a much larger missile-defense system that could neutralize much of its own nuclear arsenal, perhaps prompting Putin to announce the development of a powerful new long-distance missile, the Topol-M.[41] Initially, Putin threatened to retarget its nuclear warheads toward Europe, but in June 2007 he took a more conciliatory approach, offering use of a radar site that Moscow rents in Azerbaijan. Putin later offered yet another radar facility in southern Russia and proposed a joint Russia-U.S.-Europe defense arrangement.[42]

Moscow welcomed the election of President Barack Obama, hoping for a reversal of Bush's moves to weaponize space. Before meeting Obama for the first time, new Russian president Dmitry Medvedev declared, "I hope that this sincere desire to open a new chapter in Russian-American cooperation will be brought into fruition."[43] In April 2010, the two presidents signed a new agreement intended to shrink their respective nuclear arsenals.

Obama canceled the Czech and Poland sites in 2009, but Moscow has not canceled its ABM program. Russia successfully tested what it called an upgraded version of its A-135 Gazelle missile in Kazakhstan in 2004. But Russian defense analyst Pavel Felgenhauer insists that it was not a new missile, but one manufactured in 1988. "The aim of the test was to determine whether the missile would still fly properly after such a long time on the shelf," Felgenhauer explained. "In the Soviet era, old missiles would simply be scrapped and replaced, but at present, Russia does not have the money or industrial capability to replace its missiles."[44] Unless Moscow can continue to upgrade—or even maintain—its space-based assets, they will cease to provide even a very small amount of national security.

EUROPEAN SPACE AGENCY

In 1987, more than a decade after its founding, the ESA announced an ambitious agenda that would make western Europe completely autonomous from the U.S. or Soviet space programs. The *Hermes* space shuttle would ferry European crew to the *Columbus* space laboratory, atop an

Ariane 5 rocket launcher. However, the budget for these programs was immense. *Hermes* alone was budgeted at $9 billion. ESA began to look for outside partners, including NASA and, after 1991, the Russian space program. It joined the multinational effort to build the *ISS* and moved toward a smaller shuttlecraft, the *Jules Verne*. The Ariane, however, has been an unqualified success and currently handles almost half of the world's commercial satellite launches.

ESA is funded through a complex system of member-state contributions. Funding is divided into two categories. All states must contribute to mandatory programs, which include administrative capacity, technology research and investment, information systems, and training. Countries are required to contribute an amount in proportion to their gross domestic product (GDP). France and Germany are the top two investors by a considerable margin. Optional programs are more project-oriented, such as Earth observation, telecommunications, space transportation, and the *ISS*. Contracts for ESA programs are proportionally distributed according to the principle of *juste retour:* the value of contracts awarded within a member state should reflect the amount of money it subscribes. If Belgium, for example, contributes 15 percent of a project's budget, it should receive 15 percent of the associated business. When national budgets are being slashed, countries can easily slash their optional program contributions. ESA as a whole was severely handicapped in the early 1990s, when Germany had to handle the enormous, unexpected costs of German reunification.

History of the European Space Agency

Following *Sputnik*, France, Germany, Italy, the United Kingdom, and other western European countries banded together to establish a joint space program. Still recovering from World War II, the countries of western Europe did not individually have the resources that would be needed to develop a space program that could rival those of the United States or the USSR. Governments considered placing space under the supervision of NATO, but in the 1960s opted instead for two independent organizations devoted to peaceful exploration. The European Launcher Development Organisation (ELDO) developed the *Europa* launch vehicle, and the European Space Research Organisation (ESRO) oversaw construction of seven satellites that were to be launched by the United States. Despite their successes, by 1973 both programs were in disarray as countries bickered over budget contributions. As a solution, the UK trade minister Michael Heseltine proposed creating an umbrella organization with its own staff to better coordinate the two strands of research.[45] The two organizations were merged in 1975, forming the ESA with 10 founding members. Headquartered in Paris,

France, the ESA currently has 18 member states and an annual budget of about €3.7 billion.

Current Program

ESA can build and launch satellites, but it must rely on Russia or the United States for rockets powerful enough for manned missions. European astronauts have trained in the United States and Russia and subsequently participated in U.S. space shuttle and several Russian Soyuz missions. In 1990, ESA established a training facility, the European Astronaut Center, in Germany.

ESA is working to build a comprehensive spaceport at its current launch center in French Guiana. The French Guiana Kourou spaceport will be equipped with Soyuz, Ariane 5, and Vega launchers, allowing it to handle missions of varying sizes and configurations. Most launches are handled by Arianespace, which is a subsidiary of the Centre National d'Etudes Spatiales (CNES), using the Ariane 5 rocket. The Vega is a newer, smaller, and more cost-efficient system, scheduled to enter service in late 2010, that can deploy multiple payloads.

Kourou's location on the equator provides a slingshot effect from launching at the fastest point of the Earth's rotation, allowing for bigger payloads without bigger rockets. "Rockets that launch closest to the equator benefit from a slingshot effect as they leave that part of the Earth's surface which is moving fastest. Soyuz, merely by blasting off from Kourou, can get a 50 percent increase in performance with no technical modifications."[46] "In recent years Arianespace has moved to expand its capabilities at Kourou, with the goal of offering a complete portfolio; that is, being able to launch using large Ariane 5, medium-sized Soyuz, and mini Vega rockets. This will give the company the ability to launch "any mass, to any orbit," according to the company fleet brochure.[47] However, Kourou will not have human spaceflight capacity; the Soyuz-propelled manned missions will continue to operate from Russia's Baikonur cosmodrome.

The ESA's Stance on the Militarization of Space

The ESA is primarily concerned with scientific and commercial applications rather than military uses. The *Hubble Space Telescope*, for example, was a joint NASA-ESA project. While the George W. Bush administration approached several European defense contractors for possible work on a missile defense system, the ESA was not involved.[48]

In line with its scientific mandate, the ESA contributed the Columbus Laboratory to the *ISS*. The lab studies microgravity conditions, and its findings will be used to develop new materials, drugs, and crops.[49] After a six-year delay, caused first by Russian funding problems over the *ISS* and the loss of the shuttle *Columbia,* the module containing the lab was launched aboard the shuttle *Atlantis* in February 2008. The Columbus Laboratory had originally been considered for the space station *Freedom,* a planned, permanently manned space station that was never completed and parts of which make up the *ISS.*[50] On March 8, 2008, Arianespace launched the *Jules Verne,* a new automated transfer vehicle (ATV), from French Guiana. The spacecraft, the largest ESA project to date, automatically docked with the *ISS* on April 3. At NASA's insistence, *Jules Verne* also employed a new escape procedure, in case of technical problems that could threaten the space station itself.[51] The craft successfully delivered five tons of supplies to the *ISS,* was loaded with 2.75 tons of waste, detached from the station, and then burned up upon reentry.

Following the success of the *Jules Verne,* the ESA ordered five additional ATVs. They could provide NASA with a non-Russian alternative route for reaching the *ISS* while awaiting the next generation of U.S. spacecraft to come on line. "When the U.S. space shuttle retires, ATV will be the largest vehicle supplying the *ISS,*" said the ESA director for human spaceflight Simonetta Di Pippo. "Considering its technological challenges, like automatic rendezvous and docking, ATV is the most sophisticated space vehicle ever built."[52] The second ATV, the *Johannes Kepler,* is scheduled to launch in February 2011. In March 2010, the ESA announced that ATV-3 would be named for the Italian space pioneer Edoardo Amaldi. The ESA plans launches every 17 months and hopes future models will be able to reenter Earth's atmosphere with cargo or be able to transport crews to and from the *ISS.*

In early 2008, Arianespace had orders for launch services totaling nearly €1 billion. These contracts are in stark contrast to Arianespace's financial situation in May 2003, "when a flat market and the disastrous failure of an Ariane vehicle in flight had brought the company to its knees."[53] ESA officials have asked the ATV contractor, Astrium, to investigate the technical demands and potential costs of upgrading the ATV to transport crew. Further confirming Europe's commitment to human spaceflight, in May 2009 the ESA announced a new class of six European astronauts.[54] The main obstacle to upgrading the ATV is financial, not technological. The ESA is also working to launch *Galileo,* a European global positioning system (GPS) that will require a constellation of at least 28 satellites. The first two satellites are scheduled to launch in 2011.

CHINA

China never reached superpower status during the cold war, but it has emerged as a major international player in the 21st century. In the 1950s, Moscow and Beijing maintained good relations, and Chinese leaders hoped their comrades would share the technology behind *Sputnik.* The USSR thought differently, and the space race became a contributing factor in the Sino-Soviet split. Relations began to worsen in the 1960s and lasted until the breakup of the Soviet Union in 1991.

In January 1956, Chairman Mao Zedong announced a campaign to catch up with world-class science and technology standards. The program "established fifty-seven priority tasks to ensure China's independence in rocket and jet technology in twelve years."[55] The first step in this effort was to buy a Soviet R-1 missile in October 1956. One year later, China and the USSR signed a "New Defense Technical Accord" whereby Moscow would supply technical information and blueprints, missile models, and even engineers to establish a rocket system in China. Bilateral cooperation collapsed with the Sino-Soviet split in August 1960. The 1,400 technicians returned to the USSR, "taking their blueprints back with them and shredding the remainder."[56] But that October, China launched its first missile Dong Feng-1. Today, China has one of the most ambitious space programs in the world, and much of the technology developed for space exploration has potential military use as well.

History of China's Space Program

China entered the space race using its own homegrown technology and launched its first satellite *Dong Fang Hong-1* (DFH1) in 1970. In 2003, China became the third country to launch a human into orbit.

China's early rocket program was directed by Qian Xuesen, a China-born, U.S.-educated pioneer of the U.S. rocket program. He was a co-founder of what became Caltech's Jet Propulsion Laboratory (JPL), a government research and development center in California. Qian was a victim of the Red Scare of the 1950s when the U.S. government accused him of being a communist sympathizer and stripped him of his security clearance. Qian was detained at Terminal Island, a prison in California, and spent five years under house arrest. He was released in 1955 in exchange for five American pilots captured during the Korean War and immediately returned to China. Barely one year later, China launched its own missile program. The space program "got its start when engineers took military rockets and stuck capsules on the tip."[57]

China has launched numerous satellites since the 1970s. Their missions—and failures—often were not announced. Western observers believe these were primarily reconnaissance and remote-sensing satellites. A few were recovered after several days in orbit, suggesting that they had taken photographs that needed to be developed. Starting in the mid-1980s, China also began launching communications, weather, remote-sensing, and scientific satellites for its own use and on a commercial basis.[58] Deng Xiaoping, who became leader after Mao's death in 1976, considered several defense industry sectors, such as satellite launches, as opportunities to spur economic growth and earn foreign currency.[59]

Current Program

China's current space program has scientific, economic, and military dimensions. Beijing is developing a manned spaceflight program, a satellite launch industry, and, apparently, antisatellite weapons capabilities. Dean Cheng, who tracks Chinese military and technology issues at the Heritage Foundation think tank, sees four purposes. First, the program promotes research and development projects, yielding advanced technologies that have potential civilian applications. High-profile space activities, such as manned missions, can draw attention to China's other space programs, including its profitable launch services. Second, like the 2008 Beijing Olympic Games, the space program showcases China's growing prestige. Third, the Chinese regime has weathered many domestic crises in the 2000s, such as melamine-tainted milk products and shoddy construction practices that compounded earthquake damage. Leaders may think that a successful national space program can strengthen the legitimacy of the ruling regime among the domestic population. Finally, the program can enhance China's military prestige, as any rocket that can launch a satellite can also potentially launch multiple warheads. "The manned program is all the things I have mentioned [above] and more," Cheng told BBC News. "It is a sign of a wealthy country—this is a luxury item. It puts China ahead of every other Asian country—significantly—in terms of space."[60] While important politically, the domestic legitimacy aspect is not discussed here.

China's human spaceflight and space weapons programs are controlled by the military. While there are, at least on paper, civilian agencies handling space policy, in reality they can all be traced back to the military. The China National Space Administration, for example, is a civilian body responsible for international agreements, technology, and the space industry.[61] China's defense budget is top secret, but officials have estimated that it spends $500 million on space programs, a fraction of NASA's $18.69 billion budget for fiscal 2010.[62]

China's space policy, for its 11th five-year plan covering 2006–11, was detailed in a document entitled, "China's Space Activities in 2006" that became available in the West.[63] Purportedly written by the China National Space Administration, the document stresses Beijing's desire for peaceful, cooperative use of outer space. The words *defense* and *security* appear only three times and the People's Liberation Army (PLA) is not mentioned at all.[64] One entire 1,606-word section of the report is devoted to international cooperation, noting that China has signed agreements on the peaceful use of outer space with Argentina, Brazil, Canada, France, Malaysia, Pakistan, Russia, Ukraine, and Europe. It also announces plans for China, Bangladesh, Indonesia, Iran, Mongolia, Pakistan, Peru, Thailand, and Turkey to create an Asia-Pacific Space Cooperation Organization based in Beijing.

China repeatedly emphasized that its space program is based on home-grown technology, a point of national pride and indication of its self-sufficiency.[65] Beijing has set an ambitious, three-step agenda for its manned spaceflight program. First, China wanted to send a human into orbit; this goal was achieved in 2003. Second, China is planning to launch a manned lunar mission by 2017. Third, China plans to deploy its *Tiangong 1* space laboratory in early 2011.[66] According to Kevin Pollpeter, a China expert at the Strategic Studies Institute, "The eventual goal is to build a space station. For them, that's become one of the trappings of being a great power."[67] But even if no explicit military use of space is stated, almost all satellite and launch technology could have military applications as well.

China can launch both satellites and manned spacecraft from its Jiuquan spaceport. There are six kinds of Chinese satellites: "recoverable remote sensing satellites, Dong Fang Hong (DFH) communication satellites, Feng Yun (FY) meteorological satellites, Shi Jian (SJ) scientific research and technological experiment satellites, Zi Yuan (ZY) Earth resources satellites, and Bei Dou navigation and positioning satellites."[68] By 2009, China had more than 50 satellites in orbit, nearly all deployed using Long March rockets; the rockets themselves have an enviable 92 percent lifetime success rate.[69]

The United States has been wary of cooperating with the Chinese on civilian and scientific space projects, fearing the inadvertent transfer of technology with potential military applications. Because the United States dominated the market in communications satellite technology in the 1980s, Washington insisted on reviewing any deal with China involving U.S.-manufactured parts. In 1988, for example, the U.S. government blocked a deal whereby China would launch two Australian satellites that contained American-made components. Washington cited evidence of Chinese missile sales to Pakistan as the reason to withhold or revoke export licenses.[70]

Washington also banned Chinese participation in the *ISS*, fearing it would provide opportunities to steal advanced U.S. weapons technologies. But as China looks to other partners, such as Russia on the human spaceflight program, Europe on the Galileo satellite navigation system, and Brazil for building remote-sensing satellites, it is becoming increasingly likely that it will be able to obtain any needed technology from these sources.[71] Beijing now plans to develop its own space station by 2022.[72]

China's manned spaceflight program has developed rapidly. As of mid-2009, China had launched three manned spaceflight missions. On October 15, 2003, the *Shenzhou 5* took Yang Liwei, China's first astronaut (taikonaut), into Earth's orbit for 21 hours. The *Shenzhou 6*, launched on October 12, 2005, carried two astronauts on a five-day mission. *Shenzhou 7*, launched on September 27, 2008, carried a three-man crew. The highlight of this mission came when astronaut Zhai Zhigang conducted his country's first space walk, in a $4.4 million Chinese-made space suit.[73]

The space walk and China's space program are enormously popular at home. Roger Lanius of the Smithsonian Air and Space Museum explained, "It is a demonstration of technological virtuosity. It's a method of showing the world they are second to none—which is a very important objective for [China]."[74] Theresa Hitchens, director of the United Nations Institute for Disarmament Research, agreed, saying the manned program "has gone a long way to proving to potential customers that their products are safe."[75]

On January 11, 2007, China shot down one of its own Feng Yun weather satellites in LEO, about 537 miles above Earth. The operation presumably used a ground-based, medium-range ballistic missile, making China only the third country to successfully shoot down an object in space, after the United States and the USSR. Beijing refused to confirm the test for nearly two weeks, infuriating other countries that had detected the huge debris field.[76]

The test seemed to sharply contradict Beijing's long support—along with Moscow—for a treaty outlawing the weaponization of space. Instead, "This is the first real escalation in the weaponization of space that we've seen in 20 years," according to Jonathan McDowell of Harvard University. Hitchens did not see a contradiction. Instead, she argued that the test might be a show of strength intended to bring Washington to the negotiating table, as the Bush administration had explicitly ruled out such a treaty in August 2006.[77]

In fact, the Chinese military initially denied that the test occurred. According to Phillip C. Saunders of the Institute for National Strategic Studies at the U.S. National Defense University and Charles D. Lutes at the National Security Council, the episode suggests either a serious lack of coordination within the Chinese security structures, a miscalculation about the fallout from the test (both literally and diplomatically), or both. The Ministry of Foreign Affairs and Ministry of Defense seem to have not known of the test ahead of time. Ultimately, a representative from the Ministry of Foreign Affairs issued a tepid statement: "This test was not directed at any country and does not constitute a threat to any country." The prime minister made a nearly identical statement.[78]

In contrast to the secrecy over the ASAT test, Beijing was perhaps too eager to announce the success of the September 27, 2008, space walk mission.

91

On September 25—before *Shenzhou 7* even left the launchpad—China's official news agency, Xinhua, posted an article describing the launch and subsequent space walk, even providing detailed dialogue between the astronaut outside his ship and his comrades still inside:

> After this order, signal lights all were switched on, various data show up on rows of screens, hundreds of technicians staring at the screens, without missing any slightest changes . . .
> "One minute to go!"
> Changjiang No.1 found the target!"
> The firm voice of the controller broke the silence of the whole ship. Now, the target is captured 12 seconds ahead of the predicted time . . .
> "The air pressure in the cabin is normal!"
> Ten minutes later, the ship disappears below the horizon. Warm clapping and excited cheering breaks the night sky, echoing across the silent Pacific Ocean.[79]

The exchange raises questions about the reliability of any official statement from Chinese authorities.

China's Stance on the Militarization of Space

China does not object to the use of space-based military assets, but it has long called for a ban on the weaponization of outer space. Beijing also favors international cooperation and collaboration in space. Along with Russia, it has jointly proposed an international treaty banning the deployment of weapons in outer space. But until the United States signs, China's leaders recognize the need for weapons and strategies that account for its relative military weakness, compared to the United States. China's ambassador to the Conference on Disarmament, Li Changhe, explained his country's reasoning:

> Military satellites involve rather complex issues and their role should not be all together negated. Therefore, the primary goal at present in our efforts to prevent the weaponization of and an arms race in outer space is to ban the testing, deployment and use of weapons, weapon systems and components in outer space.[80]

China's opposition to space weapons has at least three dimensions. First, it fears that any weapons the United States might deploy in outer space could become part of a missile defense system. Second, a deployed missile defense system could negate China's relatively small nuclear arsenal, forcing it into an expensive arms race. Third, China fears that such weapons could increase Washington's efforts at global dominance. The fact that many recent U.S. war games have used China as the fictional enemy is hardly reassuring.[81]

At the same time, China's neighbors are wary of Beijing's growing space prowess, knowing that the same launchers that launched Chinese astronauts could also launch Chinese warheads. One South Korean think tank, the Korea Institute for Defense Analyses, has urged Seoul to muster a united front against China that includes other concerned states, such as Vietnam, India, and Mongolia.[82] Chinese leaders, in contrast, appear worried that a joint disaster management agreement between Japan and India could lead to an anti-Beijing alliance.[83]

Beijing's successful use of a direct-ascent ASAT weapon to destroy a satellite in January 2007 immediately changed the security calculation for countries with objects in LEO. The event released more than 2,500 chunks of debris into a heavily populated zone. In addition to weather satellites, this zone contains at least 125 civilian and military satellites owned by many countries. The United States, for example, has reconnaissance, remote sensing, surveillance, and weather satellites at this altitude. These vital assets are potentially threatened in two ways: They could be targeted by Chinese ASAT weapons, and they could be damaged by the debris cloud left by the destroyed satellite.[84] Saunders and Lutes report that the test "appears to be part of a larger Chinese ASAT program that includes ground-based lasers and jamming of satellite signals."[85]

Furthermore, Saunders and Lutes also cite reports that not only is China developing ground-based lasers that could blind satellites and interfere with transmissions, they focused one of these lasers on a U.S. satellite in September 2006.[86] However, other experts, such as a panel from the Union of Concerned Scientists (UCS), analyzed these reports and offered an alternative, and much more benign, scenario—the laser belonged to one of China's satellite laser-ranging stations, which are used by countries around the world to track satellites.[87] China may also be developing microsatellites and a co-orbital ASAT weapon.

Taiwan considered the 2007 satellite interception to be "an aggressive act by the Chinese side . . . Taiwan stands out to be the first country that might have to suffer if a future conflict were to erupt between China and other countries."[88] In one plausible scenario, China could attack Taiwan and then jam or blind U.S. space-based military assets, delaying the arrival of U.S. forces until Taiwan surrendered. The ASAT strategy could work equally well against Taiwan's Formosat satellites, which are also deployed in LEO. This scenario is not merely academic, because the United States is obligated to defend the island by the U.S. Taiwan Relations Act of 1972.[89]

The United States, Japan, and Australia also voiced concern that China's actions could trigger a space arms race. The Japanese parliament overturned a ban on using its space program for military purposes.[90] In the United States, the dismay is not solely over the strategic implications. Each successful Chinese mission takes it closer to its stated goal of a manned lunar mission by 2017. A draft of a speech that former NASA administrator Michael Griffin was to make before Congress in 2008 contained the line, "The bare fact of this accomplishment will have an enormous, and not fully predictable, effect on global perceptions of U.S. leadership in the world." The line was later deleted.[91] China, for its part, is wary of an alliance between the space programs of Japan and India, fearing it might challenge China's current leading role in Asia. But that is not the only troubling aspect. According to the journalist Peter J. Brown, "China is concerned about the general effort of the U.S. during the Bush Administration to form a Japanese-Indian alliance to contain China. They are more concerned about what this implies about U.S. intentions rather than what it implies about the intentions of the Japanese or the Indians, particularly as it concerns space."[92]

Whether or not China deploys weapons in space may be difficult to confirm. Information about the Chinese space program is so highly classified that many top government officials are not informed of ongoing activities. It is difficult even to get physically near the launch site in Jiuquan, and the first manned spaceflights were not announced until they had been successfully completed.[93] The official response to the international uproar that followed Beijing's destruction of its own aging weather satellite suggests "a lack of internal coordination, perhaps due to security compartmentation."[94] Some analysts even believe that the PLA "either is acting alone in this matter or has such influence or acts with such little supervision that it can take significant actions without notifying other government organizations or even the top Chinese leadership."[95] Thus weaponization could signal or trigger an internal power struggle. Any of China's neighbors would feel threatened by Chinese weapons deployed in space.

At the same time, China appears to be pursuing a national security strategy based on asymmetric power. U.S. space weapons experts believe that China's policy targets certain perceived vulnerabilities in U.S. defense. Specifically, "Chinese strategic analysts are well aware of increasing U.S. military dependence on space-based assets; ASAT weapons can potentially exploit this vulnerability and reduce the American ability to operate in the western Pacific. . . . ASAT weapons are a logical and relatively inexpensive response to U.S. military dominance, which rests heavily on space capabilities."[96]

EMERGING PLAYERS

Japan

Japan has become a major player both in the satellite business and in space exploration despite the fact that its efforts to pursue military applications were long blocked by a 1969 parliamentary ban on military use of space. Japanese courts interpreted the provision very narrowly, ruling that it meant nonmilitary; that is, potential military assets, such as communication satellites were not allowed.[97] Tokyo skirted the issue in 1998 to permit the deployment of information-gathering satellites, as long as they were not controlled by the military. Two reconnaissance satellites were launched in 2003 and more in 2007. Two satellites were lost on October 29, 2003, when the H-2A rocket failed.

The Japanese parliament voted to abandon this rule in 2008. Under the Basic Space Law the threshold would become "nonaggressive" instead of "nonmilitary."[98] A new Strategic Headquarters for Space Policy, chaired by the president, would now oversee Japan's space policy. The president of the Japanese Aerospace Exploration Agency, Keiji Tachikawa, explained that the Basic Space Law has five planks: building prosperity; contributing to national security; promoting diplomacy; developing industries; and investing in national dreams and in the next generation, through projects such as planetary exploration and human space activity.[99]

Leaders hope that the changes in the space law will expand Japan's satellite market, as it will not be excluded from the military-related projects. In 2007, former education, science, and technology minister Takeo Kawamura warned that "[i]f the current state of affairs is left unattended, Japan is doomed to be outdone by China and India and fall into the ranks of underdeveloped countries as far as the space industry is concerned."[100]

HISTORY OF MILITARY POLICY IN JAPAN

When Japan surrendered at the end of World War II, the Allies drafted a new constitution to prevent a replay of Japan's militant history. The emperor was no longer considered to be a god, rights to privacy and individual liberty were introduced, the powers of the police were severely curtailed, the secret police were abolished, and Tokyo was stripped of an army. Article Nine of the Japanese Constitution of 1947 renounced the country's right to wage war. Specifically, it stated that Japan "forever renounces the threat or use of force as a means of settling international disputes" and "land, sea, and air forces . . . will never be maintained." Today, Japan has no military, aside from about 250,000 members of the Self-Defense Forces. The National Police

Agency has minimal intelligence-gathering capabilities. Instead, Japan has largely relied on the United States for national security. The two countries signed mutual cooperation treaties in 1952 and 1960.

Following the September 11, 2001, terrorist attacks against the United States, Japan began to take a greater role in international security issues. Tokyo had been criticized for not participating in the 1990–91 Gulf War, and Prime Minister Junichiro Koizumi was not going to allow a repeat with the War on Terror. He introduced a seven-point plan into parliament that, after much debate, allowed three Japanese Aegis cruisers to offer logistical support to U.S. forces in Afghanistan and to establish a refugee camp and hospital in Pakistan. Three Japanese destroyers and other small ships accompanied U.S. forces to the Middle East in late September 2001, where they would carry out support operations, such as refueling.[101] The country sent several hundred support troops to Iraq and in 2009 dispatched warships to deal with the pirates operating along Somalia's coastline.[102] Tokyo also agreed to provide humanitarian relief and share intelligence with the U.S. and coalition countries.[103]

Closer to home, Japan has long feared an attack from North Korea, and these fears have only increased with Pyongyang's nuclear program and missile testing. Indeed, Tokyo, partly in response to North Korea firing a missile in its direction in 1998, budgeted $58 million for the Japan Aerospace Exploration Agency (JAXA) to begin immediate research on a reconnaissance satellite.[104] The Japanese government also turned to the United States for protection. Japan could also mobilize its satellite program.

In December 2002, Washington assured Tokyo that the U.S. Aegis destroyers deployed near Japan could intercept medium-range missiles launched by North Korea.[105] Four years later, U.S.-Japanese cooperation intensified when the two countries agreed to jointly produce an interceptor missile. In June 2006, the United States moved its X-Band radar from the U.S. military base in Misawa to a Japanese Self-Defense Agency base to potentially detect incoming North Korean missiles.[106] Japan moved to further increase its protection in December 2007 when it upgraded its four Aegis destroyers to full Aegis ballistic missile defense capability.

North Korea's April 2009 missile test entered Japanese air space and prompted Japan to put the nation and military on high alert, while officials warned that they might even shoot the missile down. The incident triggered yet another round of domestic calls for Japan to abandon its pacifism and take action. However, Hideshi Takesada, executive director of the Japanese Ministry of Defense National Institute for Defense Studies, argues that the current constitution would permit the country to deploy cruise missiles and satellite-controlled alert systems as defensive weapons.[107]

Global Perspectives

CURRENT PROGRAM

Japan's space system is used more for commerce and science than security, although it does have intelligence-gathering satellites. The country launched its first two spy satellites in 2003 in reaction to North Korean missile tests. In November 2009, Japan launched a more sophisticated imaging satellite to replace one that had reached the end of its planned five-year life. The new satellite reportedly has a resolution of tens of centimeters while the older equipment was around one meter.[108]

JAXA is responsible for developing and launching satellites, especially ones for scientific research and communications. However, the company suffered a series of failed launches in the 2000s, damaging its efforts to enter the commercial launch market. While most launches have been for commercial satellites, a probe was sent to the Moon in 1990, and the agency has overseen a series of astronomy programs, focusing on X-ray technology, solar observations, Halley's comet, and other subjects. In February 2007, it launched its fourth spy satellite, using a Japanese-engineered H-2A rocket, followed in May 2010 by the launch of the *Akatsuki*, billed as the "first interplanetary weather satellite," en route to Venus.[109]

JAPAN'S STANCE ON THE MILITARIZATION OF SPACE

Given Japan's history of conquest in Asia, any effort for the country to expand its military capacity with long-range missiles or, especially, nuclear weapons would substantially alter the balance of power in the region. North Korea and perhaps China might see the need for preemptive attacks against Japanese military assets. While North Korea does not yet have the technology necessary for such precision air strikes, China likely does. However, Tokyo has consistently opposed proliferation, both on Earth and in outer space.

For now, the United States is providing Japan with missile defense technology to detect and intercept incoming missiles, presumably launched by North Korea. Should the Japanese government decide to pursue missile defense—or offense—on its own, it would mean a major reorientation of the country's postwar commitment to pacifism and could further destabilize Asia.

North Korea

The Democratic People's Republic of Korea (DPRK) is a tightly closed country run by a highly eccentric dictator Kim Jong Il. It is hostile toward its neighbors, and it can be difficult to determine if its belligerent statements are mere boasts or grounded in facts. It claims to have nuclear weapons, advanced satellites, and powerful missiles. The United States, China, and Japan use their own space-based assets to monitor the situation in North Korea.

The Korean peninsula has a long history of competing claims. Japan occupied the Korean peninsula from 1910 to 1945; after Japan surrendered at the end of World War II, the Allies agreed to divide Korea into two countries with the line of separation set at the 38th parallel. While the United States occupied South Korea, the Soviet Union occupied North Korea. On September 9, 1948, Kim Il Sung established a communist regime in North Korea, ruled by the Korean Workers' Party. With Soviet backing, North Korea invaded South Korea in 1950 in an attempt to reunify the peninsula. The United States joined the conflict to defend South Korea; a cease-fire was negotiated in 1953 and remains in effect.

Kim Il Sung established a harsh dictatorship based on the ideology of *juche*, self-reliance and independence. Pyongyang has sought to distance itself from both Moscow and Beijing and maintains a belligerent stance toward South Korea, Japan, and the United States. The reclusive regime rules with pervasive propaganda and harsh internment and reeducation camps for dissidents, and it manipulates the economy to maintain an extensive military.[110] Faulty economic planning, flooding, and poor harvests have led to widespread starvation. Kim Il Sung died in July 1994, and his eldest son, Kim Jong Il, succeeded him. The younger Kim reportedly suffered a stroke in August 2008, setting off rumors about the succession.

In the past, the DPRK regime received subsidized food, energy, and equipment from the Soviet Union and China. These funds were cut off after the dissolution of the USSR in late 1991, triggering a precipitous drop in the food supply. The government responded with rationing and even launched a campaign of "Let's Eat Two Meals a Day." Even with aid from the World Food Program, between 1994 and 1998 some 2 to 3 million people died of starvation, while millions more suffered permanent damage to their health. The famine severely undermined popular support for the Kim regime, which responded by spending even more funds toward military buildup. As in China, North Korea's leaders may use successful military and space programs to reinforce their domestic legitimacy.

HISTORY OF MISSILE PROGRAMS IN NORTH KOREA

North Korea has tested a nuclear device twice and is developing and producing ballistic missiles of increasing range. The next logical step would be to mount a nuclear weapon atop one of its Taepdong missiles, and a successfully deployed reconnaissance satellite could further facilitate targeting. The country is believed to have enough fuel for five to 12 nuclear weapons.

North Korea's missile program was initiated in the 1980s when it began to manufacture SCUD short-range missiles capable of hitting South Korea. The SCUD prototype and documentation came from Egypt in 1976, "in return for its support against Israel in the Yom Kippur War."[111] By 1984, North Korea manufactured both SCUDs and the medium-range Nodong missile. Since then the North Korean program has expanded to include the longer-range Taepdong missile, which is a combination of SCUD and Nodong technologies. Additional technology may also have been derived from a 1950s-era Soviet submarine-launched missile that likely was illegally sold to Pyongyang by a Russian arms trader around the time the USSR collapsed.[112]

Pyongyang announced that it successfully launched a Taepdong-1 ballistic missile on August 31, 1998. The government also claimed to have successfully deployed North Korea's first home-built satellite, *Kwangmyongsong-1*, which reportedly played "patriotic music" as it orbited the Earth.[113] However, the truth of this statement could not be confirmed by outsiders. International observers never detected the satellite or its transmissions, which North Korea stated were over the same frequency used by CB radios, and they believe a misfire in the third stage of the rocket prevented the satellite from being successfully placed into orbit. "The launch probably had multiple purposes: to serve as an advertisement for the country's missile technology and to serve as a bargaining chip to win concessions from the U.S."[114] A test of the Taepdong-2 ballistic missile in July 2006 failed shortly after launch. Japan roundly condemned the test and asked the UN to impose sanctions on the DPRK. When the DPRK tested a nuclear weapon on October 9, 2006, Tokyo went back to the UN, which issued UN Security Council Resolution 1718, and this time sanctions were imposed.

CURRENT PROGRAM

In April 2009, Pyongyang announced that not only had it launched a second long-range missile, it had carried a Kwangmyongsong-2 satellite that was successfully placed into orbit. Newspapers claimed Kim had personally observed the launch and declared, "It is a striking demonstration of the might of our Juche-oriented science and technology."[115] Again, experts monitoring the launch declared it a failure. Like the *Kwangmyongsong-1* satellite launched previously, this satellite also supposedly played patriotic music. Scientists at Harvard University and the U.S. Northern Command (USNORTHCOM) declared that the three-stage rocket failed to fully engage and that if the satellite was really playing patriotic tunes, as the North Koreans insisted, it was doing so on the Pacific Ocean floor.

North Korea's claims still alarmed Japan, South Korea, and the United States, as the launch could conceivably be a test run for a long-range missile powerful enough to deliver a nuclear weapon. The UN Security Council called an emergency session to discuss an appropriate response. After three hours, the participants had not reached a conclusion, saying they needed more proof that the launch had taken place. But U.S. ambassador Susan E. Rice dismissed this logic, saying, "We think that what was launched is not the issue; the fact that there was a launch using ballistic missile technology is itself a clear violation" of negotiated limits.[116] China, however, downplayed the incident, saying that like any country North Korea has a right to launch satellites. "Our position," explained China's ambassador to the UN, Yesui Zhang, "is that all countries concerned should show restraint and refrain from taking actions that might lead to increased tensions."

On April 24, 2009, the UN Security Council voted to sanction three state companies, Korea Mining Development Trading Corp., Tanchon Commercial Bank, and Korea Ryongbong General Corp., as well as to strengthen a trade ban on missile-relevant technology. The United States had previously sanctioned the same three firms for providing Iran, Yemen, and Pakistan with missile technology, but this was the first time the UN sanctioned firms.[117]

Pyongyang responded angrily, saying:

> Such sanctions can never work on the DPRK which has been subject to all sorts of sanctions and blockade by the hostile forces for the past scores of years.
>
> What is serious is the fact that the UNSC [United Nations Security Council] has set out in directly jeopardizing the security of the country and the nation, the supreme interests of the DPRK, though it had already wantonly infringed the sovereignty of a sovereign state, pursuant to the U.S. moves.[118]

The main fallout from the incident is likely to be within the DPRK's missile program. "It's got to be embarrassing," said the Massachusetts Institute of Technology missile expert Geoffrey E. Forden. "I can imagine heads flying if [Kim] finds out the satellite didn't fly into orbit."[119]

One month later, on May 23, North Korea detonated a second nuclear device and tested three short-range missiles. The United Nations imposed even tougher sanctions, including a new resolution "calling on states, for the first time, to seize banned weapons and technology from the North that are found aboard ships on the high seas."[120] The North Korean Ministry of Foreign Affairs responded defiantly, "It makes no difference to North Korea whether its nuclear status is recognized or not. It has become an absolutely impossible option for North Korea to even think about giving up its nuclear weapons."

NORTH KOREA'S STANCE ON THE MILITARIZATION OF SPACE

The North Korean government argues that it needs nuclear weapons and missile technology to deter outside attacks; South Korea, its archrival, has the capability to produce nuclear weapons and is protected by the defense policy of the United States, which has stated that nuclear weapons might be used against North Korea if it attacked South Korea. North Korea has also used military technology for economic gains and as a bargaining chip in international negotiations. It claims the right, like any other country, to launch satellites into space. In March 2009, the country joined the 1976 Convention on Registration of Objects Launched into Outer Space.

North Korea's dogged pursuit of nuclear materials and long-range missiles has made it an international outcast. President George W. Bush included the DPRK in his axis of evil and designated it as a state sponsor of terrorism for its weapons sales to Syria and Libya. China has always maintained a special relationship with North Korea, but the DPRK's nuclear and missile tests in 2006 finally caused Beijing to back away from the Kim regime. By 2009, however, the chill in China-DPRK relations appeared to thaw, as the two countries marked "60 years of friendship."[121] Beijing has also taken the lead in coaxing Pyongyang back to the Six-Party Talks, an international forum to discuss North Korea's nuclear program that began in August 2003 with the United States, China, and North Korea, later joined by South Korea, Japan, and Russia.

During these talks, both Japan and South Korea have threatened to strengthen their military capacity if North Korea continues to pursue ballistic missiles. Seoul has raised the possibility of developing long-range missiles, which would require U.S. approval. When it became clear that North Korea planned a missile test in April 2009, Japan threatened to shoot it down and put its military on high alert, preparing Patriot missile batteries and dispatching ships with sea-to-air destroyers.[122] Although it is not likely to have sufficient technology for missiles capable of reaching the United States, Washington and much of Europe fear that North Korea would sell its missiles to third parties or even non-state actors such as terrorists for economic gains.

As with most events in North Korea, missile launches have been extremely secretive, and they are not broadcast live. Proclamations of success are issued after launches and are often at odds with external scientific evaluations. However, as required by the 1976 Registration Convention, Pyongyang notified the International Maritime Organization ahead of its April 2009 rocket test, providing the expected location of debris.[123]

Given North Korea's insular nature, it is difficult to gauge popular attitudes toward the nuclear and missile programs, but launches have become national events, timed to coincide with major holidays both in North Korea and the United States. For example, the 1998 test came just days before the 50th anniversary of the establishment of North Korea. Other tests were seemingly scheduled to coincide with U.S. holidays, including July 4, 2006, and July 4, 2009, as well as Memorial Day weekend in May 2009.[124]

Since most North Korean ballistic missile tests have failed, its launch technology clearly lags behind other countries and private launch services. According to the arms control expert Jeffrey G. Lewis, "It's not unusual to have a series of failures at the beginning of a missile program. But they don't test enough to develop confidence that they're getting over the problems. . . . Given that both versions of the Taepdong-2 have failed now, we have very little confidence in the reliability of the system."[125] However, David Wright of the UCS predicts that a successful ballistic missile of this type could carry a 2,200-pound warhead for 3,700 miles—far enough to reach Alaska.[126] So far, North Korea's efforts to create nuclear bombs and missiles have apparently produced little, but should reality ever match the rhetoric, it will destabilize much of Asia.

South Korea

South Korea is investing heavily in space technology. After launching 11 satellites from other countries, South Korea developed its own launch vehicle with help from Russia. So far, the country has attempted twice, in August 2009 and June 2010, to launch a rocket into space, each time unsuccessfully. In April 2008, South Korea sent its first astronaut into space; it plans to have a completely domestic-built satellite by 2018 and to build and launch a lunar probe by 2025. Although South Korea insists that its interest in space technology is for purely scientific and peaceful purposes, the technology it develops can be used both for military and commercial purposes. South Korea also has the capability to produce nuclear weapons, although it abandoned its nuclear weapons program in the 1970s.

HISTORY OF SPACE PROGRAMS IN SOUTH KOREA

South Korea began a national satellite and space program in 1989 with the establishment of the Korea Aerospace Research Institute (KARI). France launched the first South Korean–made satellite in August 1992, and more followed using French and other contract launch services. However, KARI sought to develop its own launch capacity. As KARI's president, Choi Dong Whan, explained in 2000, "We don't want to depend on other countries anymore. We pay a lot of money to outside countries and we have to wait for launching time which is dependent on their schedule and wastes our time and budget."[127]

Thus in 2000, the South Korean government shifted KARI's goals from developing satellites to be launched by other countries to acquiring its own launch capabilities. Announcing the new mission, then president

Kim Dae Jung declared that South Korea would build "a working satellite-launching facility by the year 2005, constructed with Korean technology and equipment." Specifically, Seoul sought to one day have a launch vehicle that could carry long-range ballistic missiles in addition to satellites.[128] Kim also predicted significant developments with high-technology research in electronics and computer software.

The shift into domestic missile production changed the region's security situation. In 1979, the South Korean government had signed an agreement with the United States to refrain from acquiring missiles with a range beyond 100 miles. But when North Korea test-fired a medium-range missile toward Japan in 1998, Seoul asked to expand the limit to 300 miles. The United States declined, fearing it could lead to missile proliferation and destabilize the Korean peninsula. Reacting to the changes at KARI, a U.S. government official explained, "One of the reasons we are not fully pleased about countries' starting up space launch programs that are not economically viable is that they could export the launching vehicle to someone else who has no intention of using it for satellites."[129] Turned down by the United States, South Korea turned to Russia, where it spent $200 million for ballistic missile technology.[130]

CURRENT PROGRAM

The Korea Space Launch Vehicle (KSLV) that South Korea developed with help from Russia consists of two stages. The main stage is a Russian-made Naro-1 rocket, while the second stage, as well as any satellite payload, is South Korea–built. The government constructed a launch facility at the Naro Space Center on an island south of Seoul in 2008. Begun in 2003, the center cost an estimated $323 million.[131]

Seoul scheduled its first domestic satellite launch for August 19, 2009, but scrubbed the mission seven minutes before liftoff. One week later, KSLV-3 launched on August 25, but a glitch at separation meant the satellite was deployed 22 miles higher than intended, overshooting its orbit. KARI was unable to communicate with the satellite.[132] A second rocket was launched on June 10, 2010, after a one-day delay, but the rocket exploded shortly after liftoff. A third launch is currently being planned.

South Korea is also moving into manned space activities. The first South Korean astronaut, Ko San, was selected in 2007 and sent to Moscow to be trained by Russians and later transported by Russia to the *ISS*.[133] The astronaut selection process was turned into a television reality series—more than 36,000 people entered. As Kim Chang Woo, head of the Ministry of Science and Technology space division explained, "We want to raise the dreams and hopes of the young generation for the space program . . . We want to make space science more popular. So that's why we opened the astronaut selection to the entire population."[134] But in March 2008 Ko was accused of improperly removing materials from the training facility and fired. Instead, the backup astronaut, bioengineer Yi So-yeon, left for a 10-day stay at the space station in April 2008.[135] She became the first Korean and the second Asian woman to fly in space.

SOUTH KOREA'S STANCE ON THE MILITARIZATION OF SPACE

While South Korea's space program has concentrated on commercial and scientific applications, the technology can also be used to launch weapons. North Korea certainly realizes the potential military applications of South Korean launchers. Pyongyang called the 2006 launch of Arirang-2, a photoreconnaissance satellite, a "grave provocative act . . . aimed at spying on the North"[136] by the Korean Central News Agency (KCNA).

North Korea accused the United States and other countries of holding a double standard, when South Korea's August 2009 satellite launch did not receive the condemnation and sanctions applied following North Korea's satellite launch only months earlier. The U.S. Department of State dismissed North Korea's complaints. "The South Koreans have developed their program in a very open and transparent way and in keeping with the international agreements that they have signed on to," said spokesman Ian Kelly. "This is in stark contrast to the example set by North Korea, which has not abided by its international agreements."[137]

South Korea's space program also calls into question continued U.S. influence on the peninsula. Washington refused to help fund Seoul's space program, so the government turned to Russia. "They did go to the Americans," said Daniel Pinkston of the International Crisis Group, "and expected them to be forthcoming and provide the technology transfers. But in fact the U.S. government, because of proliferation concerns, has blocked them."[138]

India

India is becoming a formidable player in the international space market. With a reputation for high-quality engineering and information technology, India has a strong presence in the international satellite market, for both satellite construction and launch services. The country is located in a region of instability, facing threats from domestic and international terrorism, long-standing tensions with Pakistan, and a growing threat from Afghanistan. Satellite reconnaissance would provide intelligence about insurgent activities in the region's mountainous terrain, while the Indian armed forces are interested in developing ASAT technology with an eye toward balancing China's increasing space activity.

HISTORY OF SPACE PROGRAMS IN INDIA

India established the Indian Space Research Organisation (ISRO) in 1969, which launched its first satellite in 1980. "From ISRO's inception, a driving objective has been delivering services such as remote education, telemedicine, weather forecasting, and disaster monitoring—all efforts aiding primarily economic development."[139] Over the past four decades, India's remote-sensing satellites have been used to "find well water in dry regions, saving the government's drill-boring program $100 million."[140] India tested its first nuclear bomb in 1974 and was punished by being excluded from NASA and European space missions.[141] As a result, New Delhi turned to Moscow for help, and in 1975 the USSR launched India's first satellite, the Aryabhata. India's first man in space, Rakesh Sharma, spent time at the Soviet *Salyut-7* space station. In time, relations between the United States and India warmed, and on June 25, 2005, Washington and New Delhi established the India–United States Conference on Space Science.

India launched an unmanned lunar probe (Chandrayaan 1) on October 22, 2008. The probe carried an international payload, including scientific instruments provided by the ESA and a water-ice-sensing radar from the United States.[142] The next phase of the lunar project, the Chandrayaan 2, developed as a joint project with Russia, is tentatively set to deploy a land rover on the Moon in 2012.[143]

CURRENT PROGRAM

India has a thriving and lucrative satellite industry led by Antrix, a division of ISRO. The country has been able to launch satellites since 1980. In 2006, Antrix earned over $500 million—half of the total ISRO budget. Most of the funding is generated through satellite services. According to the Indian Space Agency, "its space program has earned a return of $2 on every dollar invested by the government."[144] India developed three launch vehicle prototypes and by 2008 had conducted 22 launch vehicle missions and 53 spacecraft missions.

Like Beijing, New Delhi repeatedly emphasized that its space program is based on home-grown technology, a point of national pride and indication of its self-sufficiency.[145] On January 10, 2007, the day before China's ASAT test, India became the fourth country to successfully retrieve a satellite after it de-orbited. This is a major benchmark in efforts to send a probe or manned mission to the Moon and back.

Also like China, India hopes to build on its satellite program by establishing a manned space program. In February 2009, ISRO announced that India would launch its own astronauts into orbit by 2015. Russia will help build the spacecraft and train Indian astronauts. Critics such as analyst Gopal Raj say the program is an expensive ploy to catch up with China. "It's becoming a national prestige issue. I am not sure what you get from astronauts in space. Even the Europeans, who are much richer, have not got manned spaceflight programs."[146]

INDIA'S STANCE ON THE MILITARIZATION OF SPACE

India was quite disturbed when China demonstrated its ASAT capability in January 2007. Following China's test, Indian prime minister Manmohan Singh joined with Russian president Vladimir Putin to call for an international treaty to maintain a "weapons free outer space." A few weeks later, President A. P. J. Kalim outlined a new vision for the Indian Air Force:

> The winner of future warfare will be warriors who can visualize the strength of the enemy, based on the current scientific and technological capabilities and develop suitable strategies for meeting any eventuality . . . I visualize the Indian Air Force of 2025 to be based on our scientific and technological competence in the development of communications satellites, high precision resource mapping satellites, missile systems, unmanned super-sonic aerial vehicles and electronics and communication systems.[147]

Writing about the "new space race" in Asia, *Newsweek* noted, "India in recent years has transformed its space program from a utilitarian affair of meteorological and communications satellites into a hyperactive project that seems designed to make a splash on the world stage."[148] In May 2010, India announced that it plans to put its first dedicated military satellite in space by March 2011. The purpose of the planned satellite is to provide enhanced reconnaissance capability to the Indian navy.

The government defines India's space security strategy in terms of "security of space assets and infrastructure, their renewal and expansion where needed, and the continuity of the operational services of its space assets."[149] India reportedly has been researching laser weapons and an exoatmospheric kill vehicle, which could be deployed in space in the future.[150]

China has begun to see India as a regional economic and military rival in Asia. India's increasing partnership with the United States has left Beijing apprehensive, as has the quadrilateral cooperation among the United States, India, Japan, and Australia.[151] Any moves India makes toward deploying weapons in space will not be well received by China or Pakistan.

Pakistan

Pakistan's space program is much less developed than India's. The national space agency, Pakistan Space and Upper Atmosphere Research Commission

(SUPARCO), was established in September 1961 and had some limited success in launching satellites. The agency was soon downgraded and quite underfunded for the next three decades. In 1990, Pakistan's first satellite, Badr-1, was launched by China. SUPARCO was revived by President Pervez Musharraf in 2005. That same year Pakistan, China, and seven other countries signed a treaty for the establishment of the Asia Pacific Space Cooperation Organization (APSCO) to promote use of space resources and space applications in the Asia-Pacific region. In May 2007, this led to a bilateral agreement between China and Pakistan meant to help rapidly develop Pakistan's satellite capabilities. China pledged to provide satellite technology, while Pakistan would allow a Chinese satellite tracking station on its territory.[152] The cooperation has led to a new generation of satellites named Paksat. Paksat IR is slated to be completed in 2010 with state of the art weather and communication technology and will replace Paksat 1, which had limited functionality. Three additional satellites are in the development stage in cooperation with China. Pakistan, a nuclear-armed country, also develops and builds ballistic missiles.

Iran

Tehran announced that it placed Omid (Hope), a telecommunications satellite, into orbit in January 2009. While Iran had hired Russia to launch its Sina-1 satellite in October 2005, Omid was launched by a Safir-2 rocket from Iran's new space center in an undisclosed location. According to President Mahmoud Ahmadinejad, the satellite would spread "monotheism, peace, and justice," to the world.[153] A second rocket, Kavoshgar-3 (Explorer), was successfully launched in February 2010.

Washington viewed Iran's increasing mastery of satellite technology with caution. "The technology that is used to get this satellite into orbit . . . is one that could also be used to propel long-range ballistic missiles," the Pentagon spokesman Geoff Morrell told reporters in Washington, while Secretary of State Hillary Rodham Clinton warned Iran to comply with United Nations and International Atomic Energy Agency guidelines on proliferation.[154] Iran's pursuit of missile and nuclear technology has raised many of the same fears surrounding North Korea's missile program.

CONCLUSION

Space has definitely become a more crowded realm in the post–cold war era. New players such as China, South Korea, India, and Iran have developed satellite, missile, and even manned space programs. While once the United States and the Soviet Union faced each other with nuclear warheads, warning satellites, and radars, now the two countries are finding common ground on many issues. However, the one thing that has not changed since *Sputnik* is international restraint on placing weapons in outer space—for now.

[1] Marc Kaufman. "U.S. Finds It's Getting Crowded Out There." *Washington Post* (9 July 2008), p. A1.

[2] Kaufman, "U.S. Finds It's Getting Crowded Out There."

[3] Everett C. Dolman. *Astropolitik: Classical Geopolitics in the Space Age.* New York: Frank Cass, 2002, p. 92.

[4] Walter A. McDougall. *The Heavens and the Earth: A Political History of the Space Age.* New York: Basic Books, 1985.

[5] McDougall, *The Heavens and the Earth,* p. 26.

[6] McDougall, *The Heavens and the Earth,* p. 34.

[7] Quoted in McDougall, *The Heavens and the Earth,* p. 35.

[8] Cited in McDougall, *The Heavens and the Earth,* p. 53.

[9] Victor Mizin. "Russian Perspectives on Space Security." In *Collective Security in Space: European Perspectives,* edited by John M. Logsdon, James Clay Moltz, and Emma S. Hinds. Washington, D.C.: George Washington University Space Policy Institute, 2007.

[10] Christoph Bluth. *Soviet Strategic Arms Policy before SALT.* New York: Cambridge University Press, 1992, pp. 52–53.

[11] Matthew Brzezinski. *Red Moon Rising: Sputnik and the Hidden Rivalries That Ignited the Space Age.* New York: Times Books, 2007, p. 148.

[12] For samples, see "Animals as Cold Warriors: Missiles, Medicine, and Man's Best Friend," U.S. National Library of Medicine. Available online. URL: http://www.nlm.nih.gov/exhibition/animals/laika.html. Accessed April 26, 2010.

[13] "Russia Opens Monument to Space Dog Laika." AP (April 11, 2008); Anatoly Zak. "The True Story of Laika the Dog." Space.com (November 3, 1999). Available online. URL: http://www.space.com/news/laika_anniversary_001103.html. Accessed April 26, 2010.

[14] Matthew Mowthorpe. *The Militarization and Weaponization of Space.* Lanham, Md.: Rowman and Littlefield, 2004, p. 57.

[15] Conventional practice refers to Soviet/Russian weapons by their U.S. or NATO designation.

[16] David Scott Yost. *Soviet Ballistic Missile Defense and the Western Alliance.* Cambridge, Mass.: Harvard University Press, 1988, p. 27.

[17] "V-1000 Griffon." FAS Special Weapons Monitor. Available online. URL: http://www.fas.org/spp/starwars/program/soviet/v-1000.htm. Accessed May 31, 2010.

[18] Mowthorpe, *The Militarization and Weaponization of Space,* p. 42.

[19] Yost, *Soviet Ballistic Missile Defense and the Western Alliance,* pp. 27–29.

[20] Giles Sparrow. *Spaceflight: The Complete Story from Sputnik to Shuttle—and Beyond.* New York: DK, 2007.

[21] Pavel Felgenhauer. "New PR for an Old Missile." In *CDI Russia Weekly* (December 14, 2004).

[22] Quoted in Michael Lemonick. "Surging Ahead: The Soviets Overtake the U.S. as the No. 1 Spacefaring Nation." *Time*, October 5, 1987, pp. 64–71.

[23] Anatoly Dobrynin. *In Confidence: Moscow's Ambassador to America's Six Cold War Presidents.* New York: Times Books, 1995, p. 319.

[24] Dobrynin, *In Confidence*, p. 393.

[25] Dobrynin, *In Confidence*, p. 534.

[26] Dobrynin, *In Confidence*, p. 529.

[27] "Gorbachev: 'It Would Have Taken a Madman to Accept' SDI Proposal." *Washington Post* (13 October 1986), p. A28; Johanna McGeary. "Sunk by Star Wars." *Time*, October 20, 1986.

[28] Celestine Bohlen. "Gorbachev Indicates Flexibility on SDI." *Washington Post* (1 December 1987), p. A1.

[29] Quoted in Lemonick, "Surging Ahead."

[30] Michael Spector. "Where Sputnik Once Soared into History, Hard Times Take Hold." *New York Times* (21 March 1995), p. C1; Ilya Kramnik. "Baikonur Space Center Marks 55th Anniversary." *Space Daily* (15 February 2010).

[31] John-Thor Dahlburg. "Ambitious Soviet Space Program Collapses with State." *Houston Chronicle* (29 March 1992), p. 26.

[32] Kathy Sawyer. "U.S. Proposes Space Merger with Russia." *Washington Post* (5 November 1993), p. A1.

[33] See information at International Launch Services Web site. Available online. URL: http://www.ilslaunch.com; Stephen Clark. "Russia Delays Angara Rocket Debut as Testing Progresses." *Spaceflight Now* (3 December 2009).

[34] Alejandro Lazo. "Russian Firm Buys McLean's International Launch." *Washington Post* (30 May 2008), p. D1.

[35] Marc Kaufman. "NASA Wary of Relying on Russia." *Washington Post* (7 March 2008), p. A1.

[36] Mizin, "Russian Perspectives on Space Security," p. 82.

[37] Mizin, "Russian Perspectives on Space Security," pp. 84–85.

[38] Mizin, "Russian Perspectives on Space Security."

[39] Mizin, "Russian Perspectives on Space Security," p. 85.

[40] Quoted in James Oberg. "An Outer-Space War of Words Escalates." MSNBC (November 10, 2006).

[41] Washington noted that the Topol was neither very new nor successful in its tests to date. "Russia Deploys Nuclear Missiles." *Los Angeles Times* (23 December 2003), p. A10.

[42] "Q&A: U.S. Missile Defense." BBC News (October 12, 2007). Available online. URL: http://news.bbc.uk/1/hi/world/europe/6720153.stm. Accessed June 6, 2010.

[43] Stephen R. Hurst. "Obama, Medvedev Vow to Rebuild U.S.-Russia Ties." *Boston Globe* (6 July 2006), p. A6.

[44] Pavel Felgenhauer. "New PR for an Old Missile." *CDI Russia Weekly* (14 December 2004).

[45] Brian Harvey. *Europe's Space Programme: To Ariane and Beyond.* New York: Springer, 2003; Roger Bonnet and Vittorio Manno. *International Cooperation in Space: The Example of the European Space Agency.* Cambridge, Mass: Harvard University Press, 1994.

[46] Jonathan Amos. "European Spaceport's Sky-High Ambition." BBC News (March 7, 2008).

[47] Arianespace Web site. Available online. URL: http://www.arianespace.com/launch-services/fleet_brochure.pdf. Accessed April 20, 2010.

[48] Anne Marie Squeo. "U.S. Considers European Role for Missile Defense Program." *Wall Street Journal* (7 June 2002), p. A4.

[49] Jonathan Amos. "A Key Moment in Europe's Space Journey." BBC News (February 8, 2008).

[50] Amos, "A Key Moment in Europe's Space Journey."

[51] Quoted in Jonathan Amos. "Huge Space Truck Races into Orbit." BBC News (March 9, 2008).

[52] "Second ATV Heading to Kourou for Launch." ESA Press Release, May 11, 2010. Available online. URL: http://www.esa.int/export/SPECIALS/ATV/SEMMYA19Y8G_0.html. Accessed June 28, 2010.

[53] Jonathan Amos. "European Spaceport's Sky-High Ambition." BBC News (March 7, 2008).

[54] ESA Prepares for the Next Generation of Human Spaceflight and Exploration by Recruiting a New Class of European Astronauts." ESA Press Release No. 12-2009. Available online. URL: http://www.esa.int/esaCP/Pr_12_2009_p_EN.html. Accessed June 28, 2010.

[55] Mowthorpe, *The Militarization and Weaponization of Space*, p. 85.

[56] Mowthorpe, *The Militarization and Weaponization of Space*, p. 86.

[57] Mary Hennock, Adam B. Kushner, and Jason Overdorf. "The New Space Race." *Newsweek*, October 20, 2008.

[58] Marcia S. Smith. "China's Space Program: A Brief Overview Including Commercial Launches of U.S.-Built Satellites." In CRS Report for Congress, September 3, 1998.

[59] Mowthorpe, *The Militarization and Weaponization of Space*, p. 91.

[60] Paul Rincon. "What's Driving China Space Efforts?" BBC News (September 25, 2008).

[61] Kevin Pollpeter. "Competing Perceptions of the U.S. and Chinese Space Programs." In *China Brief* (10 January 2007).

[62] "China Confirms Satellite Downed." BBC News (January 23, 2007).

[63] University of Mississippi School of Law. "China's Space Activities in 2006." Available online. URL: http://www.spacelaw.olemiss.edu/library/space/China/whitepapers/2006-whitepaper.pdf. Accessed May 29, 2010.

[64] Pollpeter, "Competing Perceptions of the U.S. and Chinese Space Programs."

[65] Hennock, Kushner, and Overdorf, "The New Space Race."

[66] Amy Klamper. "Official Details 11-Year Path to Developing China's Own Space Station." Space News. Available online. URL: http://www.spacenews.com/civil/100414-path-china-space-station.html. Accessed May 29, 2010.

Global Perspectives

[67] Ben Blanchar. "China Counts Down Minutes to Blast-off." Reuters (September 25, 2008).

[68] Pollpeter, "Competing Perceptions of the U.S. and Chinese Space Programs."

[69] Union of Concerned Scientists. "UCS Satellite Database." Available online. URL: http://www.ucsusa.org/nuclear_weapons_and_global_security/space_weapons/technical_issues/ucs-satellite-database.html. Accessed May 29, 2010.

[70] Smith, "China's Space Program."

[71] John C. Baker and Kevin Pollpeter. "A Future for U.S.-China Space Cooperation?" In *RAND Commentary* (13 December 2004).

[72] Amy Klamper, "Official Details 11-Year Path to Developing China's Own Space Station."

[73] However, Zhai's rescue buddy wore a Russian-made suit. Emma Graham-Harrison. "China's Spacewalk Astronauts Return as Heroes." Reuters (September 28, 2008).

[74] "Lift-off for China Space Mission." BBC News (September 25, 2008).

[75] Hennock, Kushner, and Overdorf, "The New Space Race."

[76] "China Confirms Satellite Downed." Phillip C. Saunders and Charles D. Lutes. "China's ASAT Test: Motivations and Implications." Washington, D.C.: Institute for National Strategic Studies, National Defense University, 2007.

[77] William J. Broad and David E. Sanger. "Flexing Muscle, China Destroys Satellite in Test." *New York Times* (19 January 2007).

[78] Edward Cody. "China Confirms Firing Missile to Destroy Satellite." *Washington Post* (24 January 2007).

[79] "China Space Mission Article Hits Web before Launch." AP (September 25, 2008).

[80] Nuclear Threat Initiative. "China's Attitude Toward Outer Space Weapons." Available online. URL:http://www.nti.org/db/china/spacepos.htm. Accessed May 29, 2010.

[81] Nuclear Threat Initiative. "China's Attitude Toward Outer Space Weapons."

[82] Scott Snyder and See-won Byun. "Year of China-DPRK Friendship; North's Rocket Fizzles." In *Comparative Connections* (April 2009).

[83] Peter J. Brown. "China Fears India-Japan Space Alliance." *Asia Times* (12 November 2008).

[84] Brown, "China Fears India-Japan Space Alliance."

[85] Brown, "China Fears India-Japan Space Alliance."

[86] Kevin Pollpeter. "Motives and Implications Behind China's ASAT Test." In *China Brief* (9 May 2007).

[87] Union of Concerned Scientists. "Satellite Laser Ranging in China." Available online. URL:http://www.ucsusa.org/nuclear_weapons_and_global_security/space_weapons/technical_issues/chinese-lasers-and-us.html. Accessed May 29, 2010.

[88] "China Confirms Satellite Downed." BBC News. (January 23, 2007).

[89] This scenario is examined in Michael O'Hanlon. *Neither Star Wars Nor Sanctuary: Constraining the Military Use of Space*. Washington, D.C.: Brookings Institution, 2004, ch. 4.

[90] Hennock, Kushner, and Overdorf, "The New Space Race."

[91] "Shooting the Moon." *Economist,* September 27, 2008.

[92] Brown, "China Fears India-Japan Space Alliance."

[93] "Shooting the Moon."

[94] Saunders and Lutes, "China's ASAT Test."

[95] Kevin Pollpeter, "Motives and Implications Behind China's ASAT Test."

[96] Saunders and Lutes, "China's ASAT Test."

[97] Setsuko Aoki. "Japanese Perspectives on Space Security". In *Collective Security in Space: Asian Perspectives,* edited by John M. Logsdon and James Clay Moltz. Washington, D.C.: George Washington University Space Policy Institute, 2008.

[98] Manuel Manriquez. "Japan's Space Law Revision: The Next Step toward Re-Militarization?" In *NTI: Issue Brief* (2008).

[99] JAXA. "2009: A New Era for Japan's Space Program." Available online. URL:http://www.jaxa.jp/article/interview/vol44/index_e.html. Accessed May 29, 2010.

[100] Manriquez, "Japan's Space Law Revision: The Next Step toward Re-Militarization?"

[101] "Japan Decides to Continue to Dispatch MSDF Vessels to the Indian Ocean in Order to Support International Efforts to Fight Against Terrorism." Ministry of Foreign Affairs of Japan press release, October 27, 2005. Available online. URL: www.jofa.go.jp.

[102] John Murphy and Yumiko Ono. "Japanese Pacifism Is Put to the Test." *Wall Street Journal* (5 April 2009).

[103] Larry Wortzel. "Joining Forces Against Terrorism: Japan's New Law Commits More Than Words to U.S. Effort." Heritage Foundation. Available online. URL: http://www.heritage.org/Research/Reports/2001/11/Joining-Forces-Against-Terrorism. Accessed April 20, 2010.

[104] Manriquez, "Japan's Space Law Revision: The Next Step toward Re-Militarization?"

[105] James Brooke. "Japan Fears North Korea; U.S. Promises Defense Shield." *New York Times* (26 December 2002), p. A6.

[106] Joseph Coleman. "U.S., Japan Expand Missile-Defense Plan." *Washington Post* (23 June 2006).

[107] Center for Defense Information. "Japan's Recent Step-up in Missile Defense." (October 10, 2003); Masako Toki. "Missile Defense in Japan." *Bulletin of the Atomic Scientists* (January 16, 2009).

[108] "New Spy Satellite Launched into Orbit." *Japan Times* (29 November 2009).

[109] "Japan Launches New Spy Satellite." BBC News (February 24, 2007); Stephen Clark. "Dawn Launch Sends Japanese Orbiter on the Way to Venus." *Spaceflight Now* (May 20, 2010).

[110] Andrew Natsios. "The Politics of Famine in North Korea." *USIP Special Report.* Washington, D.C.: U.S. Institute of Peace, 1999.

[111] "North Korea's Missile Programme." BBC News (May 27, 2009).

[112] Joby Warrick and R. Jeffrey Smith. "U.S.-Russian Team Deems Missile Shield in Europe Ineffective." *Washington Post* (19 May 2009).

[113] "North Korea: Military Programs." James Martin Center for Nonproliferation Studies. Available online. URL: http://cns.miis.edu/space/nkorea.mil.htm.

[114] Federation of American Scientists. "North Korea Space Guide." *World Space Guide* (1998).

[115] John M. Galionna. "North Korea's Rocket Didn't Reach Orbit, but Kim's in Another World." *Los Angeles Times* (6 April 2009).

[116] Helene Cooper and David E. Sanger. "Obama Seizes on Missile Launch in Seeking Nuclear Cuts." *New York Times* (6 April 2009).

[117] Colum Lynch. "U.N. Sanctions 3 N. Korean Firms over Missile Launch." *Washington Post* (25 April 2009).

[118] Council on Foreign Relations. Available online. URL: http://www.cfr.org/publication/19254/statement_by_north_korea_on_nuclear_tests_april_2009.html?breadcrumb=%2Fregion%2F276%2Fnorth_korea. Accessed April 26, 2010.

[119] William J. Broad. "North Korean Missile Launch Was a Failure, Experts Say." *New York Times* (6 April 2009).

[120] Blaine Harden. "North Korea Says It Will Start Enriching Uranium." *Washington Post* (14 June 2009).

[121] Snyder and Byun, "Year of China-DPRK Friendship; North's Rocket Fizzles."

[122] Murphy and Ono, "Japanese Pacifism Is Put to the Test."

[123] Broad, "North Korean Missile Launch Was a Failure, Experts Say."

[124] Choe Sang-hun. "Defying U.S., N. Korea Fires Barrage of Missiles." *New York Times* (4 July 2009).

[125] Broad, "North Korean Missile Launch Was a Failure, Experts Say."

[126] Broad, "North Korean Missile Launch Was a Failure, Experts Say."

[127] Calvin Sims. "South Korea Plans to Begin Rocket Program." *New York Times* (15 January 2000), p. A3.

[128] Sims, "South Korea Plans to Begin Rocket Program."

[129] Quoted in Sims, "South Korea Plans to Begin Rocket Program."

[130] R. Jeffrey Smith and Stella Kim. "S. Korea Launch Raises Questions." *New York Times* (18 August 2009), p. A8.

[131] "Boost for South Korea's Space Program." *Asia Times* (June 6, 2007).

[132] Evan Ramstad "Korean Rocket Overshoots Orbit." *Wall Street Journal* (26 August 2009), p. A10.

[133] John Sudworth. "South Korea Buys into Space Dream." BBC News (November 11, 2007).

[134] Sudworth, "South Korea Buys into Space Dream."

[135] Hyung-Jin Kim. "South Korea Switches Astronauts After Violations." *USA Today* (10 March 2008).

[136] KCNA News. "S. Korea's Launch of Spy Satellite under Fire." Available online. URL: http://www.kcna.co.jp/. Accessed April 20, 2010.

[137] Ramstad, "Korean Rocket Overshoots Orbit."

[138] Quoted in Sudworth, "South Korea Buys into Space Dream."

[139] Jessica Guiney. "India's Space Ambitions: Headed Toward Space War?" CDI Policy Brief (May 2008).

[140] Guiney, "India's Space Ambitions: Headed Toward Space War?"

[141] Guiney, "India's Space Ambitions: Headed Toward Space War?"

[142] "India's Flag Is on the Moon." Available online. URL: www.spacetoday.org/india/IndiaMoonFlights.html. Accessed April 20, 2010.

[143] "India's Flag Is on the Moon."

[144] "India's Flag Is on the Moon."

[145] "India's Flag Is on the Moon."

[146] Randeep Ramesh. "India to Launch Its First Astronauts into Space by 2015." *Guardian* (23 February 2009).

[147] "IAF Will Be Model for World by 2025: Kalam." *India Times* (7 March 2007).

[148] Hennock, Kushner, and Overdorf, "The New Space Race."

[149] Dipankar Banerjee. "Indian Perspectives on Regional Space Security." In *Collective Security in Space: Asian Perspectives*, edited by John M. Logsdon and James Clay Moltz. Washington, D.C.: George Washington University Space Policy Institute, 2008.

[150] Peter B. de Selding. "India Developing Means to Destroy Satellite." *Space News* (4 January 2010).

[151] Banerjee, "Indian Perspectives on Regional Space Security."

[152] Syed Fazi-e-Haider. "China, Pakistan Cooperate in Space." *Asia Times* (26 April 2007).

[153] "Iran Launches Homegrown Satellite." BBC News (February 2, 2009).

[154] Borzou Daragahi. "Iran Satellite Launch Raises Alarm in West." *Los Angeles Times* (4 February 2009), p. A1.

PART II

Primary Sources

4

United States Documents

National Aeronautics and Space Act of 1958 (excerpt)

Released a few months after Sputnik, *this document firmly declares that the U.S. space program is for peaceful, scientific purposes only. To emphasize this orientation, the Act created a civilian administrative body—NASA.*

DECLARATION OF POLICY AND PURPOSE

Sec. 102. (a) The Congress hereby declares that it is the policy of the United States that activities in space should be devoted to peaceful purposes for the benefit of all mankind.

(b) The Congress declares that the general welfare and security of the United States require that adequate provision be made for aeronautical and space activities. The Congress further declares that such activities shall be the responsibility of, and shall be directed by, a civilian agency exercising control over aeronautical and space activities sponsored by the United States, except that activities peculiar to or primarily associated with the development of weapons systems, military operations, or the defense of the United States (including the research and development necessary to make effective provision for the defense of the United States) shall be the responsibility of, and shall be directed by, the Department of Defense; and that determination as to which such agency has responsibility for and direction of any such activity shall be made by the President in conformity with section 201 (e).

(c) The aeronautical and space activities of the United States shall be conducted so as to contribute materially to one or more of the following objectives:

(1) The expansion of human knowledge of phenomena in the atmosphere and space;

(2) The improvement of the usefulness, performance, speed, safety, and efficiency of aeronautical and space vehicles;

(3) The development and operation of vehicles capable of carrying instruments, equipment, supplies and living organisms through space;

(4) The establishment of long-range studies of the potential benefits to be gained from, the opportunities for, and the problems involved in the utilization of aeronautical and space activities for peaceful and scientific purposes.

(5) The preservation of the role of the United States as a leader in aeronautical and space science and technology and in the application thereof to the conduct of peaceful activities within and outside the atmosphere.

(6) The making available to agencies directly concerned with national defenses of discoveries that have military value or significance, and the furnishing by such agencies, to the civilian agency established to direct and control nonmilitary aeronautical and space activities, of information as to discoveries which have value or significance to that agency;

(7) Cooperation by the United States with other nations and groups of nations in work done pursuant to this Act and in the peaceful application of the results, thereof; and

(8) The most effective utilization of the scientific and engineering resources of the United States, with close cooperation among all interested agencies of the United States in order to avoid unnecessary duplication of effort, facilities, and equipment.

[. . .]

TITLE II—COORDINATION OF AERONAUTICAL AND SPACE ACTIVITIES NATIONAL AERONAUTICS AND SPACE COUNCIL

[. . .]

NATIONAL AERONAUTICS AND SPACE ADMINISTRATION

Sec. 202. (a) There is hereby established the National Aeronautics and Space Administration (hereinafter called the "Administration"). The Administration shall be headed by an Administrator, who shall be appointed from civilian life by the President by and with the advice and consent of the Senate, and shall receive compensation at the rate of $22,500 per annum. Under the supervision and direction of the President, the Administrator shall be responsible for the exercise of all powers and the discharge of all duties of the Administration, and shall have authority and control over all personnel and activities, thereof.

(b) There shall be in the Administration a Deputy Administrator, who shall be appointed from civilian life by the President by and with the advice and consent of the Senate, shall receive compensation of $21,500 per annum, and shall perform such duties and exercise such powers as the Administrator may prescribe. The Deputy Administrator shall act for, and exercise the powers of, the Administrator during his absence or disability.

(c) The Administrator and the Deputy Administrator shall not engage in any other business, vocation, or employment while serving as such.

Source: NASA History Division. "Key Documents in the History of Space Policy." Available online. URL: http://history.nasa.gov/spdocs.html#1950s. Accessed April 26, 2010.

"Urgent National Needs" Speech by President John F. Kennedy (May 25, 1961) (excerpt)

On April 12, 1961, Soviet cosmonaut Yuri Gagarin became the first man to orbit the Earth. Six weeks later, President John F. Kennedy came to Capitol Hill to rally the American population to meet the challenges of the cold war. In the most famous portion of his speech, Kennedy declared, "this nation should commit itself to achieving the goal, before this decade is out, of landing a man on the moon and returning him safely to the Earth."

Finally, if we are to win the battle that is now going on around the world between freedom and tyranny, the dramatic achievements in space which occurred in recent weeks should have made clear to us all, as did the Sputnik in 1957, the impact of this adventure on the minds of men everywhere, who are attempting to make a determination of which road they should take. Since early in my term, our efforts in space have been under review. With the advice of the Vice President, who is Chairman of the National Space Council, we have examined where we are strong and where we are not, where we may succeed and where we may not. Now it is time to take longer strides—time for a great new American enterprise—time for this nation to take a clearly leading role in space achievement, which in many ways may hold the key to our future on earth.

I believe we possess all the resources and talents necessary. But the facts of the matter are that we have never made the national decisions or marshalled the national resources required for such leadership. We have never specified long-range goals on an urgent time schedule, or managed our resources and our time so as to insure their fulfilment.

Recognizing the head start obtained by the Soviets with their large rocket engines, which gives them many months of lead-time, and recognizing the likelihood that they will exploit this lead for some time to come in still more impressive successes, we nevertheless are required to make new efforts on our own. For while we cannot guarantee that we shall one day be first, we can guarantee that any failure to make this effort will make us last. We take an additional risk by making it in full view of the world, but as shown by the feat of astronaut Shepard, this very risk enhances our stature when we are successful. But this is not merely a race. Space is open to us now; and our eagerness to share its meaning is not governed by the efforts of others. We go into space because whatever mankind must undertake, free men must fully share.

I therefore ask the Congress, above and beyond the increases I have earlier requested for space activities, to provide the funds which are needed to meet the following national goals:

First, I believe that this nation should commit itself to achieving the goal, before this decade is out, of landing a man on the moon and returning him safely to the earth. No single space project in this period will be more impressive to mankind, or more important for the long-range exploration of space; and none will be so difficult or expensive to accomplish. We propose to accelerate the development of the appropriate lunar spacecraft. We propose to develop alternate liquid and solid fuel boosters, much larger than any now being developed, until certain which is superior. We propose additional funds for other engine development and for unmanned explorations— explorations which are particularly important for one purpose which this nation will never overlook: the survival of the man who first makes this daring flight. But in a very real sense, it will not be one man going to the moon—if we make this judgment affirmatively, it will be an entire nation. For all of us must work to put him there.

Secondly, an additional $23 million, together with $7 million already available, will accelerate development of the Rover nuclear rocket. This gives promise of some day providing a means for even more exciting and ambitious exploration of space, perhaps beyond the moon, perhaps to the very end of the solar system itself.

Third, an additional $50 million will make the most of our present leadership, by accelerating the use of space satellites for world-wide communications.

Fourth, an additional $75 million—of which $53 million is for the Weather Bureau—will help give us at the earliest possible time a satellite system for world-wide weather observation.

Let it be clear—and this is a judgment which the Members of the Congress must finally make—let it be clear that I am asking the Congress and the country to accept a firm commitment to a new course of action, a course which will last for many years and carry very heavy costs: $531 million in fiscal '62—an estimated seven to nine billion dollars additional over the next five years. If we are to go only half way, or reduce our sights in the face of difficulty, in my judgment it would be better not to go at all.

Now this is a choice which this country must make, and I am confident that under the leadership of the Space Committees of the Congress, and the Appropriating Committees, that you will consider the matter carefully.

It is a most important decision that we make as a nation. But all of you have lived through the last four years and have seen the significance of space and the adventures in space, and no one can predict with certainty what the ultimate meaning will be of mastery of space.

I believe we should go to the moon. But I think every citizen of this country as well as the Members of the Congress should consider the matter carefully in making their judgment, to which we have given attention over many weeks and months, because it is a heavy burden, and there is no sense in agreeing or desiring that the United States take an affirmative position in outer space, unless we are prepared to do the work and bear the burdens to make it successful. If we are not, we should decide today and this year.

This decision demands a major national commitment of scientific and technical manpower, materiel and facilities, and the possibility of their diversion from other important activities where they are already thinly spread. It means a degree of dedication, organization and discipline which have not always characterized our research and development efforts. It means we cannot afford undue work stoppages, inflated costs of material or talent, wasteful interagency rivalries, or a high turnover of key personnel.

New objectives and new money cannot solve these problems. They could in fact, aggravate them further—unless every scientist, every engineer, every serviceman, every technician, contractor, and civil servant gives his personal pledge that this nation will move forward, with the full speed of freedom, in the exciting adventure of space.

Source: John F. Kennedy. "Special Message to the Congress on Urgent National Needs." May 25, 1961. Available online. URL: http://www.jfklibrary.org/Historical+Resources/Archives/Reference+Desk/Speeches/JFK/003POF03National Needs05251961.htm. Accessed May 20, 2010.

Kennedy Address Before the 18th General Assembly of the United Nations (September 20, 1963) (excerpt)

In his final appearance at the United Nations, President John F. Kennedy reminded listeners that the world had been at the brink of nuclear war one year earlier, with the Cuban Missile Crisis. Now he reached out to Moscow, proposing a joint U.S.-USSR lunar mission.

We meet again in the quest for peace.

Twenty-four months ago, when I last had the honor of addressing this body, the shadow of fear lay darkly across the world. The freedom of West Berlin was in immediate peril. Agreement on a neutral Laos seemed remote. The mandate of the United Nations in the Congo was under fire. The financial outlook for this organization was in doubt. Dag Hammarskjold was dead. The doctrine of troika was being pressed in his place, and atmospheric tests had been resumed by the Soviet Union.

Those were anxious days for mankind—and some men wondered aloud whether this organization could survive. But the 16th and 17th General Assemblies achieved not only survival but progress. Rising to its responsibility, the United Nations helped reduce the tensions and helped to hold back the darkness.

Today the clouds have lifted a little so that new rays of hope can break through. The pressures on West Berlin appear to be temporarily eased.

Political unity in the Congo has been largely restored. A neutral coalition in Laos, while still in difficulty, is at least in being. The integrity of the United Nations Secretariat has been reaffirmed. A United Nations Decade of Development is under way. And, for the first time in 17 years of effort, a specific step has been taken to limit the nuclear arms race.

I refer, of course, to the treaty to ban nuclear tests in the atmosphere, outer space, and under water—concluded by the Soviet Union, the United Kingdom, and the United States—and already signed by nearly 100 countries. It has been hailed by people the world over who are thankful to be free from the fears of nuclear fallout, and I am confident that on next Tuesday at 10:30 o'clock in the morning it will receive the overwhelming endorsement of the Senate of the United States.

The world has not escaped from the darkness. The long shadows of conflict and crisis envelop us still. But we meet today in an atmosphere of rising hope, and at a moment of comparative calm. My presence here today is not a sign of crisis, but of confidence. I am not here to report on a new threat to the peace or new signs of war. I have come to salute the United Nations and to show the support of the American people for your daily deliberations.

For the value of this body's work is not dependent on the existence of emergencies—nor can the winning of peace consist only of dramatic victories. Peace is a daily, a weekly, a monthly process, gradually changing opinions, slowly eroding old barriers, quietly building new structures. And however undramatic the pursuit of peace, that pursuit must go on.

Today we may have reached a pause in the cold war—but that is not a lasting peace. A test ban treaty is a milestone—but it is not the millennium. We have not been released from our obligations—we have been given an opportunity. And if we fail to make the most of this moment and this momentum—if we convert our new-found hopes and understandings into new walls and weapons of hostility—if this pause in the cold war merely leads to its renewal and not to its end—then the indictment of posterity will rightly point its finger at us all. But if we can stretch this pause into a period of cooperation—if both sides can now gain new confidence and experience in concrete collaborations for peace—if we can now be as bold and farsighted in the control of deadly weapons as we have been in their creation—then surely this first small step can be the start of a long and fruitful journey.

123

The task of building the peace lies with the leaders of every nation, large and small. For the great powers have no monopoly on conflict or ambition. The cold war is not the only expression of tension in this world—and the nuclear race is not the only arms race. Even little wars are dangerous in a nuclear world. The long labor of peace is an undertaking for every nation— and in this effort none of us can remain unaligned. To this goal none can be uncommitted.

The reduction of global tension must not be an excuse for the narrow pursuit of self-interest. If the Soviet Union and the United States, with all of their global interests and clashing commitments of ideology, and with nuclear weapons still aimed at each other today, can find areas of common interest and agreement, then surely other nations can do the same—nations caught in regional conflicts, in racial issues, or in the death throes of old colonialism. Chronic disputes which divert precious resources from the needs of the people or drain the energies of both sides serve the interests of no one—and the badge of responsibility in the modern world is a willingness to seek peaceful solutions.

It is never too early to try; and it's never too late to talk; and it's high time that many disputes on the agenda of this Assembly were taken off the debating schedule and placed on the negotiating table.

The fact remains that the United States, as a major nuclear power, does have a special responsibility in the world. It is, in fact, a threefold responsibility— a responsibility to our own citizens; a responsibility to the people of the whole world who are affected by our decisions; and to the next generation of humanity. We believe the Soviet Union also has these special responsibilities—and that those responsibilities require our two nations to concentrate less on our differences and more on the means of resolving them peacefully. For too long both of us have increased our military budgets, our nuclear stockpiles, and our capacity to destroy all life on this hemisphere—human, animal, vegetable—without any corresponding increase in our security.

Our conflicts, to be sure, are real. Our concepts of the world are different. No service is performed by failing to make clear our disagreements. A central difference is the belief of the American people in the self-determination of all people.

We believe that the people of Germany and Berlin must be free to reunite their capital and their country.

We believe that the people of Cuba must be free to secure the fruits of the revolution that have been betrayed from within and exploited from without.

In short, we believe that all the world—in Eastern Europe as well as Western, in Southern Africa as well as Northern, in old nations as well as new—that people must be free to choose their own future, without discrimination or dictation, without coercion or subversion.

These are the basic differences between the Soviet Union and the United States, and they cannot be concealed. So long as they exist, they set limits to agreement, and they forbid the relaxation of our vigilance. Our defense around the world will be maintained for the protection of freedom—and our determination to safeguard that freedom will measure up to any threat or challenge.

But I would say to the leaders of the Soviet Union, and to their people, that if either of our countries is to be fully secure, we need a much better weapon than the H-bomb—a weapon better than ballistic missiles or nuclear submarines—and that better weapon is peaceful cooperation.

We have, in recent years, agreed on a limited test ban treaty, on an emergency communications link between our capitals, on a statement of principles for disarmament, on an increase in cultural exchange, on cooperation in outer space, on the peaceful exploration of the Antarctic, and on temporing last year's crisis over Cuba.

I believe, therefore, that the Soviet Union and the United States, together with their allies, can achieve further agreements—agreements which spring from our mutual interest in avoiding mutual destruction.

There can be no doubt about the agenda of further steps. We must continue to seek agreements on measures which prevent war by accident or miscalculation. We must continue to seek agreements on safeguards against surprise attack, including observation posts at key points. We must continue to seek agreement on further measures to curb the nuclear arms race, by controlling the transfer of nuclear weapons, converting fissionable materials to peaceful purposes, and banning underground testing, with adequate inspection and enforcement. We must continue to seek agreement on a freer flow of information and people from East to West and West to East.

We must continue to seek agreement, encouraged by yesterday's affirmative response to this proposal by the Soviet Foreign Minister, on an arrangement to keep weapons of mass destruction out of outer space. Let us get our negotiators back to the negotiating table to work out a practicable arrangement to this end.

In these and other ways, let us move up the steep and difficult path toward comprehensive disarmament, securing mutual confidence through mutual verification, and building the institutions of peace as we dismantle the engines of war. We must not let failure to agree on all points delay agreements where agreement is possible. And we must not put forward proposals for propaganda purposes.

Finally, in a field where the United States and the Soviet Union have a special capacity—in the field of space—there is room for new cooperation, for further joint efforts in the regulation and exploration of space. I include among these possibilities a joint expedition to the moon. Space offers no problems of sovereignty; by resolution of this Assembly, the members of the United Nations have foresworn any claim to territorial rights in outer space or on celestial bodies, and declared that international law and the United Nations Charter will apply. Why, therefore, should man's first flight to the moon be a matter of national competition? Why should the United States and the Soviet Union, in preparing for such expeditions, become involved in immense duplications of research, construction, and expenditure? Surely we should explore whether the scientists and astronauts of our two countries—indeed of all the world—cannot work together in the conquest of space, sending someday in this decade to the moon not the representatives of a single nation, but the representatives of all of our countries.

[. . .]

The United Nations cannot survive as a static organization. Its obligations are increasing as well as its size. Its Charter must be changed as well as its customs. The authors of that Charter did not intend that it be frozen in perpetuity. The science of weapons and war has made us all, far more than 18 years ago in San Francisco, one world and one human race, with one common destiny. In such a world, absolute sovereignty no longer assures us of absolute security. The conventions of peace must pull abreast and then ahead of the inventions of war. The United Nations, building on its successes and learning from its failures, must be developed into a genuine world security system.

But peace does not rest in charters and covenants alone. It lies in the hearts and minds of all people. And if it is cast out there, then no act, no pact, no treaty, no organization can hope to preserve it without the support and the wholehearted commitment of all people. So let us not rest all our hopes on parchment and on paper; let us strive to build peace, a desire for peace, a willingness to work for peace, in the hearts and minds of all our people. I believe that we can. I believe the problems of human destiny are not beyond the reach of human beings.

Two years ago I told this body that the United States had proposed, and was willing to sign, a limited test ban treaty. Today that treaty has been signed. It will not put an end to war. It will not remove basic conflicts. It will not secure freedom for all. But it can be a lever, and Archimedes, in explaining the principles of the lever, was said to have declared to his friends: "Give me a place where I can stand—and I shall move the world."

My fellow inhabitants of this planet: Let us take our stand here in this Assembly of nations. And let us see if we, in our own time, can move the world to a just and lasting peace.

Source: John F. Kennedy. "Address before the 18th General Assembly of the United Nations." September 20, 1963. Available online. URL: http://www.jfklibrary.org/Historical+Resources/Archives/Reference+Desk/Speeches/JFK/003POF03_18thGeneralAssembly09201963.htm. Accessed May 20, 2010.

Report of the Space Task Group, 1969 (excerpt)

Two months after Apollo 11 landed on the Moon, a presidential task group issued a report recommending a reusable shuttle and a manned mission to Mars no later than 1981. President Richard Nixon had established the panel on February 13, 1969, and it was led by Vice President Spiro Agnew. The report outlined the wide-ranging impact of the Apollo mission on science, technology, and national security, as well as the national pride and international prestige it had generated. The report declared the United States the winner in the "race to the moon," but pointed out, "Achievement of the Apollo goal resulted in a new feeling of 'oneness' among men everywhere. It inspired a common sense of victory that can provide the basis for new initiatives for international cooperation."

CONCLUSIONS AND RECOMMENDATIONS

The Space Task Group in its study of future directions in space, with recognition of the many achievements culminating in the successful flight of Apollo 11, views these achievements as only a beginning to the long-term exploration and use of space by man. We see a major role for this Nation in proceeding from the initial opening of this frontier to its exploitation for the benefit of mankind, and ultimately to the opening of new regions of space to access by man.

We have found increasing interest in the exploitation of our demonstrated space expertise and technology for the direct benefit of mankind in such areas as earth resources, communications, navigation, national security, science and technology, and international participation. We have concluded that the space program for the future must include increased emphasis upon space applications.

We have also found strong and wide-spread personal identification with the manned flight program, and with the outstanding men who have participated as astronauts in this program. We have concluded that a forward-looking space program for the future for this Nation should include continuation of manned spaceflight activity. Space will continue to provide new challenges to satisfy the innate desire of man to explore the limits of his reach.

We have surveyed the important national resource of skilled program managers, scientists, engineers, and workmen who have contributed so much to the success the space program has enjoyed. This resource together with industrial capabilities, government, and private facilities and growing expertise in space operations are the foundation upon which we can build.

We have found that this broad foundation has provided us with a wide variety of new and challenging opportunities from which to select our future directions. We have concluded that the Nation should seize these new opportunities, particularly to advance science and engineering, international relations, and enhance the prospects for peace.

We have found questions about national priorities, about the expense of manned flight operations, about new goals in space which could be interpreted as a "crash program." Principal concern in this area relates to decisions about a manned mission to Mars. We conclude that NASA has the demonstrated organizational competence and technology base, by virtue of the Apollo success and other achievements, to carry out a successful program to land man on Mars within 15 years. There are a number of precursor activities necessary before such a mission can be attempted. These activities can proceed without developments specific to a Manned

Mars Mission-but for optimum benefit should be carried out with the Mars mission in mind. We conclude that a manned Mars mission should be accepted as a long-range goal for the space program. Acceptance of this goal would not give the manned Mars mission overriding priority relative to other program objectives, since options for decision on its specific date are inherent in a balanced program. Continuity of other unmanned exploration and applications efforts during periods of unusual budget constraints should be supported in all future plans.

[. . .]

THE POST-APOLLO SPACE PROGRAM:
DIRECTIONS FOR THE FUTURE
I. INTRODUCTION

With the successful flight of Apollo 11, man took his first step on a heavenly body beyond his own planet. As we look into the distant future it seems clear that this is a milestone—a beginning—and not an end to the exploration and use of space. Success of the Apollo program has been the capstone to a series of significant accomplishments for the United States in space in a broad spectrum of manned and unmanned exploration missions and in the application of space techniques for the benefit of man. In the short span of twelve years man has suddenly opened an entirely new dimension for his activity.

In addition, the national space program has made significant contributions to our national security, has been a political instrument of international value, has produced new science and technology, and has given us not only a national pride of accomplishment, but has offered a challenge and example for other national endeavors.

The Nation now has the demonstrated capability to move on to new goals and new achievements in space in all of the areas pioneered during the decade of the sixties. In each area of space exploration what seemed impossible yesterday has become today's accomplishment. Our horizons and our competence have expanded to the point that we can consider unmanned missions to any region in our solar system; manned bases in earth orbit, lunar orbit or on the surface of the Moon; manned missions to Mars; space transportation systems that carry their payloads into orbit and then return and land as a conventional jet aircraft; reusable nuclear-powered rockets for space operations; remotely controlled roving science vehicles on the Moon or on Mars; and application of space capability to a variety of services of benefit to man here on earth. Our opportunities are great and we have a broad spectrum of choices available to us. It remains only to chart the course and to set the pace of progress in this new dimension for man.

The Space Task Group, established under the chairmanship and direction of the Vice President (Appendices A and B), has examined the spectrum of new opportunities available in space, values and benefits from space activities, casts and resource implications of future options, and international aspects of the space program. A great wealth of data has been made available to the Task Group, including reports from the National Aeronautics and Space Administration and the Department of Defense reflecting very extensive planning and review activities, a detailed report from the President's Science Advisory Committee, views from [2] members of Congress, the National Academy of Sciences Space Science Board, and the American Institute of Aeronautics and Astronautics. In addition, a series of individual reports from a special group of distinguished citizens who were asked for their personal recommendations on the future course of the space program were of considerable value to the Task Group. This broad range of material was considered and evaluated as part of the Task Group deliberations. This report presents in summary form the views of the Space Task Group on the Nation's future directions in space.

Background

Twelve years ago, when the first artificial Earth satellite was placed into orbit, most of the world's population was surprised and stunned by on achievement so new and foreign to human experience. Today people of all nations are familiar with satellites, orbits, the concept of zero 'g', manned operations in space, and a host of other aspects characteristic of this new age—the age of space exploration.

The United States has carried out a diversified program during these early years in space, requiring innovation in many fields of science, technology, and the human and social sciences. The Nation's effort has been interdisciplinary, drawing successfully upon a synergistic combination of human knowledge, management experience, and production know-how to bring this National to a position of leadership in space.

Space activities have become a part of our national agenda.

We now have the benefit of twelve years of space activity and our leadership position as background for our examination of future directions in space.

National Priorities

By its very nature, the exploration and exploitation of space is a costly undertaking and must compete for funds with other national or individual enterprises. Now that the national goal of manned lunar landing has been achieved, discussion of future space goals has produced increasing pressures for reexamination of, and possible changes in, our national priorities.

Many believe that funds spent for the space program contribute less to our national economic growth and social well-being than funds allocated for other programs such as health, education urban affairs, or revenue sharing. Others believe that funds spent for space exploration will ultimately return great economic and social benefits not now foreseen. These divergent views will persist and must be recognized in making decisions on future space activities.

The Space Task Group has not attempted to reconcile these differences. Neither have we attempted to classify the space program in a hierarchy of national priorities. The Space Task Group has identified major technical and scientific challenges in space in the belief that returns will accrue to the society that takes up those challenges.

Values and Benefits

The magnitude of predicted great economic and social benefits from space activities cannot be precisely determined. Nevertheless, there should be a recognition that significant direct benefits have been realized as a result of space investments, particularly from applications programs, as a long-term result of space science activities, DOD space activities, and advancing technology. These direct benefits are only part of the total set of benefits from the space program, many of which are very difficult to quantify and therefore are not often given adequate consideration when costs and benefits from space activities are weighed or assessed in relation to other national programs.

National Resource

In the eleven years since its creation, NASA has provided the Nation with a broad capability for a wide variety of space activity, and has successfully completed a series of challenging tasks culminating in the first manned lunar landing. These accomplishments have involved rapid increases to peak annual expenditures of almost $6 billion and a peak civil service and contractor work force of 420,000 people. Expenditures for NASA have subsequently dropped over the last three years from this peak to the present level of about $4 billion and supporting manpower has dropped to about 190,000 people.

In addition to NASA space activity, the DOD has developed and operated space systems satisfying unique military requirements. Spending for military space grew rapidly in the early sixties and has increased gradually during the past few years to approximately $2 billion per year.

The Nation's space program has fostered the growth of a valuable reservoir of highly trained, competent engineers, managers, skilled workmen and scientists within government, industry and universities. The climactic achievement of Apollo 11 is tribute to their capability.

This resource together with supporting facilities, technology and organizational entities capable of complex management tasks grew and matured during the 1960's largely in response to the stimulation of Apollo, and if it is to be maintained, needs a new focus for its future.

Manned Space Flight
There has been universal personal identification with the astronauts and a high degree of interest in manned space activities which reached a peak both nationally and internationally with Apollo. The manned flight program permits vicarious participation by the man-in-the-street in exciting, challenging, and dangerous activity. Sustained high interest, judged in the light of current experience, however, is related to availability of new tasks and new mission activity—new challenges for man in space. The presence of man in space, in addition to its effect upon public interest in space activity, can also contribute to mission success by enabling man to exercise his unique capabilities, and thereby enhance mission reliability, flexibility, ability to react to unpredicted conditions, and potential for exploration.

While accomplishments related to man in space have prompted the greatest acclaim for our Nation's space activities, there has been increasing public reaction over the large investments required to conduct the manned flight program. Scientists have been particularly vocal about these high costs and problems encountered in performing science experiments as part of Apollo, a highly engineering oriented program in its early phases.

Much of the negative reaction to manned spaceflight, therefore, will diminish if costs for placing and maintaining man in space are reduced and opportunities for challenging new missions with greater emphasis upon science return are provided.

Science and Applications
Although high public interest has resided with manned spaceflight, the Nation has also enjoyed a successful and highly productive science and applications program.

The list of more achievements in space science is great, ranging from our first exploratory orbital flights resulting in discoveries about the Earth and its environment to the most recent Mariner missions to the vicinity of Mars producing new data about our neighbor planet.

Both optical and radio astronomy have been stimulated by the opening of new regions of the electromagnetic spectrum and new fields of interest have been uncovered—notably in the high energy X-ray and gamma-ray regions. Astronomy is advancing rapidly at present, partly with the aid of observations from space, and a deeper understanding of the nature and structure of the universe is emerging. In planetary exploration, we have a unique opportunity to pursue a number of the major questions man has asked about his relation to the universe. What is the history of the formation and evolution of the solar system? Are there clues to the origin of life? Does life exist elsewhere in the solar system?

In the life sciences, questions about the effect of zero 'g' upon living systems, demands of long-duration spaceflight upon our understanding of man and his interaction or response to his environment, both physiologically and psychologically, promise new insights into the understanding of complex living systems.

International Aspects
Achievement of the Apollo goal resulted in a new feeling of "oneness" among men everywhere. It inspired a common sense of victory that can provide the basis for new initiatives for international cooperation.

The U.S. and the USSR have widely been portrayed as in a "race to the Moon" or as vying over leadership in space. In a sense, this has been on accurate reflection of one of the several strong motivations for U.S. space program decisions over the previous decade.

MILITARIZATION OF SPACE

Now with the successes of Apollo, of the Mariner 6 and 7 Mars flybys, of communications and meteorology applications, the U.S. is at the peak of its prestige and accomplishments in space. For the short term, the race with the Soviets has been won. In reaching our present position, one of the great strengths of the U.S. space program has been its open nature, and the broad front of solid achievement in science and applications that has accompanied the highly successful manned flight program.

The attitude of the American people has gradually been changing and public frustration over Soviet accomplishments in space, an important force in support of the Nation's acceptance of the lunar landing in 1961, is not now present. Today, new Soviet achievements are not likely to have the effect of those in the post. Nevertheless, the Soviets have continued development of capability for future achievements and dramatic missions of high political impact are possible. There is no sign of retrenchment or withdrawal by the Soviets from the public arena of space activity despite launch vehicle and spacecraft failures and the pre-emptive effect of Apollo 11.

The landing on the Moon has captured the imagination of the world. It is now abundantly clear to the man in the street, as well as to the political leaders of the world, that mankind now has at his service a new technological capability, an important characteristic of which is that its applicability transcends national boundaries. If we retain the identification of the world with our space program, we have on opportunity for significant political effects on nations and peoples and on their relationships to each other, which in the long run may be quite profound.

III. GOALS AND OBJECTIVES

Goals

NASA has outlined plans that would include a manned Mars mission in 1981 with the development decision on a Mars Excursion Module in FY 1974, if the Nation were to accept this commitment. Such a program would result in maximum stimulation of our technology and creation of new capability. There are many precursor activities that will be required before a manned Mars mission is attempted, such as detailed study of biomedical aspects, both physiological and psychological, of flights lasting 500–600 days, unmanned reconnaissance of the planets, creation of highly reliable life support systems, power supplies, and propulsion capability adequate for the rigors of such a voyage and reliable enough to support man. Decision to proceed with a 1981 mission would require early attention to these precursor activities.

While launch of a manned Mars exploration mission appears achievable as early as 1981, it can also be accomplished at any one of the roughly biennial launch opportunities following this date, provided essential precursor activities have been carried out.

Thus, the understanding that we are ultimately going to explore the planets with man provides a shaping function for the post-Apollo space program. However, in a balanced program containing other goals and objectives, this focus should not assume over-riding priority and cause sacrifice of other important activity in times of severe budget constraints. Flexibility in program content and options for decision on the specific date for a manned Mars mission are inherent in this understanding.

The Space Task Group, in response to the President's request for a "Coordinated program and budget proposal," has therefore chosen this balanced program as that plan best calculated to meet the Nation's needs for direction of its future space activity. In reaching this conclusion we have considered international and domestic influences, weighed and placed in perspective science and engineering development, exploration and application of space, manned and unmanned approaches to space missions, and have appraised interagency influences. Discussion of the principal objectives which describe this balanced program follows.

Program Objectives

Elements of the balanced program recommended by the Space Task Group can be identified within the following set of program objectives which define major emphases for future space activity:

- Application of space technology to the direct benefit of mankind
- Operation of military space systems to enhance national defense
- Exploration of the solar system and beyond
- Development of new capabilities for operating in space
- International participation and cooperation

[. . .]

Source: NASA. "Report of the Space Task Group, 1969." Available online. URL: http://www.hq.nasa.gov/office/pao/History/taskgrp.html. Accessed May 20, 2010.

President Richard Nixon Announces the Space Shuttle Program (January 5, 1972)

Drawing upon the recommendations in the 1969 report of the Space Task Group, President Richard M. Nixon announced the creation of the space shuttle program. The first shuttle, Columbia, *launched nine years later. The shuttle was a completely new prototype. Previous spacecraft had been consumed during their missions, but the space shuttle was envisioned as a low-cost way to fully explore and exploit space. Most important, Nixon's statement makes no mention of a Mars mission, the primary recommendation of the task group.*

I have decided today that the United States should proceed at once with the development of an entirely new type of space transportation system designed to help transform the space frontier of the 1970's into familiar territory, easily accessible for human endeavor in the 1980's and '90's.

This system will center on a space vehicle that can shuttle repeatedly from Earth to orbit and back. It will revolutionize transportation into near space, by routinizing it. It will take the astronomical costs out of astronautics. In short, it will go a long way toward delivering the rich benefits of practical space utilization and the valuable spinoffs from space efforts into the daily lives of Americans and all people.

The new year 1972 is a year of conclusion for America's current series of manned flights to the Moon. Much is expected from the two remaining Apollo missions—in fact, their scientific results should exceed the return from all the earlier flights together. Thus they will place a fitting capstone on this vastly successful undertaking. But they also bring us to an important decision point—a point of assessing what our space horizons are as Apollo ends, and of determining where we go from here.

In the scientific arena, the past decade of experience has taught us that spacecraft are an irreplaceable tool for learning about our near-Earth space environment, the Moon, and the planets, besides being an important aid to our studies of the Sun and stars. In utilizing space to meet needs on Earth, we have seen the tremendous potential of satellites for international communications and world-wide weather forecasting. We are gaining the capability to use satellites as tools in global monitoring and management of nature resources, in agricultural applications, and in pollution control. We can foresee their use in guiding airliners across the oceans and in bringing TV education to wide areas of the world.

However, all these possibilities, and countless others with direct and dramatic bearing on human betterment, can never be more than fractionally realized so long as every single trip from Earth to orbit remains a matter of special effort and staggering expense. This is why commitment to the Space Shuttle program is the right step for America to take, in moving out from our present beach-head in the sky to achieve a real working presence in space—because the Space Shuttle will give us routine access to space by sharply reducing costs in dollars and preparation time.

The new system will differ radically from all existing booster systems, in that most of this new system will be recovered and used again and again—up to 100 times. The resulting economies may bring operating costs down as low as one-tenth of those present launch vehicles.

The resulting changes in modes of flight and re-entry will make the ride safer, and less demanding for the passengers, so that men and women with work to do in space can "commute" aloft, without having to spend years in training for the skills and rigors of old-style spaceflight. As scientists and technicians are actually able to accompany their instruments into space, limiting boundaries between our manned and unmanned space programmes will disappear. Development of new space applications will be able to proceed much faster. Repair or servicing of satellites in space will become possible, as will delivery of valuable payloads from orbit back to Earth.

The general reliability and versatility which the Shuttle system offers seems likely to establish it quickly as the workhorse of our whole space effort, taking the place of all present launch vehicles except the very smallest and very largest.

NASA and many aerospace companies have carried out extensive design studies for the Shuttle. Congress has reviewed and approved this effort. Preparation is now sufficient for us to commence the actual work of construction with full confidence of success. In order to minimize technical and economic risks, the space agency will continue to take a cautious evolutionary approach in the development of this new system. Even so, by moving ahead at this time, we can have the Shuttle in manned flight by 1978, and operational a short time later.

It is also significant that this major new national enterprise will engage the best efforts of thousands of highly skilled workers and hundreds of contractor firms over the next several years. The amazing 'technology explosion' that has swept this country in the years since we ventured into space should remind us that robust activity in the aerospace industry is healthy for everyone—not just in jobs and income, but in the extension of our capabilities in every direction. The continued pre-eminence of America and American industry in the aerospace field will be an important part of the Shuttle's 'payload'.

Views of the Earth from space have shown us how small and fragile our home planet truly is. We are learning the imperatives of universal brotherhood and global ecology learning to think and act as guardians of one tiny blue and green island in the trackless oceans of the Universe. This new program will give more people more access to the liberating perspectives of space, even as it extends our ability to cope with physical challenges of Earth and broadens our opportunities for international cooperation in low-cost, multi-purpose space missions.

'We must sail sometimes with the wind and sometimes against it', said Oliver Wendell Holmes, 'but we must sail, and not drift, nor lie at anchor'. So with man's epic voyage into space—a voyage the United States of America has led and still shall lead.

STATEMENT BY DR. FLETCHER, NASA ADMINISTRATOR.

As indicated in the President's statement, the studies by NASA and the aerospace industry of the Space Shuttle have now reached the point where the decision can be made to proceed into actual development of the Space Shuttle vehicle. The decision to proceed, which the President has now approved, is consistent with the plans presented to and approved by the Congress in NASA's FY 1972 budget.

The decision by the President is a historic step in the nation's space program—it will change the nature of what man can do in space. By the end of this decade the nation will have the means of getting men and equipment to and from space routinely, on a moment's notice if necessary, and at a small fraction of today's cost. This will be done within the framework of a useful total space program of science, exploration, and applications at approximately the present overall level of the space budget.

The Space Shuttle will consist of an aircraft-like orbiter, about the size of a DC-9. It will be capable of carrying into orbit and back again to Earth useful payloads up to 15 ft. in diameter by 60 ft. long, and weighing up to 65,000 lb.

Fuel for the orbiter's liquid-hydrogen / liquid-oxygen engines will be carried in an external tank that will be jettisoned in orbit.

The orbiter will be launched by an unmanned booster.

The orbiter can operate in space for about a week. The men onboard will be able to launch, service, or recover unmanned spacecraft; perform experiments and other useful operations in Earth orbit; and farther in the future resupply with men and equipment space modules which themselves have been brought to space by the Space Shuttle. When each mission has been completed, the Space Shuttle will return to Earth and land on a runway like an airplane.

There are four main reasons why the Space Shuttle is important and is the right step in manned spaceflight and the US space program.

- The Shuttle is the only meaningful new manned space program which can be accomplished on a modest budget;
- It is needed to make space operations less complex and less costly;
- It is needed to do useful things, and
- It will encourage greater international participation in spaceflight.
- On the basis of today's decision, NASA will proceed as follows;

This spring we will issue a request for prospective contractors. This summer we will place Space Shuttle under contract and development work will start. Between now and about the end of February, NASA and our contractors will focus their study efforts on technical areas where further detailed information is required before the requests for contractor proposals can be issued. These areas include comparisons of pressure-fed liquid and solid rocket motor options for the booster stage.

Source: NASA History Division. "President Nixon's 1972 Announcement of the Space Shuttle." Available online. URL: http://history.nasa.gov/stsnixon.htm. Accessed May 20, 2010.

National Security Decision Directive Number 42, "National Space Policy" (July 4, 1982)

The Reagan White House conducted its own review of U.S. space policy, culminating in National Security Decision Directive Number 42 (NSDD-42). *Official policy continued to view the space shuttle program as the "major factor" in U.S. efforts and noted its dual role for defense and scientific purposes. NSDD-42 explicitly states that the United States will oppose any effort to ban space-based military or intelligence applications.*

I. INTRODUCTION AND PRINCIPLES

This directive establishes national policy to guide the conduct of United States space program and related activities; it supersedes Presidential Directives 37, 42, and 54, as well as National Security Decision Directive 8. This directive is consistent with and augments the guidance contained in existing directives, executive orders, and law. The decisions outlined in this directive provide the broad framework and the basis for the commitments necessary for the conduct of United States space programs.

The Space Shuttle is to be a major factor in the future evolution of United States space programs. It will continue to foster cooperation between the national security and civil efforts to ensure efficient and effective use of national resources. Specifically, routine use of the manned Space Shuttle will provide the opportunity to understand better and evaluate the role of man in space, to increase the utility of space programs, and to expand knowledge of the space environment.

The basic goals of United States space policy are to: (a) strengthen the security of the United States; (b) maintain United States space leadership; (c) obtain economic and scientific benefits through the exploitation of space; (d) expand United States private-sector investment and involvement in civil space and space-related activities; (e) promote international cooperative activities that are in the national interest; and (f) cooperate with other nations in maintaining the freedom of space for all activities that enhance the security and welfare of mankind.

[2] The United States space program shall be conducted in accordance with the following basic principles:

A. The United States is committed to the exploration and use of outer space by all nations for peaceful purposes and for the benefit of all mankind. [Sentence deleted during declassification review]
B. The United States rejects any claims to sovereignty by any nation over outer space or celestial bodies, or any portion thereof, and rejects any limitations on the fundamental right to acquire data from space.
C. The United States considers the space systems of any nation to be national property with the right of passage through the operations in

space without interference. Purposeful interference with space systems shall be viewed as infringement upon sovereign rights.

D. The United States encourages domestic commercial exploration of space capabilities, technology, and systems for national economic benefit. These activities must be consistent with national security concerns, treaties, and international agreements.

E. The United States will conduct international cooperative space-related activities that achieve sufficient scientific, political, economic, or national security benefits for the nation.

F. [Paragraph deleted in declassification review]

G. The United States Space Transportation System (STS) is the primary space launch system for both national security and civil government missions. STS capabilities and capacities shall be developed to meet appropriate national needs and shall be available to authorized users— domestic and foreign, commercial, and governmental.

H. The United States will pursue activities in space in support of its right of self-defense.

I. The United States will continue to study space arms control options. The United States will consider verifiable and equitable arms control measures that would ban or otherwise limit testing and deployment of specific weapons systems should those measures be compatible with United States national security. The United States will oppose arms control concepts or legal regimes that seek general prohibitions on the military or intelligence use of space.

II. SPACE TRANSPORTATION SYSTEM

The Space Transportation System (STS) is composed of the Space shuttle, associated upper stages, and related facilities. The following policies shall govern the development and operation of the STS:

A. The STS is a vital element of the United States space program and is the primary space launch system for both United States national security and civil government missions. The STS will be afforded the degree of survivability and security protection required for a critical national space resource.

B. The first priority of the STS program is to make the system fully operational and cost-effective in providing routine access to space.

C. The United States is fully committed to maintaining world leadership in space transportation with an STS capacity sufficient to meet appropriate national needs. The STS program requires sustained commitments by all affected departments and agencies. The United States will continue to

141

develop the STS through the National Aeronautics and Space Administration (NASA) in cooperation with the Department of Defense (DoD). Enhancements of STS operational capability, upper stages, and efficient methods of deploying and retrieving payloads should be pursued as national requirements are defined.

D. United States Government spacecraft should be designed to take advantage of the unique capabilities of the STS. The completion of transition to the Shuttle should occur as expeditiously as practical.

[4]E. [Paragraph deleted in declassification review]

F. Expandable launch vehicle operations shall be continued by the United States Government until the capabilities of the STS are sufficient to meet its needs and obligations. Unique national security considerations may dictate developing special-purpose launch capabilities.

G. For the near-term, the STS will continue to be managed and operated in an institutional arrangement consistent with the current NASA/DoD Memoranda of Understanding. Responsibility will remain in NASA for operational control of the STS for civil missions and in the DoD for operational control of the STS for national security missions. Mission management is the responsibility of the mission agency. As the STS operations mature, options will be considered for possible transition to a different institutional structure.

H. Major changes to STS program capabilities will require Presidential approval.

III. CIVIL SPACE PROGRAM

The United States shall conduct civil space programs to expand knowledge of the Earth, its environment, the solar system, and the universe; to develop and promote selected civil applications of space technology; to preserve the United States leadership in critical aspects of space science, applications, and technology; and to further United States domestic and foreign policy objectives. Consistent with the National Aeronautics and Space Act, the following policies shall govern the conduct of the civil space program.

A. Science, Applications, and Technology: United States Government civil programs shall continue a balanced strategy of research, development, operations, and exploration for science, applications, and technology. The key objectives of these programs are to:

(1) Preserve the United States preeminence in critical major space activities to enable continued exploitation and exploration of space.

[5] (2) Conduct research and experimentation to expand understanding of: (a) astrophysical phenomena and the origin and evolution of the universe, through long-term astrophysical observation; (b) the Earth, its environment, and its dynamic relation with the Sun; (c) the origin and evolution of the solar system, through solar, planetary, and lunar sciences and exploration; and (d) the space environment and technology required to advance knowledge in the biological sciences.

(3) Continue to explore the requirements, operational concepts, and technology associated with permanent space facilities.

(4) Conduct appropriate research and experimentation in advanced technology and systems to provide a basis for future civil space applications.

B. Private Sector Participation: The United States Government will provide a climate conducive to expanded private sector investment and involvement in civil space activities, with due regard to public safety and national security. Private sector space activities will be authorized and supervised or regulated by the government to the extent required by treaty and national security.

C. International Cooperation: United States cooperation in international civil activities will:

(1) Support the public, nondiscriminatory direct readout of data from Federal civil systems to foreign ground stations and provision of data to foreign users under specified conditions.

(2) Continue cooperation with other nations by conducting joint scientific and research programs that yield sufficient benefits to the United States in areas such as access to foreign scientific and technological expertise, and access to foreign research and development facilities, and that serve other national goals. All international space ventures must be consistent with United States technology-transfer policy.

D. Civil Operational Remote Sensing: Management of Federal civil operational remote sensing is the responsibility of the Department of Commerce. The Department of Commerce will: (a) aggregate Federal needs for civil operational remote sensing to be met by either the private sector or the Federal government; (b) identify needed civil operational system research and development objectives; and (c) in coordination with other departments or agencies, provide for regulation of private-sector operational remote sensing systems.

[6] [Page deleted in declassification review]

[7] [Page deleted in declassification review]

[8] [Paragraph deleted in declassification review]

    (1) The fact that the United States conducts satellite photoreconnaissance for peaceful purposes, including intelligence collection and the monitoring of arms control agreements, is unclassified. The fact that such photoreconnaissance includes a near-real-time capability and is used to provide defense related information for indications and warning is also unclassified. All other details, facts and products concerning the national foreign intelligence space program are subject to appropriate classification and security controls.

    (2) [Paragraph deleted in declassification review]

VI. INTER-SECTOR RESPONSIBILITIES

[Paragraphs A-F deleted in declassification review]

[9] G. The United States Government will maintain and coordinate separate national security and civil operational space systems when differing needs of the sectors dictate.

VII. IMPLEMENTATION

Normal interagency coordinating mechanisms will be employed to the maximum extent possible to implement the policies enunciated in this directive. To provide a forum to all Federal agencies for their policy views, to review and advise on proposed changes to national space policy, and to provide for orderly and rapid referral of space policy issues to the President for decisions as necessary, a Senior Interagency Group (SIG) on Space shall be established. The SIG (Space) will be chaired by the Assistant to the President for National Security Affairs and will include the Deputy or Under Secretary of State, Deputy or Under Secretary of Defense, Deputy or Under Secretary of Commerce, Director of Central Intelligence, Chairman of the Joint Chiefs of Staff, Director of the Arms Control and Disarmament Agency, and the [10] Administrator of the National Aeronautics and Space Administration. Representatives of the Office of Management and Budget and the Office of Science and Technology Policy will be include as observers. Other agencies or departments will participate based on the subjects to be addressed.

Source: NASA. "National Security Decision Directive Number 42, 'National Space Policy,' July 4, 1982." Available online. URL: http://www.hq.nasa.gov/office/pao/History/nsdd-42.html. Accessed March 19, 2010.

President Reagan's Statement on the International Space Station from State of the Union Address (January 25, 1984)

President Reagan used his 1984 annual State of the Union address to set a new goal for the U.S. space program—the establishment of an international space station within a decade. NASA would be able to invite other countries to participate in the endeavor. He predicted an overburdened space transportation system and announced that the Department of Transportation would jumpstart a commercial launch service.

Our second great goal is to build on America's pioneer spirit . . . and that's to develop that frontier. A sparkling economy spurs initiatives, sunrise industries and makes older ones more competitive.

Nowhere is this more important than our next frontier: space. Nowhere do we so effectively demonstrate our technological leadership and ability to make life better on Earth. The Space Age is barely a quarter of a century old. But already we've pushed civilization forward with our advances in science and technology. Opportunities and jobs will multiply as we cross new thresholds of knowledge and reach deeper into the unknown.

Our progress in space, taking giant steps for all mankind, is a tribute to American teamwork and excellence. Our finest minds in government, industry and academia have all pulled together. And we can be proud to say: We are first; we are the best; and we are so because we're free.

America has always been greatest when we dared to be great. We can reach for greatness again. We can follow our dreams to distant stars, living and working in space for peaceful, economic, and scientific gain. Tonight, I am directing NASA to develop a permanently manned space station and to do it within a decade.

A space station will permit quantum leaps in our research in science, communications, and in metals and lifesaving medicines which could be manufactured only in space. We want our friends to help us meet these challenges and share in their benefits. NASA will invite other countries to participate so we can strengthen peace, build prosperity, and expand freedom for all who share our goals.

Just as the oceans opened up a new world for clipper ships and Yankee traders, space holds enormous potential for commerce today. The market for space transportation could surpass our capacity to develop it. Companies interested in putting payloads into space must have ready access to private sector launch services. The Department of Transportation will help an expendable launch services industry to get off the ground. We'll soon implement a number of executive initiatives, develop proposals to ease regulatory constraints, and, with NASA's help, promote private sector investment in space.

Source: NASA. Excerpts of President Reagan's State of the Union Address, January 25, 1984. Available online. URL: http://history.nasa.gov/reagan84.htm. Accessed March 19, 2010.

President Reagan's Reaction to the *Challenger* Explosion (January 28, 1986)

U.S. president Ronald Reagan had planned to make his State of the Union address on the evening of January 28, 1986, but when the space shuttle Challenger exploded on liftoff that morning, killing all seven astronauts aboard, he canceled that speech. In an era when space travel had become almost commonplace, he reminded everyone of the dangers inherent in space exploration.

Ladies and Gentlemen, I'd planned to speak to you tonight to report on the state of the Union, but the events of earlier today have led me to change those plans. Today is a day for mourning and remembering. Nancy and I are pained to the core by the tragedy of the shuttle *Challenger*. We know we share this pain with all of the people of our country. This is truly a national loss.

Nineteen years ago, almost to the day, we lost three astronauts in a terrible accident on the ground. But, we've never lost an astronaut in flight; we've never had a tragedy like this. And perhaps we've forgotten the courage it took for the crew of the shuttle; but they, the *Challenger* Seven, were aware of the dangers, but overcame them and did their jobs brilliantly. We mourn seven heroes: Michael Smith, Dick Scobee, Judith Resnik, Ronald McNair, Ellison Onizuka, Gregory Jarvis, and Christa McAuliffe. We mourn their loss as a nation together.

For the families of the seven, we cannot bear, as you do, the full impact of this tragedy. But we feel the loss, and we're thinking about you so very much. Your loved ones were daring and brave, and they had that special grace, that special spirit that says, 'Give me a challenge and I'll meet it with joy.' They had a hunger to explore the universe and discover its truths. They wished to serve, and they did. They served all of us.

We've grown used to wonders in this century. It's hard to dazzle us. But for twenty-five years the United States space program has been doing just that. We've grown used to the idea of space, and perhaps we forget that we've only just begun. We're still pioneers. They, the members of the *Challenger* crew, were pioneers.

And I want to say something to the schoolchildren of America who were watching the live coverage of the shuttle's takeoff. I know it is hard to understand, but sometimes painful things like this happen. It's all part of the process of exploration and discovery. It's all part of taking a chance and expanding man's horizons. The future doesn't belong to the fainthearted; it belongs to the brave. The *Challenger* crew was pulling us into the future, and we'll continue to follow them.

I've always had great faith in and respect for our space program, and what happened today does nothing to diminish it. We don't hide our space program. We don't keep secrets and cover things up. We do it all up front and in public. That's the way freedom is, and we wouldn't change it for a minute. We'll continue our quest in space. There will be more shuttle flights and more shuttle crews and, yes, more volunteers, more civilians, more teachers in space. Nothing ends here; our hopes and our journeys continue. I want to add that I wish I could talk to every man and woman who works for NASA or who worked on this mission and tell them: "Your dedication and professionalism have moved and impressed us for decades. And we know of your anguish. We share it."

There's a coincidence today. On this day 390 years ago, the great explorer Sir Francis Drake died aboard ship off the coast of Panama. In his lifetime the great frontiers were the oceans, and a historian later said, 'He lived by the sea, died on it, and was buried in it.' Well, today we can say of the *Challenger* crew: Their dedication was, like Drake's, complete.

The crew of the space shuttle *Challenger* honored us by the manner in which they lived their lives. We will never forget them, nor the last time we saw them, this morning, as they prepared for the journey and waved good-bye and 'slipped the surly bonds of earth' to 'touch the face of God.'

Thank you.

Source: The History Place Great Speeches Collection. "President Ronald Reagan on the Challenger Disaster." Available online. URL: http://www.historyplace.com/speeches/reagan-challenger.htm. Accessed March 19, 2010.

Rumsfeld Commission, Final Report (2001) (excerpts)

Congress, the Department of Defense, and the Central Intelligence Agency appointed a blue-ribbon panel to assess how space projects affect national security. The Commission to Assess United States National Security Space Management and Organizations was headed by Donald Rumsfeld, a former secretary of defense who was about to return to the job under President George W. Bush. The panel is thus frequently referred to as the Rumsfeld Commission. The panel's final report warned of U.S. vulnerability to attacks from space, forecasting a space Pearl Harbor. Commission members recommended that the United States should better protect its space-based assets and prevent a space-based attack on U.S. soil.

REPORT OF THE COMMISSION TO ASSESS UNITED STATES NATIONAL SECURITY SPACE MANAGEMENT AND ORGANIZATIONS PURSUANT TO PUBLIC LAW 106-65 JANUARY 11, 2001

II. EXECUTIVE SUMMARY
A. Conclusions of the Commission

The Commission was directed to assess the organization and management of space activities in support of U.S. national security. Members of the Commission were appointed by the chairmen and ranking minority members of the House and Senate Armed Services Committees and by the Secretary of Defense in consultation with the Director of Central Intelligence.

The Commission unanimously concluded that the security and well being of the United States, its allies and friends depend on the nation's ability to operate in space.

Therefore, it is in the U.S. national interest to:

- Promote the peaceful use of space.
- Use the nation's potential in space to support its domestic, economic, diplomatic and national security objectives.
- Develop and deploy the means to deter and defend against hostile acts directed at U.S. space assets and against the uses of space hostile to U.S. interests.

The pursuit of U.S. national interests in space requires leadership by the President and senior officials. The Commission recommends an early review and, as appropriate, revision of the national space policy. The policy should provide direction and guidance for the departments and agencies of the U.S. Government to:

- Employ space systems to help speed the transformation of the U.S. military into a modern force able to deter and defend against evolving threats directed at the U.S. homeland, its forward deployed forces, allies and interests abroad and in space.
- Develop revolutionary methods of collecting intelligence from space to provide the President the information necessary for him to direct the nation's affairs, manage crises and resolve conflicts in a complex and changing international environment.
- Shape the domestic and international legal and regulatory environment for space in ways that ensure U.S. national security interests and enhance the competitiveness of the commercial sector and the effectiveness of the civil space sector.
- Promote government and commercial investment in leading edge technologies to assure that the U.S. has the means to master operations in space and compete in international markets.
- Create and sustain within the government a trained cadre of military and civilian space professionals.

The U.S. Government is increasingly dependent on the commercial space sector to provide essential services for national security operations. Those services include satellite communications as well as images of the earth useful to government officials, intelligence analysts and military commanders. To assure the United States remains the world's leading space-faring nation, the government has to become a more reliable consumer of U.S. space products and services and should:

- Invest in technologies to permit the U.S. Government to field systems one generation ahead of what is available commercially to meet unique national security requirements.
- Encourage the U.S. commercial space industry to field systems one generation ahead of international competitors.

The relative dependence of the U.S. on space makes its space systems potentially attractive targets. Many foreign nations and non-state entities are pursuing space-related activities. Those hostile to the U.S. possess, or can acquire on the global market, the means to deny, disrupt or destroy U.S. space systems by attacking satellites in space, communications links to and from the ground or ground stations that command the satellites and process their data. Therefore, the U.S. must develop and maintain intelligence collection capabilities and an analysis approach that will enable it to better understand the intentions and motivations as well as the capabilities of potentially hostile states and entities.

An attack on elements of U.S. space systems during a crisis or conflict should not be considered an improbable act. If the U.S. is to avoid a "Space Pearl Harbor" it needs to take seriously the possibility of an attack on U.S. space systems. The nation's leaders must assure that the vulnerability of the United States is reduced and that the consequences of a surprise attack on U.S. space assets are limited in their effects. The members of this Commission have, together, identified five matters of key importance that we believe need attention quickly from the top levels of the U.S. Government. We have drawn these conclusions from six months of assessing U.S. national security space activities, including 32 days of meetings with 77 present and former senior officials and knowledgeable private sector representatives. These five matters—our unanimous conclusions—are:

First, the present extent of U.S. dependence on space, the rapid pace at which this dependence is increasing and the vulnerabilities it creates, all demand that U.S. national security space interests be recognized as a top national security priority. The only way they will receive this priority is through specific guidance and direction from the very highest government levels. Only the President has the authority, first, to set forth the national space policy, and then to provide the guidance and direction to senior officials, that together are needed to ensure that the United States remains the world's leading space-faring nation. Only Presidential leadership can ensure the cooperation needed from all space sectors—commercial, civil, defense and intelligence.

Second, the U.S. Government—in particular, the Department of Defense and the Intelligence Community—is not yet arranged or focused to meet the national security space needs of the 21st century. Our growing dependence on space, our vulnerabilities in space and the burgeoning opportunities from space are simply not reflected in the present institutional arrangements. After examining a variety of organizational approaches, the Commission concluded that a number of disparate space activities should promptly be merged, chains of command adjusted, lines of communication opened and policies modified to achieve greater responsibility and accountability. Only then can the necessary trade-offs be made, the appropriate priorities be established and the opportunities for improving U.S. military and intelligence capabilities be realized. Only with senior-level leadership, when properly managed and with the right priorities will U.S. space programs both deserve and attract the funding that is required.

Third, U.S. national security space programs are vital to peace and stability, and the two officials primarily responsible and accountable for those programs are the Secretary of Defense and the Director of Central Intelligence. Their relationship is critical to the development and deployment of the space capabilities needed to support the President in war, in crisis and also in peace. They must work closely and effectively together, in partnership, both to set and maintain the course for national security space programs and to resolve the differences that arise between their respective bureaucracies. Only if they do so will the armed forces, the Intelligence Community and the National Command Authorities have the information they need to pursue our deterrence and defense objectives successfully in this complex, changing and still dangerous world.

Fourth, we know from history that every medium—air, land and sea—has seen conflict. Reality indicates that space will be no different. Given this virtual certainty, the U.S. must develop the means both to deter and to defend against hostile acts in and from space. This will require superior space capabilities. Thus far, the broad outline of U.S. national space policy is sound, but the U.S. has not yet taken the steps necessary to develop the needed capabilities and to maintain and ensure continuing superiority.

Finally, investment in science and technology resources—not just facilities, but people—is essential if the U.S. is to remain the world's leading spacefaring nation. The U.S. Government needs to play an active, deliberate role in expanding and deepening the pool of military and civilian talent in science, engineering and systems operations that the nation will need. The government also needs to sustain its investment in enabling and breakthrough technologies in order to maintain its leadership in space.

B. Space: Today and the Future

With the dramatic and still accelerating advances in science and technology, the use of space is increasing rapidly. Yet, the uses and benefits of space often go unrecognized. We live in an information age, driven by needs for precision, accuracy and timeliness in all of our endeavors—personal, business and governmental. As society becomes increasingly mobile and global, reliance on the worldwide availability of information will increase. Space-based systems, transmitting data, voice and video, will continue to play a critical part in collecting and distributing information. Space is also a medium in which highly valuable applications are being developed and around which highly lucrative economic endeavors are being built.

1. A New Era of Space
The first era of the space age was one of experimentation and discovery. Telstar, Mercury and Apollo, Voyager and Hubble, and the Space Shuttle taught Americans how to journey into space and allowed them to take the first tentative steps toward operating in space while enlarging their knowledge of the universe. We are now on the threshold of a new era of the space age, devoted to mastering operations in space.

The Role for Space
Space-based technology is revolutionizing major aspects of commercial and social activity and will continue to do so as the capacity and capabilities of satellites increase through emerging technologies. Space enters homes, businesses, schools, hospitals and government offices through its applications for transportation, health, the environment, telecommunications, education, commerce, agriculture and energy. Much like highways and airways, waterlines and electric grids, services supplied from space are already an important part of the U.S. and global infrastructures.

Space-related capabilities help national leaders to implement American foreign policy and, when necessary, to use military power in ways never before possible. Because of space capabilities, the U.S. is better able to sustain and extend deterrence to its allies and friends in our highly complex international environment.

In the coming period, the U.S. will conduct operations to, from, in and through space in support of its national interests both on the earth and in space. As with national capabilities in the air, on land and at sea, the U.S. must have the capabilities to defend its space assets against hostile acts and to negate the hostile use of space against U.S. interests.

Intelligence collected from space remains essential to U.S. national security. It is essential to the formulation of foreign and defense policies, the capacity of the President to manage crises and conflicts, the conduct of military operations and the development of military capabilities to assure the attainment of U.S. objectives. The Department of Defense and the Intelligence Community are undertaking substantial and expensive programs to replace virtually their entire inventory of satellites over the next decade or so. These programs are estimated to cost more than $60 billion during this period.

Opportunities in space are not limited to the United States. Many countries either conduct or participate in space programs dedicated to a variety of tasks, including communications and remote sensing. The U.S. will be tested over time by competing programs or attempts to restrict U.S. space activities through international regulations.

Toward the Future

Mastering near-earth space operations is still in its early stages. As mastery over operating in space is achieved, the value of activity in space will grow. Commercial space activity will become increasingly important to the global economy. Civil activity will involve more nations, international consortia and non-state actors. U.S. defense and intelligence activities in space will become increasingly important to the pursuit of U.S. national security interests.

The Commissioners appreciate the sensitivity that surrounds the notion of weapons in space for offensive or defensive purposes. They also believe, however, that to ignore the issue would be a disservice to the nation. The Commissioners believe the U.S. Government should vigorously pursue the capabilities called for in the National Space Policy to ensure that the President will have the option to deploy weapons in space to deter threats to and, if necessary, defend against attacks on U.S. interests.

2. Vulnerabilities and Threats

Space systems are vulnerable to a range of attacks that could disrupt or destroy the ground stations, launch systems or satellites on orbit. The political, economic and military value of space systems makes them attractive targets for state and non-state actors hostile to the United States and its interests. In order to extend its deterrence concepts and defense capabilities to space, the U.S. will require development of new military capabilities for operation to, from, in and through space. It will require, as well, engaging U.S. allies and friends, and the international community, in a sustained effort to fashion appropriate "rules of the road" for space.

Assessing the Threat Environment

The U.S. is more dependent on space than any other nation. Yet, the threat to the U.S. and its allies in and from space does not command the attention it merits from the departments and agencies of the U.S. Government charged with national security responsibilities. Consequently, evaluation of the threat to U.S. space capabilities currently lacks priority in the competition for collection and analytic resources. Failure to develop credible threat analyses could have serious consequences for the United States. It could leave the U.S. vulnerable to surprises in space and could result in deferred decisions on developing space-based capabilities due to the lack of a validated, well-understood threat. The ability to restrict or deny freedom of access to and operations in space is no longer limited to global military powers. Knowledge of space systems and the means to counter them is increasingly available on the international market. The reality is that there are many extant capabilities to deny, disrupt or physically destroy space systems and the ground facilities that use and control them. Examples include denial and deception, interference with satellite systems, jamming satellites on orbit, use of microsatellites for hostile action and detonation of a nuclear weapon in space.

Reducing Vulnerability

As harmful as the loss of commercial satellites or damage to civil assets would be, an attack on intelligence and military satellites would be even more serious for the nation in time of crisis or conflict. As history has shown—whether at Pearl Harbor, the killing of 241 U.S. Marines in their barracks in Lebanon or the attack on the USS *Cole* in Yemen—if the U.S. offers an inviting target, it may well pay the price of attack. With the growing commercial and national security use of space, U.S. assets in space and on the ground offer just such targets. The U.S. is an attractive candidate for a "Space Pearl Harbor." The warning signs of U.S. vulnerability include:

- In 1998, the Galaxy IV satellite malfunctioned, shutting down 80 percent of U.S. pagers, as well as video feeds for cable and broadcast transmissions. It took weeks in some cases to fully restore satellite service.
- In early 2000, the U.S. lost all information from a number of its satellites for three hours when computers in ground stations malfunctioned.
- In July 2000, the Xinhua news agency reported that China's military is developing methods and strategies for defeating the U.S. military in a high-tech and space-based future war.

The signs of vulnerability are not always so clear as those described above and therefore are not always recognized. Hostile actions against space systems can reasonably be confused with natural phenomena. Space debris or solar activity can "explain" the loss of a space system and mask unfriendly actions or the potential thereof. Such ambiguity and uncertainty could be fatal to the successful management of a crisis or resolution of a conflict. They could lead to forbearance when action is needed or to hasty action when more or better information would have given rise to a broader and more effective set of response options.

There are a number of possible crises or conflicts in which the potential vulnerability of national security space systems would be worrisome. For example:

- Efforts to identify and strike terrorist strongholds and facilities in advance of or in retaliation for terrorist attacks on U.S. forces or citizens abroad, or on the U.S. homeland or that of its allies.
- Conflict in the Taiwan Straits, in which the U.S. attempts to deter escalation through the conduct of military operations while seeking to bring it to a favorable end through diplomatic measures.
- War in the Middle East, posing a threat to U.S. friends and allies in the region and calling for a rapid political and military response to threats by an aggressor to launch ballistic missiles armed with weapons of mass destruction.

That U.S. space systems might be threatened or attacked in such contingencies may seem improbable, even reckless. However, as political economist Thomas Schelling has pointed out, "There is a tendency in our planning to confuse the unfamiliar with the improbable. The contingency we have not considered looks strange; what looks strange is thought improbable; what is improbable need not be considered seriously." Surprise is most often not a lack of warning, but the result of a tendency to dismiss as reckless what we consider improbable.

History is replete with instances in which warning signs were ignored and change resisted until an external, "improbable" event forced resistant bureaucracies to take action. The question is whether the U.S. will be wise enough to act responsibly and soon enough to reduce U.S. space vulnerability. Or whether, as in the past, a disabling attack against the country and its people—a "Space Pearl Harbor"—will be the only event able to galvanize the nation and cause the U.S. Government to act.

We are on notice, but we have not noticed.

Source: Report of the Commission to Assess United States National Security Space Management and Organization. Available online. URL: www.dod.mil/pubs/spaceintro.pdf. Accessed March 19, 2010.

U.S. Withdrawal from ABM Treaty
(December 13, 2001) (Excerpts)

The United States and Soviet Union signed a bilateral Anti-Ballistic Missile Treaty in 1972. Both countries agreed to renounce ABM systems beyond a single location on each side. But following the cold war, the United States faced many potential enemies, especially non-state actors such as terrorists. Washington and Moscow could now cooperate on efforts to eliminate such threats. Following on the heels of the 9/11 terrorist attacks, the Bush administration announced that it would withdraw from the ABM Treaty in order to deploy weapons to defend from attacks on multiple fronts.

ABM TREATY FACT SHEET

Statement by the Press Secretary

Announcement of Withdrawal from the ABM Treaty

The circumstances affecting U.S. national security have changed fundamentally since the signing of the ABM Treaty in 1972. The attacks against the U.S. homeland on September 11 vividly demonstrate that the threats we face today are far different from those of the Cold War. During that era, now fortunately in the past, the United States and the Soviet Union were locked in an implacably hostile relationship. Each side deployed thousands of nuclear weapons pointed at the other. Our ultimate security rested largely on the grim premise that neither side would launch a nuclear attack because doing so would result in a counter-attack ensuring the total destruction of both nations.

Today, our security environment is profoundly different. The Cold War is over. The Soviet Union no longer exists. Russia is not an enemy, but in fact is increasingly allied with us on a growing number of critically important issues. The depth of United States-Russian cooperation in counterterrorism is both a model of the new strategic relationship we seek to establish and a foundation on which to build further cooperation across the broad spectrum of political, economic and security issues of mutual interest.

Today, the United States and Russia face new threats to their security. Principal among these threats are weapons of mass destruction and their delivery means wielded by terrorists and rogue states. A number of such states are acquiring increasingly longer-range ballistic missiles as instruments of blackmail and coercion against the United States and its friends and allies.

156

United States Documents

The United States must defend its homeland, its forces and its friends and allies against these threats. We must develop and deploy the means to deter and protect against them, including through limited missile defense of our territory.

Under the terms of the ABM Treaty, the United States is prohibited from defending its homeland against ballistic missile attack. We are also prohibited from cooperating in developing missile defenses against long-range threats with our friends and allies. Given the emergence of these new threats to our national security and the imperative of defending against them, the United States is today providing formal notification of its withdrawal from the ABM Treaty. As provided in Article XV of that Treaty, the effective date of withdrawal will be six months from today.

At the same time, the United States looks forward to moving ahead with Russia in developing elements of a new strategic relationship.

- In the inter-related area of offensive nuclear forces, we welcome President Putin's commitment to deep cuts in Russian nuclear forces, and reaffirm our own commitment to reduce U.S. nuclear forces significantly.
- We look forward to continued consultations on how to achieve increased transparency and predictability regarding reductions in offensive nuclear forces.
- We also look forward to continued consultations on transparency, confidence building, and cooperation on missile defenses, such as joint exercises and potential joint development programs.
- The United States also plans to discuss with Russia ways to establish regular defense planning talks to exchange information on strategic force issues, and to deepen cooperation on efforts to prevent and deal with the effects of the spread of weapons of mass destruction and their means of delivery.

The United States intends to expand cooperation in each of these areas and to work intensively with Russia to further develop and formalize the new strategic relationship between the two countries.

The United States believes that moving beyond the ABM Treaty will contribute to international peace and security. We stand ready to continue our active dialogue with allies, China, and other interested states on all issues associated with strategic stability and how we can best cooperate to meet the threats of the 21st century. We believe such a dialogue is in the interest of all states.

Source: The White House. "U.S. Withdrawal from ABM Treaty (December 13, 2001)" Available online. URL: http://georgewbush-whitehouse.archives.gov/news/releases/2001/12/20011213-2.html. Accessed May 20, 2010.

U.S. COMMERCIAL REMOTE SENSING POLICY, NSPD 27 (April 25, 2003) (Excerpt)

With the U.S. government increasingly relying on commercial satellite providers to meet government remote sensing needs, President George W. Bush directed the government to work with commercial firms to maintain the U.S. lead in advanced satellite technologies. The directive also specifies that the government may step in to regulate the activities of commercial providers if it is in the interest of national security.

FACT SHEET

The President authorized a new national policy on April 25, 2003 that establishes guidance and implementation actions for commercial remote sensing space capabilities. This policy supersedes Presidential Decision Directive 23, U.S. Policy on Foreign Access to Remote Sensing Space Capabilities, dated 9 March 1994. This fact sheet provides a summary of the new policy.

I. SCOPE AND DEFINITIONS

This policy provides guidance for: (1) the licensing and operation of U.S. commercial remote sensing space systems; (2) United States Government use of commercial remote sensing space capabilities; (3) foreign access to U.S. commercial remote sensing space capabilities; and (4) government-to-government intelligence, defense, and foreign policy relationships involving U.S. commercial remote sensing space capabilities.

For the purposes of this document:

- "Remote sensing space capabilities" refers to all remote sensing space systems, technology, components, products, data, services, and related information. In this context, "space system" consists of the spacecraft, the mission package(s), ground stations, data links, and associated command and control facilities and may include data processing and exploitation hardware and software; and

- "Commercial remote sensing space capabilities" refers to privately owned and operated space systems licensed under the Land Remote Sensing Policy Act of 1992, their technology, components, products, data, services, and related information, as well as foreign systems whose products and services are sold commercially.

No legal rights or remedies, or legally enforceable causes of action are created or intended to be created by this policy. Officers of the United States and those agents acting on their behalf implementing this policy shall do so in a manner consistent with applicable law.

II. POLICY GOAL

The fundamental goal of this policy is to advance and protect U.S. national security and foreign policy interests by maintaining the nation's leadership in remote sensing space activities, and by sustaining and enhancing the U.S. remote sensing industry. Doing so will also foster economic growth, contribute to environmental stewardship, and enable scientific and technological excellence.

In support of this goal, the United States Government will:

- Rely to the maximum practical extent on U.S. commercial remote sensing space capabilities for filling imagery and geospatial needs for military, intelligence, foreign policy, homeland security, and civil users;
- Focus United States Government remote sensing space systems on meeting needs that can not be effectively, affordably, and reliably satisfied by commercial providers because of economic factors, civil mission needs, national security concerns, or foreign policy concerns;
- Develop a long-term, sustainable relationship between the United States Government and the U.S. commercial remote sensing space industry;
- Provide a timely and responsive regulatory environment for licensing the operations and exports of commercial remote sensing space systems; and
- Enable U.S. industry to compete successfully as a provider of remote sensing space capabilities for foreign governments and foreign commercial users, while ensuring appropriate measures are implemented to protect national security and foreign policy.

III. BACKGROUND

Vital national security, foreign policy, economic, and civil interests depend on the United States ability to remotely sense Earth from space. Toward these ends, the United States Government develops and operates highly capable remote sensing space systems for national security purposes, to satisfy civil mission needs, and to provide important public services. United States national security systems are valuable assets because of their high quality data collection, timeliness, volume, and coverage that provide a near real-time capability for regularly monitoring events around the world. United States civil remote sensing systems enable such activities as research on local, regional, and global change, and support services and data products for weather, climate, and hazard response, and agricultural, transportation, and infrastructure planning.

A robust U.S. commercial remote sensing space industry can augment and potentially replace some United States Government capabilities and can contribute to U.S. military, intelligence, foreign policy, homeland security, and civil objectives, as well as U.S. economic competitiveness. Continued development and advancement of U.S. commercial remote sensing space capabilities also is essential to sustaining the nation's advantage in collecting information from space. Creating a robust U.S. commercial remote sensing industry requires enhancing the international competitiveness of the industry.

IV. LICENSING AND OPERATION GUIDELINES FOR PRIVATE REMOTE SENSING SPACE SYSTEMS

To support the goals of this policy, U.S. companies are encouraged to build and operate commercial remote sensing space systems whose operational capabilities, products, and services are superior to any current or planned foreign commercial systems. However, because of the potential value of its products to an adversary, the operation of a U.S. commercial remote sensing space system requires appropriate security measures to address U.S. national security and foreign policy concerns. In such cases, the United States Government may restrict operations of the commercial systems in order to limit collection and/or dissemination of certain data and products, e.g., best resolution, most timely delivery, to the United States Government, or United States Government approved recipients.

On a case-by-case basis, the United States Government may require additional controls and safeguards for U.S. commercial remote sensing space systems potentially including them as conditions for United States Government use of those capabilities. These controls and safeguards shall include, but not be limited to: (1) the unique conditions associated with United States Government use of commercial remote sensing space systems; and (2) satellite, ground station, and communications link protection measures to allow the United States Government to rely on these systems. The United States Government also may condition the operation of U.S. commercial remote sensing space systems to ensure appropriate measures are implemented to protect U.S. national security and foreign policy interests.

V. UNITED STATES GOVERNMENT USE OF COMMERCIAL REMOTE SENSING SPACE CAPABILITIES

To support the goals of this policy, the United States Government shall utilize U.S. commercial remote sensing space capabilities to meet imagery and geospatial needs. Foreign commercial remote sensing space capabilities, including but not limited to imagery and geospatial products and services, may be integrated in United States Government imagery and geospatial architectures, consistent with national security and foreign policy objectives.

VI. FOREIGN ACCESS TO U.S. COMMERCIAL REMOTE SENSING SPACE CAPABILITIES

It is in U.S. national security, foreign policy, and economic interests that U.S. industry compete successfully as providers of remote sensing space products and capabilities to foreign governments and foreign commercial users. Therefore, license applications for U.S. commercial remote sensing space exports shall be considered favorably to the extent permitted by existing law, regulations and policy when such exports support these interests.

The United States Government will consider remote sensing exports on a case-by-case basis. These exports will continue to be licensed pursuant to the United States Munitions List or the Commerce Control List, as appropriate, and in accordance with existing law and regulations.

The following guidance will also apply, when considering license applications for remote sensing exports:

VII. GOVERNMENT-TO-GOVERNMENT INTELLIGENCE, DEFENSE, AND FOREIGN RELATIONSHIPS

The United States Government will use U.S. commercial remote sensing space capabilities to the maximum extent practicable to foster foreign partnerships and cooperation, and foreign policy objectives, consistent with the goals of this policy and with broader national security objectives. Proposals for new partnerships regarding remote sensing that would raise questions about United States Government competition with the private sector shall be submitted for interagency review. In general, the United States Government should not pursue such partnerships if they would compete with the private sector, unless there is a compelling national security or foreign policy reason for doing so.

VIII. IMPLEMENTATION ACTIONS

Implementation of this directive will be within the overall policy and resource guidance of the President and subject to the availability of appropriations. Agencies have been directed to complete a series of specific implementation actions within 120 days from the date of this directive.

Source: Federation of American Scientists. "Official Intelligence-Related Documents." Available online. URL: http://www.fas.org/irp/offdocs/nspd/remsens.html. Accessed May 21, 2010.

National Policy on Ballistic Missile Defense (May 20, 2003)

Before the terrorist attacks of September 11, 2001, U.S. security policy had primarily focused on threats from other countries. But 9/11 confirmed that the United States was also threatened by non-state actors, such as terrorist groups. Consequently, President George W. Bush called upon the development and deployment of missile defense systems, which would provide flexibility to defend against a number of targets.

NATIONAL POLICY ON BALLISTIC MISSILE DEFENSE
FACT SHEET
FOR IMMEDIATE RELEASE
OFFICE OF THE PRESS SECRETARY
MAY 20, 2003

Restructuring our defense and deterrence capabilities to correspond to emerging threats remains one of the Administration's highest priorities, and the deployment of missile defenses is an essential component of this broader effort.

Changed Security Environment
As the events of September 11 demonstrated, the security environment is more complex and less predictable than in the past. We face growing threats

from weapons of mass destruction (WMD) in the hands of states or non-state actors, threats that range from terrorism to ballistic missiles intended to intimidate and coerce us by holding the U.S. and our friends and allies hostage to WMD attack.

Hostile states, including those that sponsor terrorism, are investing large resources to develop and acquire ballistic missiles of increasing range and sophistication that could be used against the United States and our friends and allies. These same states have chemical, biological, and/or nuclear weapons programs. In fact, one of the factors that make long-range ballistic missiles attractive as a delivery vehicle for weapons of mass destruction is that the United States and our allies lack effective defenses against this threat.

The contemporary and emerging missile threat from hostile states is fundamentally different from that of the Cold War and requires a different approach to deterrence and new tools for defense. The strategic logic of the past may not apply to these new threats, and we cannot be wholly dependent on our capability to deter them. Compared to the Soviet Union, their leaderships often are more risk prone. These are leaders that also see WMD as weapons of choice, not of last resort. Weapons of mass destruction are their most lethal means to compensate for our conventional strength and to allow them to pursue their objectives through force, coercion, and intimidation.

Deterring these threats will be difficult. There are no mutual understandings or reliable lines of communication with these states. Our new adversaries seek to keep us out of their region, leaving them free to support terrorism and to pursue aggression against their neighbors. By their own calculations, these leaders may believe they can do this by holding a few of our cities hostage. Our adversaries seek enough destructive capability to blackmail us from coming to the assistance of our friends who would then become the victims of aggression.

Some states are aggressively pursuing the development of weapons of mass destruction and long-range missiles as a means of coercing the United States and our allies. To deter such threats, we must devalue missiles as tools of extortion and aggression, undermining the confidence of our adversaries that threatening a missile attack would succeed in blackmailing us. In this way, although missile defenses are not a replacement for an offensive response capability, they are an added and critical dimension of contemporary deterrence. Missile defenses will also help to assure allies and friends, and to dissuade countries from pursuing ballistic missiles in the first instance by undermining their military utility.

National Missile Defense Act of 1999

On July 22, 1999, the National Missile Defense Act of 1999 (Public Law 106-38) was signed into law. This law states, "It is the policy of the United States to deploy as soon as is technologically possible an effective National Missile Defense system capable of defending the territory of the United States against limited ballistic missile attack (whether accidental, unauthorized, or deliberate) with funding subject to the annual authorization of appropriations and the annual appropriation of funds for National Missile Defense." The Administration's program on missile defense is fully consistent with this policy.

Missile Defense Program

At the outset of this Administration, the President directed his Administration to examine the full range of available technologies and basing modes for missile defenses that could protect the United States, our deployed forces, and our friends and allies. Our policy is to develop and deploy, at the earliest possible date, ballistic missile defenses drawing on the best technologies available.

The Administration has also eliminated the artificial distinction between "national" and "theater" missile defenses.

The defenses we will develop and deploy must be capable of not only defending the United States and our deployed forces, but also friends and allies; The distinction between theater and national defenses was largely a product of the ABM Treaty and is outmoded. For example, some of the systems we are pursuing, such as boost-phase defenses, are inherently capable of intercepting missiles of all ranges, blurring the distinction between theater and national defenses; and The terms "theater" and "national" are interchangeable depending on the circumstances, and thus are not a meaningful means of categorizing missile defenses. For example, some of the systems being pursued by the United States to protect deployed forces are capable of defending the entire national territory of some friends and allies, thereby meeting the definition of a "national" missile defense system.

Building on previous missile defense work, over the past year and a half, the Defense Department has pursued a robust research, development, testing, and evaluation program designed to develop layered defenses capable of intercepting missiles of varying ranges in all phases of flight. The testing regimen employed has become increasingly stressing, and the results of recent tests have been impressive.

United States Documents

Fielding Missile Defenses

In light of the changed security environment and progress made to date in our development efforts, the United States plans to begin deployment of a set of missile defense capabilities in 2004. These capabilities will serve as a starting point for fielding improved and expanded missile defense capabilities later.

We are pursuing an evolutionary approach to the development and deployment of missile defenses to improve our defenses over time. The United States will not have a final, fixed missile defense architecture. Rather, we will deploy an initial set of capabilities that will evolve to meet the changing threat and to take advantage of technological developments. The composition of missile defenses, to include the number and location of systems deployed, will change over time.

In August 2002, the Administration proposed an evolutionary way ahead for the deployment of missile defenses. The capabilities planned for operational use in 2004 and 2005 will include ground-based interceptors, sea-based interceptors, additional Patriot (PAC-3) units, and sensors based on land, at sea, and in space. In addition, the United States will work with allies to upgrade key early-warning radars as part of our capabilities.

Under our approach, these capabilities may be improved through additional measures such as:

Deployment of additional ground- and sea-based interceptors, and Patriot (PAC-3) units; Initial deployment of the THAAD and Airborne Laser systems; Development of a family of boost-phase and midcourse hit-to-kill interceptors based on sea-, air-, and ground-based platforms; Enhanced sensor capabilities; and Development and testing of space-based defenses.

The Defense Department will begin to implement this approach and will move forward with plans to deploy a set of initial missile defense capabilities beginning in 2004.

Cooperation with Friends and Allies

Because the threats of the 21st century also endanger our friends and allies around the world, it is essential that we work together to defend against these threats. Missile defense cooperation will be a feature of U.S. relations with close, long-standing allies, and an important means to build new relationships with new friends like Russia. Consistent with these goals:

The U.S. will develop and deploy missile defenses capable of protecting not only the United States and our deployed forces, but also friends and allies; We will also structure the missile defense program in a manner that encourages industrial participation by friends and allies, consistent with overall U.S. national security; and We will also promote international missile defense cooperation, including within bilateral and alliance structures such as NATO.

As part of our efforts to deepen missile defense cooperation with friends and allies, the United States will seek to eliminate impediments to such cooperation. We will review existing policies and practices governing technology sharing and cooperation on missile defense, including U.S. export control regulations and statutes, with this aim in mind.

The goal of the Missile Technology Control Regime (MTCR) is to help reduce the global missile threat by curbing the flow of missiles and related technology to proliferators. The MTCR and missile defenses play complementary roles in countering the global missile threat. The United States intends to implement the MTCR in a manner that does not impede missile defense cooperation with friends and allies.

Conclusion
The new strategic challenges of the 21st century require us to think differently, but they also require us to act. The deployment of effective missile defenses is an essential element of the United States' broader efforts to transform our defense and deterrence policies and capabilities to meet the new threats we face. Defending the American people against these new threats is the Administration's highest priority.

Source: The White House. "National Missile Defense Act." Available online. URL: http://georgewbush-whitehouse.archives.gov/news/releases/2003/05/20030520-15.html. Accessed May 20, 2010.

President Bush Announces New Vision for Space Exploration Program (January 14, 2004)

Ten days after two U.S. rovers landed on Mars, President George W. Bush announced a major change in the human flight program, calling for completion of the International Space Station, *a return to the Moon, and, reviving the recommendations of the 1969 Space Task Group, a manned mission to Mars. He pledged to restart the space shuttle program, which had been frozen following the* Columbia *disaster, and announced creation of a new transport spacecraft, the* Crew Exploration Vehicle.

Today's Presidential Action

- Today, President Bush announced a new vision for the Nation's space exploration program. The President committed the United States to a long-term human and robotic program to explore the solar system, starting with a return to the Moon that will ultimately enable future exploration of Mars and other destinations.
- The President's vision affirms our Nation's commitment to manned space exploration. It gives NASA a new focus and clear objectives. It will be affordable and sustainable while maintaining the highest levels of safety.
- The benefits of space technology are far-reaching and affect the lives of every American. Space exploration has yielded advances in communications, weather forecasting, electronics, and countless other fields. For example, image processing technologies used in lifesaving CAT Scanners and MRIs trace their origins to technologies engineered for use in space.

Background on Today's Presidential Action

America's history is built on a desire to open new frontiers and to seek new discoveries. Exploration, like investments in other Federal science and technology activities, is an investment in our future. President Bush is committed to a long-term space exploration program benefiting not only scientific research, but also the lives of all Americans. The exploration vision also has the potential to drive innovation, development, and advancement in the aerospace and other high-technology industries. The President's vision for exploration will not require large budget increases in the near term. Instead, it will bring about a sustained focus over time and a reorientation of NASA's programs.

- NASA spends, and will continue to spend, less than 1 percent of the Federal budget. Our Nation's investment in space is reasonable for a tremendously promising program of discovery and exploration that historically has resulted in concrete benefits as well as inspiring Americans and people throughout the world.

President Bush's Vision for U.S. Space Exploration

The President's plan for steady human and robotic space exploration is based on the following goals:

- First, America will complete its work on the International Space Station by 2010, fulfilling our commitment to our 15 partner countries. The United States will launch a re-focused research effort on board the International Space Station to better understand and overcome the effects of human spaceflight on astronaut health, increasing the safety of future space missions.

- To accomplish this goal, NASA will return the Space Shuttle to flight consistent with safety concerns and the recommendations of the Columbia Accident Investigation Board. The Shuttle's chief purpose over the next several years will be to help finish assembly of the Station, and the Shuttle will be retired by the end of this decade after nearly 30 years of service.

- Second, the United States will begin developing a new manned exploration vehicle to explore beyond our orbit to other worlds—the first of its kind since the Apollo Command Module. The new spacecraft, the Crew Exploration Vehicle, will be developed and tested by 2008 and will conduct its first manned mission no later than 2014. The Crew Exploration Vehicle will also be capable of transporting astronauts and scientists to the International Space Station after the Shuttle is retired.

- Third, America will return to the Moon as early as 2015 and no later than 2020 and use it as a stepping stone for more ambitious missions. A series of robotic missions to the Moon, similar to the Spirit Rover that is sending remarkable images back to Earth from Mars, will explore the lunar surface beginning no later than 2008 to research and prepare for future human exploration. Using the Crew Exploration Vehicle, humans will conduct extended lunar missions as early as 2015, with the goal of living and working there for increasingly extended periods.

- The extended human presence on the Moon will enable astronauts to develop new technologies and harness the Moon's abundant resources to allow manned exploration of more challenging environments. An extended human presence on the Moon could reduce the costs of further exploration, since lunar-based spacecraft could escape the Moon's lower gravity using less energy at less cost than Earth-based vehicles. The experience and knowledge gained on the Moon will serve as a foundation for human missions beyond the Moon, beginning with Mars.

- NASA will increase the use of robotic exploration to maximize our understanding of the solar system and pave the way for more ambitious manned missions. Probes, landers, and similar unmanned vehicles will serve as trailblazers and send vast amounts of knowledge back to scientists on Earth.

Key Points on the President's FY 2005 Budget

- The funding added for exploration will total $12 billion over the next five years. Most of this added funding for new exploration will come from reallocation of $11 billion that is currently within the five-year total NASA budget of $86 billion.
- In the Fiscal Year (FY) 2005 budget, the President will request an additional $1 billion to NASA's existing five-year plan, or an average of $200 million per year.
- From 1992 to 2000, NASA's budget decreased by a total of 5 percent. Since the year 2000, NASA's budget has increased by approximately 3 percent per year.
- From the current 2004 level of $15.4 billion, the President's proposal will increase NASA's budget by an average of 5 percent per year over the next three years, and at approximately 1 percent or less per year for the two years after those.

President's Commission on the Implementation of U.S. Space Exploration Policy

To ensure that NASA maintains a sense of focus and direction toward accomplishing this new mission, the President has directed NASA Administrator Sean O'Keefe to review all current spaceflight and exploration and direct them toward the President's goals. The President also formed a Commission on the Implementation of U.S. Space Exploration Policy to advise NASA on the long-term implementation of the President's vision.

Space Technology Affects the Lives of Every American
More than 1,300 NASA and other U.S. space technologies have contributed to U.S. industry, improving our quality of life and helping save lives.

- Image processing used in CAT Scanners and MRI technology in hospitals worldwide came from technology developed to computer-enhanced pictures of the Moon for the Apollo programs.

- Kidney dialysis machines were developed as a result of a NASA-developed chemical process, and insulin pumps were based on technology used on the Mars Viking spacecraft.
- Programmable Heart Pacemakers were first developed in the 1970s using NASA satellite electrical systems.
- Fetal heart monitors were developed from technology originally used to measure airflow over aircraft wings.
- Surgical probes used to treat brain tumors in children resulted from special lighting technology developed for plant growth experiments on Space Shuttle missions.
- Infrared hand-held cameras used to observe blazing plumes from the Shuttle have helped firefighters point out hot spots in brush fires.
- Satellite communications allow news organizations to provide live, on-the-spot broadcasting from anywhere in the world; families and businesses to stay in touch using cellphone networks; and the simple pleasures of satellite TV and radio, and the convenience of ATMs across the country and around the world.

Source: White House Fact Sheet. Available online. URL: http://history.nasa.gov/SEP%20Press%20Release.htm. Accessed May 20, 2010.

U.S. Space-Based Position, Navigation, and Timing Policy, NSPD-39 (December 15, 2004) (excerpt)

Similar to National Security Presidential Directive 27, NSPD-39 details the relationship between the U.S. government and private firms regarding the U.S. global positioning system (GPS). The primary goal is to ensure that the government has uninterrupted access to continually upgraded GPS information, designating it a segment of critical infrastructure, and the ability to prevent hostile use of the GPS network. The full document contains a lengthy breakdown of agency roles and responsibilities.

FACT SHEET

The President authorized a new national policy on December 8, 2004 that establishes guidance and implementation actions for space-based positioning, navigation, and timing programs, augmentations, and activities for U.S. national and homeland security, civil, scientific, and commercial purposes. This policy supersedes Presidential Decision Directive/National Science and

Technology Council-6, U.S. Global Positioning System Policy, dated March 28, 1996.

I. Scope and Definitions

This policy provides guidance for: (1) development, acquisition, operation, sustainment, and modernization of the Global Positioning System and U.S.-developed, owned and/or operated systems used to augment or otherwise improve the Global Positioning System and/or other space-based positioning, navigation, and timing signals; (2) development, deployment, sustainment, and modernization of capabilities to protect U.S. and allied access to and use of the Global Positioning System for national, homeland, and economic security, and to deny adversaries access to any space-based positioning, navigation, and timing services; and (3) foreign access to the Global Positioning System and United States Government augmentations, and international cooperation with foreign space-based positioning, navigation, and timing services, including augmentations.

For purposes of this document:

- "Interoperable" refers to the ability of civil U.S. and foreign space-based positioning, navigation, and timing services to be used together to provide better capabilities at the user level than would be achieved by relying solely on one service or signal;
- "Compatible" refers to the ability of U.S. and foreign space-based positioning, navigation, and timing services to be used separately or together without interfering with each individual service or signal, and without adversely affecting navigation warfare; and
- "Augmentation" refers to space and/or ground-based systems that provide users of space-based positioning, navigation, and timing signals with additional information that enables users to obtain enhanced performance when compared to the un-augmented space-based signals alone. These improvements include better accuracy, availability, integrity, and reliability, with independent integrity monitoring and alerting capabilities for critical applications.

II. Background

Over the past decade, the Global Positioning System has grown into a global utility whose multi-use services are integral to U.S. national security, economic growth, transportation safety, and homeland security, and are an essential element of the worldwide economic infrastructure. In the year 2000, the United States recognized the increasing importance of the

Global Positioning System to civil and commercial users by discontinuing the deliberate degradation of accuracy for non-military signals, known as Selective Availability. Since that time, commercial and civil applications of the Global Positioning System have continued to multiply and their importance has increased significantly. Services dependent on Global Positioning System information are now an engine for economic growth, enhancing economic development, and improving safety of life, and the system is a key component of multiple sectors of U.S. critical infrastructure.

While the growth in civil and commercial applications continues, the positioning, navigation, and timing information provided by the Global Positioning System remains critical to U.S. national security, and its applications are integrated into virtually every facet of U.S. military operations. United States and allied military forces will continue to rely on the Global Positioning System military services for positioning, navigation, and timing services.

The continuing growth of services based on the Global Positioning System presents opportunities, risks, and threats to U.S. national, homeland, and economic security. The widespread and growing dependence on the Global Positioning System of military, civil, and commercial systems and infrastructures has made many of these systems inherently vulnerable to an unexpected interruption in positioning, navigation, and/or timing services. In addition, whether designed for military capabilities or not, all positioning, navigation, and timing signals from space and their augmentations provide inherent capabilities that can be used by adversaries, including enemy military forces and terrorist groups. Finally, emerging foreign space-based positioning, navigation, and timing services could enhance or undermine the future utility of the Global Positioning System.

The United States must continue to improve and maintain the Global Positioning System, augmentations, and backup capabilities to meet growing national, homeland, and economic security requirements, for civil requirements, and to meet commercial and scientific demands. In parallel, we must continue to improve capabilities to deny adversary access to all space-based positioning, navigation, and timing services, particularly including services that are openly available and can be readily used by adversaries and/or terrorists to threaten the security of the United States. In addition, the diverse requirements for and multiple applications of space-based positioning, navigation, and timing services require stable yet adaptable policies and management mechanisms. The existing management mechanisms for the Global Positioning System and its augmentations must be modified to accommodate a multi-use approach to program planning, resource allocation, system development, and operations. Therefore, the United States Government must improve the policy and management framework governing the Global Positioning System and its augmentations to support their continued ability to meet increasing and varied domestic and global requirements.

III. Goals and Objectives

The fundamental goal of this policy is to ensure that the United States maintains space-based positioning, navigation, and timing services, augmentation, back-up, and service denial capabilities that: (1) provide uninterrupted availability of positioning, navigation, and timing services; (2) meet growing national, homeland, economic security, and civil requirements, and scientific and commercial demands; (3) remain the pre-eminent military space-based positioning, navigation, and timing service; (4) continue to provide civil services that exceed or are competitive with foreign civil space-based positioning, navigation, and timing services and augmentation systems; (5) remain essential components of internationally accepted positioning, navigation, and timing services; and (6) promote U.S. technological leadership in applications involving space-based positioning, navigation, and timing services. To achieve this goal, the United States Government shall:

- Provide uninterrupted access to U.S. space-based global, precise positioning, navigation, and timing services for U.S. and allied national security systems and capabilities through the Global Positioning System, without being dependent on foreign positioning, navigation, and timing services;
- Provide on a continuous, worldwide basis civil space-based, positioning, navigation, and timing services free of direct user fees for civil, commercial, and scientific uses, and for homeland security through the Global Positioning System and its augmentations, and provide open, free access to information necessary to develop and build equipment to use these services;
- Improve capabilities to deny hostile use of any space-based positioning, navigation, and timing services, without unduly disrupting civil and commercial access to civil positioning, navigation, and timing services outside an area of military operations, or for homeland security purposes;
- Improve the performance of space-based positioning, navigation, and timing services, including more robust resistance to interference for, and consistent with, U.S. and allied national security

purposes, homeland security, and civil, commercial, and scientific users worldwide;

- Maintain the Global Positioning System as a component of multiple sectors of the U.S. Critical Infrastructure, consistent with Homeland Security Presidential Directive-7, Critical Infrastructure Identification, Prioritization, and Protection, dated December 17, 2003;

- Encourage foreign development of positioning, navigation, and timing services and systems based on the Global Positioning System. Seek to ensure that foreign space-based positioning, navigation, and timing systems are interoperable with the civil services of the Global Positioning System and its augmentations in order to benefit civil, commercial, and scientific users worldwide. At a minimum, seek to ensure that foreign systems are compatible with the Global Positioning System and its augmentations and address mutual security concerns with foreign providers to prevent hostile use of space-based positioning, navigation, and timing services; and

- Promote the use of U.S. space-based positioning, navigation, and timing services and capabilities for applications at the Federal, State, and local level, to the maximum practical extent.

IV. Management of Space-Based Positioning, Navigation, and Timing Services

This policy establishes a permanent National Space-Based Positioning, Navigation, and Timing Executive Committee.

The Executive Committee shall make recommendations to its member Departments and Agencies, and to the President through the representatives of the Executive Office of the President. In addition, the Executive Committee will advise and coordinate with and among the Departments and Agencies responsible for the strategic decisions regarding policies, architectures, requirements, and resource allocation for maintaining and improving U.S. space-based positioning, navigation, and timing infrastructures, including the Global Positioning System, its augmentations, security for these services, and relationships with foreign positioning, navigation, and timing services. Specifically, the Executive Committee shall:

- Ensure that national security, homeland security, and civil requirements receive full and appropriate consideration in the decision-making process and facilitate the integration and deconfliction of these requirements for space-based positioning, navigation, and timing capabilities, as required;

- Coordinate individual Departments' and Agencies' positioning, navigation, and timing program plans, requirements, budgets, and policies, and assess the adequacy of funding and schedules to meet validated requirements in a timely manner;
- Ensure that the utility of civil services exceeds, or is at least equivalent to, those routinely provided by foreign space-based positioning, navigation, and timing services;
- Promote plans to modernize the U.S. space-based positioning, navigation, and timing infrastructure, including: (1) development, deployment, and operation of new and/or improved national security and public safety services when required and to the maximum practical extent; and (2) determining the apportionment of requirements between the Global Positioning System and its augmentations, including consideration of user equipment;
- Review proposals and provide recommendations to the Departments and Agencies for international cooperation, as well as spectrum management and protection issues; and
- Establish a space-based Positioning, Navigation, and Timing Advisory Board. The board shall be comprised of experts from outside the United States Government, and shall be chartered as a Federal Advisory Committee.

V. Foreign Access to U.S. Space-based Positioning, Navigation, and Timing Capabilities

Any exports of U.S. positioning, navigation, and timing capabilities covered by the United States Munitions List or the Commerce Control List will continue to be licensed pursuant to the International Traffic in Arms Regulations or the Export Administration Regulations, as appropriate, and in accordance with all existing laws and regulations.

As a general guideline, export of civil or other non-United States Munitions List space-based positioning, navigation and timing capabilities that are currently available or are planned to be available in the global marketplace will continue to be considered favorably. Exports of sensitive or advanced positioning, navigation, and timing information, systems, technologies, and components will be considered on a case-by-case basis in accordance with existing laws and regulations, as well as relevant national security and foreign policy goals and considerations. In support of such reviews, the Secretary of State, in consultation with the Secretaries of Defense, Commerce, and Energy, the Administrator of the National Aeronautics and

Space Administration, and the Director of Central Intelligence, shall modify and maintain the Sensitive Technology List directed in U.S. Commercial Remote Sensing Space Policy, dated April 25, 2003, including those technology items or areas deemed sensitive for positioning, navigation and timing applications. The Secretaries of State and Commerce shall use the list in the evaluation of requests for exports.

Source: Federation of American Scientists. "Official Intelligence-Related Documents." Available online. URL: http://www.fas.org/irp/offdocs/nspd/nspd-39.htm. Accessed May 20, 2010.

Overview of the DoD Fiscal 2010 Budget Proposal (April 6, 2009) (excerpt)

The first military budget of the Obama administration drastically changed defense priorities. Programs for intelligence, surveillance, drone aircraft, and sensors increased. However, the transformational satellite, part of a secure communications network, was canceled, while the missile defense program was cut by $1.2 billion and the multiple-kill vehicle and second airborne laser programs were eliminated.

The fiscal 2010 budget proposal was crafted to achieve four principal objectives:

1. Reaffirming and strengthening the nation's commitment to care for the all-volunteer force, America's greatest strategic asset.
2. Reshaping DoD programs to institutionalize and enhance capabilities to fight the wars the U.S. is engaged in today and the scenarios the nation will most likely face in the years ahead, while also providing a hedge against other risks and contingencies.
3. Beginning a fundamental overhaul of the DoD's approach to procurement, acquisition, and contracting.
4. Providing the necessary resources to support the troops in the field.

Reshape the Force
Efforts to put defense bureaucracies on a war footing in recent years have revealed flaws in the way the department operates. Its institutions were created and organized to prepare for conventional conflicts with modern armies, navies, and air forces. The existing organization has not provided adequate institutional support for today's warfighters and their needs. Today's warfighters require steady, long-term funding and a bureaucratic constituency

similar to conventional modernization programs. The fiscal 2010 budget will begin a process of change needed to fully support today's warfighters.

ISR. The budget will increase intelligence, surveillance, and reconnaissance (ISR) support for today's warfighter by adding nearly $2 billion to the base budget. This will include:

- 50 Predator-class unmanned aerial orbits by 2011. This capability, which has been in high demand in Iraq and Afghanistan, will now be permanently funded in the base budget. It will eventually result in a 62 percent increase in capability over the current level and 127 percent compared to a year ago.
- An increase in manned ISR capabilities, such as the turbo-prop aircraft deployed so successfully as part of Task Force Odin in Iraq.
- Research and development on a number of ISR enhancements and experimental platforms optimized for today's battlefield.

Change How We Buy
To maintain America's technological and conventional edge, DoD must make a dramatic change in the way military equipment is acquired. Reform involves the following:

Transformational Satellite. The $19 billion Transformational (TSAT) program will be terminated. Instead, DoD will purchase two more Advanced Extremely High Frequency satellites.

Missile Defense. The fiscal 2010 budget will reduce the Missile Defense Agency (MDA) program by $1.2 billion, leaving a fiscal 2010 request of $7.8 billion for MDA:

- The program will be restructured to focus on the rogue state and theater missile threat.
- Ground-based interceptors in Alaska will not be increased as planned, but research and development will be funded to improve existing capabilities to defend against long-range rogue missile threats.
- The second airborne laser prototype aircraft will be canceled due to affordability and technology problems, keeping the existing aircraft as a technology demonstration effort.
- The Multiple Kill Vehicle (MKV) program will be terminated because of significant technical challenges.

Source: Department of Defense Press Release. (April 6, 2009). Available online. URL: http://www.defense.gov/news/2010%20Budget%20Proposal.pdf. Accessed May 20, 2010.

A "Phased, Adaptive Approach" for Missile Defense in Europe (September 17, 2009) (excerpt)

One of the key elements in President George W. Bush's national security policy was the need for the United States to deploy a missile defense system, and Washington withdrew from the Anti-Ballistic Missile Treaty to pursue this project. Bush ordered components to be installed in Poland and the Czech Republic, a move that angered Moscow. The Obama administration canceled those deployment plans in favor of the existing Aegis weapon system.

FACT SHEET ON U.S. MISSILE DEFENSE POLICY
A "PHASED, ADAPTIVE APPROACH" FOR MISSILE DEFENSE IN EUROPE

President Obama has approved the recommendation of Secretary of Defense Gates and the Joint Chiefs of Staff for a phased, adaptive approach for missile defense in Europe. This approach is based on an assessment of the Iranian missile threat, and a commitment to deploy technology that is proven, cost-effective, and adaptable to an evolving security environment.

Starting around 2011, this missile defense architecture will feature deployments of increasingly-capable sea- and land-based missile interceptors, primarily upgraded versions of the Standard Missile-3 (SM-3), and a range of sensors in Europe to defend against the growing ballistic missile threat from Iran. This phased approach develops the capability to augment our current protection of the U.S. homeland against long-range ballistic missile threats, and to offer more effective defenses against more near-term ballistic missile threats. The plan provides for the defense of U.S. deployed forces, their families, and our Allies in Europe sooner and more comprehensively than the previous program, and involves more flexible and survivable systems.

The Secretary of Defense and the Joint Chiefs of Staff recommended to the President that he revise the previous Administration's 2007 plan for missile defense in Europe as part of an ongoing comprehensive review of our missile defenses mandated by Congress. Two major developments led to this unanimous recommended change:

- New Threat Assessment: The intelligence community now assesses that the threat from Iran's short- and medium-range ballistic missiles is developing more rapidly than previously projected, while the threat of potential Iranian intercontinental ballistic missile (ICBM) capabilities has been slower to develop than previously estimated. In the near-term, the greatest missile threats from Iran will be to U.S. Allies and partners, as well as to U.S. deployed personnel—military and civilian and their accompanying families in the Middle East and in Europe.

- Advances in Capabilities and Technologies: Over the past several years, U.S. missile defense capabilities and technologies have advanced significantly. We expect this trend to continue. Improved interceptor capabilities, such as advanced versions of the SM-3, offer a more flexible, capable, and cost-effective architecture. Improved sensor technologies offer a variety of options to detect and track enemy missiles.

These changes in the threat as well as our capabilities and technologies underscore the need for an adaptable architecture. This architecture is responsive to the current threat, but could also incorporate relevant technologies quickly and cost-effectively to respond to evolving threats. Accordingly, the Department of Defense has developed a four-phased, adaptive approach for missile defense in Europe. While further advances of technology or future changes in the threat could modify the details or timing of later phases, current plans call for the following:

- Phase One (in the 2011 timeframe)—Deploy current and proven missile defense systems available in the next two years, including the sea-based Aegis Weapon System, the SM-3 interceptor (Block IA), and sensors such as the forward-based Army Navy/Transportable Radar Surveillance system (AN/TPY-2), to address regional ballistic missile threats to Europe and our deployed personnel and their families;
- Phase Two (in the 2015 timeframe)—After appropriate testing, deploy a more capable version of the SM-3 interceptor (Block IB) in both sea- and land-based configurations, and more advanced sensors, to expand the defended area against short- and medium-range missile threats;
- Phase Three (in the 2018 timeframe)—After development and testing are complete, deploy the more advanced SM-3 Block IIA variant currently under development, to counter short-, medium-, and intermediate-range missile threats; and
- Phase Four (in the 2020 timeframe)—After development and testing are complete, deploy the SM-3 Block IIB to help better cope with medium- and intermediate-range missiles and the potential future ICBM threat to the United States.

Throughout all four phases, the United States also will be testing and updating a range of approaches for improving our sensors for missile defense. The new distributed interceptor and sensor architecture also does not require a single, large, fixed European radar that was to be located in the Czech Republic; this approach also uses different interceptor technology than the previous program, removing the need for a single field of 10 ground-based interceptors in Poland. Therefore, the Secretary of Defense recommended that the United States no longer plan to move forward with that architecture.

The Czech Republic and Poland, as close, strategic and steadfast Allies of the United States, will be central to our continued consultations with NATO Allies on our defense against the growing ballistic missile threat.

The phased, adaptive approach for missile defense in Europe:

- Sustains U.S. homeland defense against long-range ballistic missile threats. The deployment of an advanced version of the SM-3 interceptor in Phase Four of the approach would augment existing ground-based interceptors located in Alaska and California, which provide for the defense of the homeland against a potential ICBM threat.
- Speeds protection of U.S. deployed forces, civilian personnel, and their accompanying families against the near-term missile threat from Iran. We would deploy current and proven technology by roughly 2011—about six or seven years earlier than the previous plan—to help defend the regions in Europe most vulnerable to the Iranian short- and medium-range ballistic missile threat.
- Ensures and enhances the protection of the territory and populations of all NATO Allies, in concert with their missile defense capabilities, against the current and growing ballistic missile threat. Starting in 2011, the phased, adaptive approach would systematically increase the defended area as the threat is expected to grow. In the 2018 timeframe, all of Europe could be protected by our collective missile defense architecture.
- Deploys proven capabilities and technologies to meet current threats. SM-3 (Block 1A) interceptors are deployed on Aegis ships today, and more advanced versions are in various stages of development. Over the past four years, we have conducted a number of tests of the SM-3 IA, and it was the interceptor used in the successful engagement of a decaying satellite in February 2008. Testing in 2008 showed that sensors we plan to field bring significant

capabilities to the architecture, and additional, planned research and development over the next few years offers the potential for more diverse and more capable sensors.

- Provides flexibility to upgrade and adjust the architecture, and to do so in a cost-effective manner, as the threat evolves. Because of the lower per-interceptor costs and mobility of key elements of the architecture, we will be better postured to adapt this set of defenses to any changes in threat.

We will work with our Allies to integrate this architecture with NATO members' missile defense capabilities, as well as with the emerging NATO command and control network that is under development. One benefit of the phased, adaptive approach is that there is a high degree of flexibility—in addition to sea-based assets, there are many potential locations for the architecture's land-based elements, some of which will be re-locatable. We plan to deploy elements in northern and southern Europe and will be consulting closely at NATO with Allies on the specific deployment options.

We also welcome Russian cooperation to bring its missile defense capabilities into a broader defense of our common strategic interests. We have repeatedly made clear to Russia that missile defense in Europe poses no threat to its strategic deterrent. Rather, the purpose is to strengthen defenses against the growing Iranian missile threat. There is no substitute for Iran complying with its international obligations regarding its nuclear program. But ballistic missile defenses will address the threat from Iran's ballistic missile programs, and diminish the coercive influence that Iran hopes to gain by continuing to develop these destabilizing capabilities.

Through the ongoing Department of Defense ballistic missile defense review, the Secretary of Defense and the Joint Chiefs of Staff will continue to provide recommendations to the President that address other aspects of our ballistic missile defense capabilities and posture around the world.

Source: White House Fact Sheet, September 17, 2009. Available online. URL: http://www.whitehouse.gov/ the_press_office/FACT-SHEET-US-Missile-Defense-Policy-A-Phased-Adaptive-Approach-for-Missile-Defense-in-Europe/. Accessed May 20, 2010.

Remarks by the President on Space Exploration in the 21st Century (April 15, 2010) (excerpts)

President Barack Obama came to the Kennedy Space Center in Florida to outline his vision of space policy to the men and women working on the space shuttle and Orion programs. He reassured that their jobs would continue as the United States deployed manned missions to asteroids and Mars.

. . . as President, I believe that space exploration is not a luxury, it's not an afterthought in America's quest for a brighter future—it is an essential part of that quest.

So today, I'd like to talk about the next chapter in this story. The challenges facing our space program are different, and our imperatives for this program are different, than in decades past. We're no longer racing against an adversary. We're no longer competing to achieve a singular goal like reaching the Moon. In fact, what was once a global competition has long since become a global collaboration. But while the measure of our achievements has changed a great deal over the past 50 years, what we do—or fail to do—in seeking new frontiers is no less consequential for our future in space and here on Earth.

So let me start by being extremely clear: I am 100 percent committed to the mission of NASA and its future. Because broadening our capabilities in space will continue to serve our society in ways that we can scarcely imagine. Because exploration will once more inspire wonder in a new generation—sparking passions and launching careers. And because, ultimately, if we fail to press forward in the pursuit of discovery, we are ceding our future and we are ceding that essential element of the American character.

[. . .]

But I also know that underlying these concerns is a deeper worry, one that precedes not only this plan but this administration. It stems from the sense that people in Washington—driven sometimes less by vision than by politics—have for years neglected NASA's mission and undermined the work of the professionals who fulfill it. We've seen that in the NASA budget, which has risen and fallen with the political winds.

But we can also see it in other ways: in the reluctance of those who hold office to set clear, achievable objectives; to provide the resources to meet those objectives; and to justify not just these plans but the larger purpose of space exploration in the 21st century.

All that has to change. And with the strategy I'm outlining today, it will. We start by increasing NASA's budget by $6 billion over the next five years, even — [applause] — I want people to understand the context of this. This is happening even as we have instituted a freeze on discretionary spending and sought to make cuts elsewhere in the budget.

[. . .]

And we will extend the life of the *International Space Station* likely by more than five years, while actually using it for its intended purpose: conducting advanced research that can help improve the daily lives of people here on Earth, as well as testing and improving upon our capabilities in space. This includes technologies like more efficient life support systems that will help reduce the cost of future missions. And in order to reach the space station, we will work with a growing array of private companies competing to make getting to space easier and more affordable.

[. . .]

In addition, as part of this effort, we will build on the good work already done on the *Orion* crew capsule. I've directed Charlie Bolden to immediately begin developing a rescue vehicle using this technology, so we are not forced to rely on foreign providers if it becomes necessary to quickly bring our people home from the International Space Station. And this Orion effort will be part of the technological foundation for advanced spacecraft to be used in future deep space missions. In fact, Orion will be readied for flight right here in this room.

Next, we will invest more than $3 billion to conduct research on an advanced "heavy lift rocket"—a vehicle to efficiently send into orbit the crew capsules, propulsion systems, and large quantities of supplies needed to reach deep space. In developing this new vehicle, we will not only look at revising or modifying older models; we want to look at new designs, new materials, new technologies that will transform not just where we can go but what we can do when we get there. And we will finalize a rocket design no later than 2015 and then begin to build it. And I want everybody to understand: That's at least two years earlier than previously planned—and that's conservative, given that the previous program was behind schedule and over budget.

At the same time, after decades of neglect, we will increase investment—right away—in other groundbreaking technologies that will allow astronauts to reach space sooner and more often, to travel farther and faster for less cost, and to live and work in space for longer periods of time more safely. That means tackling major scientific and technological challenges. How do we shield astronauts from radiation on longer missions? How do we harness resources on distant worlds? How do we supply spacecraft with energy needed for these far-reaching journeys? These are questions that we can answer and will answer. And these are the questions whose answers no doubt will reap untold benefits right here on Earth.

So the point is what we're looking for is not just to continue on the same path—we want to leap into the future; we want major breakthroughs; a transformative agenda for NASA.

[. . .]

The bottom line is nobody is more committed to manned spaceflight, to human exploration of space than I am. But we've got to do it in a smart way, and we can't just keep on doing the same old things that we've been doing and thinking that somehow is going to get us to where we want to go.

Some have said, for instance, that this plan gives up our leadership in space by failing to produce plans within NASA to reach low Earth orbit, instead of relying on companies and other countries. But we will actually reach space faster and more often under this new plan, in ways that will help us improve our technological capacity and lower our costs, which are both essential for the long-term sustainability of spaceflight. In fact, through our plan, we'll be sending many more astronauts to space over the next decade.

There are also those who criticized our decision to end parts of Constellation as one that will hinder space exploration below [sic] low Earth orbit. But it's precisely by investing in groundbreaking research and innovative companies that we will have the potential to rapidly transform our capabilities—even as we build on the important work already completed, through projects like Orion, for future missions. And unlike the previous program, we are setting a course with specific and achievable milestones.

Early in the next decade, a set of crewed flights will test and prove the systems required for exploration beyond low Earth orbit. And by 2025, we expect new spacecraft designed for long journeys to allow us to begin the first-ever crewed missions beyond the Moon into deep space. So we'll start—we'll start by sending astronauts to an asteroid for the first time in history. By the mid-2030s, I believe we can send humans to orbit Mars and return them safely to Earth. And a landing on Mars will follow. And I expect to be around to see it.

[. . .]

Now, I understand that some believe that we should attempt a return to the surface of the Moon first, as previously planned. But I just have to say pretty bluntly here: We've been there before.

[. . .]

There's a lot more of space to explore, and a lot more to learn when we do. So I believe it's more important to ramp up our capabilities to reach—and operate at—a series of increasingly demanding targets, while advancing our technological capabilities with each step forward. And that's what this strategy does. And that's how we will ensure that our leadership in space is even stronger in this new century than it was in the last.

[. . .]

I'm proposing a $40 million initiative led by a high-level team from the White House, NASA, and other agencies to develop a plan for regional economic growth and job creation. And I expect this plan to reach my desk by August 15th. It's an effort that will help prepare this already skilled workforce for new opportunities in the space industry and beyond.

So this is the next chapter that we can write together here at NASA. We will partner with industry. We will invest in cutting-edge research and technology. We will set far-reaching milestones and provide the resources to reach those milestones. And step by step, we will push the boundaries not only of where we can go but what we can do.

Fifty years after the creation of NASA, our goal is no longer just a destination to reach. Our goal is the capacity for people to work and learn and operate and live safely beyond the Earth for extended periods of time, ultimately in ways that are more sustainable and even indefinite. And in fulfilling this task, we will not only extend humanity's reach in space—we will strengthen America's leadership here on Earth.

[. . .]

Now, little more than 40 years ago, astronauts descended the nine-rung ladder of the lunar module called Eagle, and allowed their feet to touch the dusty surface of the Earth's only Moon. This was the culmination of a daring and perilous gambit—of an endeavor that pushed the boundaries of our knowledge, of our technological prowess, of our very capacity as human beings to solve problems. It wasn't just the greatest achievement in NASA's history—it was one of the greatest achievements in human history.

And the question for us now is whether that was the beginning of something or the end of something. I choose to believe it was only the beginning.

[. . .]

Source: White House press release. Available online. URL: http://www.whitehouse.gov/the-press-office/remarks-president-space-exploration-21st-century. Accessed May 20, 2010.

National Space Policy of the United States of America (June 28, 2010) (excerpts)

The Obama White House issued a new national space policy on June 28, 2010, encouraging the use of commercial sources whenever possible, including launches of U.S. payloads. The policy also stresses the importance of diplomacy, peaceful uses of space, making space available to all responsible parties, and strengthening stability in space including measures to mitigate orbital debris.

NATIONAL SPACE POLICY
of the
UNITED STATES OF AMERICA

June 28, 2010
[. . .]

The space age began as a race for security and prestige between the superpowers. The opportunities were boundless, and the decades that followed have seen a radical transformation in the way we live our daily lives, in large part due to our use of space. Space systems have taken us to other celestial bodies and extended humankind's horizons back in time to the very first moments of the universe and out to the galaxies at its far reaches. Satellites contribute to increased transparency and stability among nations and provide a vital communications path for avoiding potential conflicts. Space systems increase our knowledge in many scientific fields, and life on Earth is far better as a result.

The utilization of space has created new markets; helped save lives by warning us of natural disaster, expediting search and rescue operations, and making recovery efforts faster and more effective; made agriculture and natural resource management more efficient and sustainable; expanded our frontiers; and provided global access to advanced medicine, weather forecasting, geospatial information, financial operations, broadband and other communications, and scores of other activities worldwide. Space systems allow people and governments around the world to see with clarity, communicate with certainty, navigate with accuracy, and operate with assurance.

The legacy of success in space and its transformation also presents new challenges. When the space age began, the opportunities to use space were limited to only a few nations, and there were limited consequences for irresponsible or unintentional behavior. Now, we find ourselves in a world where the benefits of space permeate almost every facet of our lives. The growth and evolution of the global economy has ushered in an ever-increasing number of nations and organizations using space. The now-ubiquitous and interconnected nature of space capabilities and the world's growing dependence on them mean that irresponsible acts in space can have damaging consequences for all of us. For example, decades of space activity have littered Earth's orbit with debris; and as the world's space-faring nations continue to increase activities in space, the chance for a collision increases correspondingly.

As the leading space-faring nation, the United States is committed to addressing these challenges. But this cannot be the responsibility of the United States alone. All nations have the right to use and explore space, but with this right also comes responsibility. The United States, therefore, calls on all nations to work together to adopt approaches for responsible activity in space to preserve this right for the benefit of future generations.

MILITARIZATION OF SPACE

From the outset of humanity's ascent into space, this Nation declared its commitment to enhance the welfare of humankind by cooperating with others to maintain the freedom of space.

The United States hereby renews its pledge of cooperation in the belief that with strengthened international collaboration and reinvigorated U.S. leadership, all nations and peoples—space-faring and space-benefiting—will find their horizons broadened, their knowledge enhanced, and their lives greatly improved.

PRINCIPLES

In this spirit of cooperation, the United States will adhere to, and proposes that other nations recognize and adhere to, the following principles:

- It is the shared interest of all nations to act responsibly in space to help prevent mishaps, misperceptions, and mistrust. The United States considers the sustainability, stability, and free access to, and use of, space vital to its national interests. Space operations should be conducted in ways that emphasize openness and transparency to improve public awareness of the activities of government, and enable others to share in the benefits provided by the use of space.
- A robust and competitive commercial space sector is vital to continued progress in space. The Untied States is committed to encouraging and facilitating the growth of a U.S. commercial space sector that supports U.S. needs, is globally competitive, and advances U.S. leadership in the generation of new markets and innovation-driven entrepreneurship.
- All nations have the right to explore and use space for peaceful purposes, and for the benefit of all humanity, in accordance with international law. Consistent with this principle, "peaceful purposes" allows for space to be used for national and homeland security activities.
- As established in international law, there shall be no national claims of sovereignty over outer space or any celestial bodies. The United States considers the space systems of all nations to have the rights of passage through, and conduct of operations in, space without interference. Purposeful interference with space systems, including supporting infrastructure, will be considered an infringement of a nation's rights.
- The United States will employ a variety of measures to help assure the use of space for all responsible parties, and, consistent with the inherent right of self-defense, deter others from interference and

188

attack, defend our space systems and contribute to the defense of allied space systems, and, if deterrence fails, defeat efforts to attack them.

GOALS

Consistent with these principles, the United States will pursue the following goals in its national space programs:

- Energize competitive domestic industries to participate in global markets and advance the development of: satellite manufacturing; satellite-based services; space launch; terrestrial applications; and increased entrepreneurship.
- Expand international cooperation on mutually beneficial space activities to: broaden and extend the benefits of space; further the peaceful use of space; and enhance collection and partnership in sharing of space-derived information.
- Strengthen stability in space through: domestic and international measures to promote safe and responsible operations in space; improved information collection and sharing for space object collision avoidance; protection of critical space systems and supporting infrastructures, with special attention to the critical interdependence of space and information systems; and strengthening measures to mitigate orbital debris.
- Increase assurance and resilience of mission-essential functions enabled by commercial, civil, scientific, and national security spacecraft and supporting infrastructure against disruption, degradation, and destruction, whether from environmental, mechanical, electronic, or hostile causes.
- Pursue human and robotic initiatives to develop innovative technologies, foster new industries, strengthen international partnerships, inspire our Nation and the world, increase humanity's understanding of the Earth, enhance scientific discovery, and explore our solar system and the universe beyond.
- Improve space-based Earth and solar observation capabilities needed to conduct science, forecast terrestrial and near-Earth space weather, monitor climate and global change, manage natural resources, and support disaster response and recovery.

All actions undertaken by departments and agencies in implementing this directive shall be within the overall resource and policy guidance provided by the President; consistent with U.S. law and regulations, treaties and other agreements to which the United States is a party, other applicable international law, U.S. national and homeland security requirements, U.S. foreign policy, and national interests; and in accordance with the Presidential Memorandum on Transparency and Open Government.

[. . .]

Source: "National Space Policy of the United States of America, June 28, 2010." Available online. URL: http://www.whitehouse.gov/sites/default/files/national_space_policy_6-28-10.pdf. Accessed July 10, 2010.

5

International Documents

In a September 25, 1961, speech to the United Nations, President John F. Kennedy called for the international community to expand the UN's jurisdiction to include the whole universe. This immediately applied existing international law to space, paving the way for agreements on activities in space. The resolution also calls for creating a launch registry.

The General Assembly,

Recognizing the common interest of mankind in furthering the peaceful uses of outer space and the urgent need to strengthen international cooperation in this important field,

Believing that the exploration and use of outer space should be only for the betterment of mankind and to the benefit of States irrespective of the stage of their economic or scientific development,

1. COMMENDS TO STATES FOR THEIR GUIDANCE IN THE EXPLORATION AND USE OF OUTER SPACE THE FOLLOWING PRINCIPLES:

(a) International law, including the Chapter of the United Nations, applies to outer space and celestial bodies;

(b) Outer space and celestial bodies are free for exploration and use by all States in conformity with international law and are not subject to national appropriation;

MILITARIZATION OF SPACE

2. INVITES THE COMMITTEE ON THE PEACEFUL USES OF OUTER SPACE TO STUDY AND REPORT ON THE LEGAL PROBLEMS WHICH MAY ARISE FROM THE EXPLORATION AND USE OF OUTER SPACE.

Section B

The General Assembly,

Believing that the United Nations should provide a focal point for international co-operation in the peaceful exploration and use of outer space,

1. *Calls upon* States launching objects into orbit or beyond to furnish information promptly to the Committee on the Peaceful Uses of Outer Space, through the Secretary-General, for the registration of launchings;
2. *Requests* the Secretary-General to maintain a public registry of the information furnished in accordance with paragraph 1 above;
3. *Requests* the Committee on the Peaceful Uses of Outer Space, as it deems appropriate, to review that report and submit its comments and recommendations to the Economic and Social Council and to the General Assembly.

[. . .]

Section D

The General Assembly,

Believing that communication by means of satellites should be available to the nations of the world as soon as practicable on a global and non-discriminatory basis;

Convinced of the need to prepare the way for the establishment of effective operational satellite communication,

1. *Notes with satisfaction* that the International Telecommunication Union plans to call a special conference in 1963 to make allocations of radio frequency bands for outer space activities;
2. *Recommends* that the International Telecommunication Union consider at that conference those aspects of space communication in which international co-operation will be required;
3. *Notes* the potential importance of communication satellites for use by the United Nations and its principal organs and specialized agencies for both operational and informational requirements;

4. *Invites* the Special Fund and the Expanded Programme of Technical Assistance, in consultation with the International Telecommunication Union, to give sympathetic consideration to requests from Member States for technical and other assistance for the survey of their communication needs and for the development of their domestic communication facilities, so that they may make effective use of space communication;

[. . .]

Source: UN Office for Outer Space Affairs. Available online. URL: http://www.oosa.unvienna.org/oosa/SpaceLaw/gares/html/gares_16_1721.html. Accessed April 6, 2010.

Treaty Banning Nuclear Weapon Tests in the Atmosphere, in Outer Space and Under Water (1963) (excerpt)

The United States exploded its first hydrogen bomb in November 1952, with the Soviet Union following in August. After realizing the destructive capacity and associated fallout of the devices, the public and the international community called for controlling the use and even testing of such weapons. The treaty process lasted eight years, with verification procedures as the principal sticking point. The original signatories were the United States, the USSR, and the United Kingdom; China and France have not signed. The treaty does not ban underground testing.

The Governments of the United States of America, the United Kingdom of Great Britain and Northern Ireland, and the Union of Soviet Socialist Republics, hereinafter referred to as the "Original Parties,"

Proclaiming as their principal aim the speediest possible achievement of an agreement on general and complete disarmament under strict international control in accordance with the objectives of the United Nations which would put an end to the armaments race and eliminate the incentive to the production and testing of all kinds of weapons, including nuclear weapons,

Seeking to achieve the discontinuance of all test explosions of nuclear weapons for all time, determined to continue negotiations to this end, and desiring to put an end to the contamination of man's environment by radioactive substances, Have agreed as follows:

ARTICLE I

1. Each of the Parties to this Treaty undertakes to prohibit, to prevent, and not to carry out any nuclear weapon test explosion, or any other nuclear explosion, at any place under its jurisdiction or control: (a) in the atmosphere; beyond its limits, including outer space; or under water, including territorial waters or high seas; or

(b) in any other environment if such explosion causes radioactive debris to be present outside the territorial limits of the State under whose jurisdiction or control such explosion is conducted. It is understood in this connection that the provisions of this subparagraph are without prejudice to the conclusion of a Treaty resulting in the permanent banning of all nuclear test explosions, including all such explosions underground, the conclusion of which, as the Parties have stated in the Preamble to this Treaty, they seek to achieve.

2. Each of the Parties to this Treaty undertakes furthermore to refrain from causing, encouraging, or in any way participating in, the carrying out of any nuclear weapon test explosion, or any other nuclear explosion, anywhere which would take place in any of the environments described, or have the effect referred to, in paragraph 1 of this Article.

[. . .]

Source: Arms Control Association. "Limited Test Ban Treaty (LTBT)." Available online. URL: http://www.armscontrol.org/documents/LTBT. Accessed April 6, 2010.

Treaty on the Principles Governing the Use of Outer Space, Including the Moon and Other Celestial Bodies "Outer Space Treaty" (1967) (excerpt)

The landmark Outer Space Treaty is the foundation of all space law. It asserts that space is to be open to all countries and cannot be claimed or occupied by any country. Article IV bans the deployment of "nuclear weapons or any other kinds of weapons of mass destruction" in outer space, either in orbit around Earth or deployed on a celestial body such as the Moon. However, it does not ban deployment of conventional weapons. This presumably means that conventional weapons, using technology such as lasers or kinetic energy, are permitted.

ARTICLE I

The exploration and use of outer space, including the moon and other celestial bodies, shall be carried out for the benefit and in the interests of all countries, irrespective of their degree of economic or scientific development, and shall be the province of all mankind.

Outer space, including the moon and other celestial bodies, shall be free for exploration and use by all States without discrimination of any kind, on a basis of equality and in accordance with international law, and there shall be free access to all areas of celestial bodies.

There shall be freedom of scientific investigation in outer space, including the moon and other celestial bodies, and States shall facilitate and encourage international co-operation in such investigation.

ARTICLE II

Outer space, including the moon and other celestial bodies, is not subject to national appropriation by claim of sovereignty, by means of use or occupation, or by any other means.

ARTICLE III

States Parties to the Treaty shall carry on activities in the exploration and use of outer space, including the moon and other celestial bodies, in accordance with international law, including the Charter of the United Nations, in the interest of maintaining international peace and security and promoting international co-operation and understanding.

ARTICLE IV

States Parties to the Treaty undertake not to place in orbit around the earth any objects carrying nuclear weapons or any other kinds of weapons of mass destruction, instal such weapons on celestial bodies, or station such weapons in outer space in any other manner.

The moon and other celestial bodies shall be used by all States Parties to the Treaty exclusively for peaceful purposes. The establishment of military bases, installations and fortifications, the testing of any type of weapons and the conduct of military manoeuvres on celestial bodies shall be forbidden. The use of military personnel for scientific research or for any other peaceful purposes shall not be prohibited. The use of any equipment or facility necessary for peaceful exploration of the moon and other celestial bodies shall also not be prohibited.

ARTICLE V

States Parties to the Treaty shall regard astronauts as envoys of mankind in outer space and shall render to them all possible assistance in the event of accident, distress, or emergency landing on the territory of another State Party or on the high seas. When astronauts make such a landing, they shall be safely and promptly returned to the State of registry of their space vehicle.

In carrying on activities in outer space and on celestial bodies, the astronauts of one State Party shall render all possible assistance to the astronauts of other States Parties.

States Parties to the Treaty shall immediately inform the other States Parties to the Treaty or the Secretary-General of the United Nations of any phenomena they discover in outer space, including the moon and other celestial bodies, which could constitute a danger to the life or health of astronauts.

ARTICLE VI

States Parties to the Treaty shall bear international responsibility for national activities in outer space, including the moon and other celestial bodies, whether such activities are carried on by governmental agencies or by non-governmental entities, and for assuring that national activities are carried out in conformity with the provisions set forth in the present Treaty. The activities of non- governmental entities in outer space, including the moon and other celestial bodies, shall require authorization and continuing supervision by the appropriate State Party to the Treaty. When activities are carried on in outer space, including the moon and other celestial bodies, by an international organization, responsibility for compliance with this Treaty shall be borne both by the international organization and by the States Parties to the Treaty participating in such organization.

ARTICLE VII

Each State Party to the Treaty that launches or procures the launching of an object into outer space, including the moon and other celestial bodies, and each State Party from whose territory or facility an object is launched, is internationally liable for damage to another State Party to the Treaty or to its natural or juridical persons by such object or its component parts on the Earth, in air space or in outer space, including the moon and other celestial bodies.

ARTICLE VIII

A State Party to the Treaty on whose registry an object launched into outer space is carried shall retain jurisdiction and control over such object, and over any personnel thereof, while in outer space or on a celestial body. Ownership of objects launched into outer space, including objects landed or constructed on a celestial body, and of their component parts, is not affected by their presence in outer space or on a celestial body or by their return to the Earth. Such objects or component parts found beyond the limits of the State Party of the Treaty on whose registry they are carried shall be returned to that State Party, which shall, upon request, furnish identifying data prior to their return.

ARTICLE IX

In the exploration and use of outer space, including the moon and other celestial bodies, States Parties to the Treaty shall be guided by the principle of co-operation and mutual assistance and shall conduct all their activities in outer space, including the moon and other celestial bodies, with due regard to the corresponding interests of all other States Parties to the Treaty. States Parties to the Treaty shall pursue studies of outer space, including the moon and other celestial bodies, and conduct exploration of them so as to avoid their harmful contamination and also adverse changes in the environment of the Earth resulting from the introduction of extra-terrestrial matter and, where necessary, shall adopt appropriate measures for this purpose. If a State Party to the Treaty has reason to believe that an activity or experiment planned by it or its nationals in outer space, including the moon and other celestial bodies, would cause potentially harmful interference with activities of other States Parties in the peaceful exploration and use of outer space, including the moon and other celestial bodies, it shall undertake appropriate international consultations before proceeding with any such activity or experiment. A State Party to the Treaty which has reason to believe that an activity or experiment planned by another State Party in outer space, including the moon and other celestial bodies, would cause potentially harmful interference with activities in the peaceful exploration and use of outer space, including the moon and other celestial bodies, may request consultation concerning the activity or experiment.

ARTICLE X

In order to promote international co-operation in the exploration and use of outer space, including the moon and other celestial bodies, in conformity with the purposes of this Treaty, the States Parties to the Treaty shall consider on a basis of equality any requests by other States Parties to the Treaty to be afforded an opportunity to observe the flight of space objects launched by those States.

The nature of such an opportunity for observation and the conditions under which it could be afforded shall be determined by agreement between the States concerned.

ARTICLE XI

In order to promote international co-operation in the peaceful exploration and use of outer space, States Parties to the Treaty conducting activities in outer space, including the moon and other celestial bodies, agree to inform the Secretary-General of the United Nations as well as the public and the international scientific community, to the greatest extent feasible and practicable, of the nature, conduct, locations and results of such activities. On receiving the said information, the Secretary-General of the United Nations should be prepared to disseminate it immediately and effectively.

ARTICLE XII

All stations, installations, equipment and space vehicles on the moon and other celestial bodies shall be open to representatives of other States Parties to the Treaty on a basis of reciprocity. Such representatives shall give reasonable advance notice of a projected visit, in order that appropriate consultations may be held and that maximum precautions may be taken to assure safety and to avoid interference with normal operations in the facility to be visited.

ARTICLE XIII

The provisions of this Treaty shall apply to the activities of States Parties to the Treaty in the exploration and use of outer space, including the moon and other celestial bodies, whether such activities are carried on by a single State Party to the Treaty or jointly with other States, including cases where they are carried on within the framework of international inter-governmental organizations.

Any practical questions arising in connexion with activities carried on by international inter-governmental organizations in the exploration and use of outer space, including the moon and other celestial bodies, shall be resolved by the States Parties to the Treaty either with the appropriate

International Documents

international organization or with one or more States members of that international organization, which are Parties to this Treaty.

[. . .]

Source: NASA History Division. "Key Documents in the History of Space Policy." Available online. URL: http://history.nasa.gov/spdocs.html#1960s. Accessed April 6, 2010.

Rescue Treaty
Agreement on the Rescue of Astronauts, the Return of Astronauts and the Return of Objects Launched into Outer Space (1968) (excerpt)

Manned spaceflight began with Yuri Gagarin's trip around Earth on April 12, 1961, and the United States quickly followed. Consequently, U.S. and Soviet leaders sought to protect their personnel and equipment. The Rescue Agreement designates astronauts and cosmonauts as "personnel of a spacecraft" and requires countries to render aid to such personnel in distress. It also requires the assisting state to notify the launching authority immediately. While most debris burns up on reentry, any debris or equipment that does fall to Earth is to be promptly returned to the appropriate authority—at the launching authority's expense.

ARTICLE 1

Each Contracting Party which receives information or discovers that the personnel of a spacecraft have suffered accident or are experiencing conditions of distress or have made an emergency or unintended landing in territory under its jurisdiction or on the high seas or in any other place not under the jurisdiction of any State shall immediately:

(a) notify the launching authority or, if it cannot identify and immediately communicate with the launching authority, immediately make a public announcement by all appropriate means of communication at its disposal;

(b) notify the Secretary-General of the United Nations, who should disseminate the information without delay by all appropriate means of communication at his disposal.

ARTICLE 2

If, owing to accident, distress, emergency or unintended landing, the personnel of a spacecraft land in territory under the jurisdiction of a Contracting

Party, it shall immediately take all possible steps to rescue them and render them all necessary assistance. It shall inform the launching authority and also the Secretary-General of the United Nations of the steps it is taking and of their progress. If assistance by the launching authority would help to effect a prompt rescue or would contribute substantially to the effectiveness of search and rescue operations, the launching authority shall co-operate with the Contracting Party with a view to the effective conduct of search and rescue operations. Such operations shall be subject to the direction and control of the Contracting Party, which shall act in close and continuing consultation with the launching authority.

ARTICLE 3

If information is received or it is discovered that the personnel of a space-craft have alighted on the high seas or in any other place not under the juris-diction of any State, those Contracting Parties which are in a position to do so shall, if necessary, extend assistance in search and rescue operations for such personnel to assure their speedy rescue. They shall inform the launch-ing authority and the Secretary-General of the United Nations of the steps they are taking and of their progress.

ARTICLE 4

If, owing to accident, distress, emergency or unintended landing, the personnel of a spacecraft land in territory under the jurisdiction of a Con-tracting Party or have been found on the high seas or in any other place not under the jurisdiction of any State, they shall be safely and promptly returned to representatives of the launching authority.

ARTICLE 5

1. Each Contracting Party which receives information or discovers that a space object or its component parts has returned to Earth in territory under its jurisdiction or on the high seas or in any other place not under the jurisdiction of any State, shall notify the launching authority and the Secretary- General of the United Nations.
2. Each Contracting Party having jurisdiction over the territory on which a space object or its component parts has been discovered shall, upon the request of the launching authority and with assistance from that authority if requested, take such steps as it finds practicable to recover the object or component parts.
3. Upon request of the launching authority, objects launched into outer space or their component parts found beyond the territorial limits of

the launching authority shall be returned to or held at the disposal of representatives of the launching authority, which shall, upon request, furnish identifying data prior to their return.

4. Notwithstanding paragraphs 2 and 3 of this article, a Contracting Party which has reason to believe that a space object or its component parts discovered in territory under its jurisdiction, or recovered by it elsewhere, is of a hazardous or deleterious nature may so notify the launching authority, which shall immediately take effective steps, under the direction and control of the said Contracting Party, to eliminate possible danger of harm.

5. Expenses incurred in fulfilling obligations to recover and return a space object or its component parts under paragraphs 2 and 3 of this article shall be borne by the launching authority.

ARTICLE 6

For the purposes of this Agreement, the term "launching authority" shall refer to the State responsible for launching, or, where an international intergovernmental organization is responsible for launching, that organization, provided that organization declares its acceptance of the rights and obligations provided for in this Agreement and a majority of the States members of that organization are Contracting Parties to this Agreement and to the Treaty on Principles Governing the Activities of States in the Exploration and Use of Outer Space, including the Moon and Other Celestial Bodies.

[. . .]

Source: UN Office for Outer Space Affairs. "Agreement on the Rescue of Astronauts, the Return of Astronauts and the Return of Objects Launched into Outer Space" Available online. URL: http://www.oosa.unvienna.org/oosa/en/ SpaceLaw/gares/html/gares_22_2345.html. Accessed April 6, 2010.

Liability Convention (1972) (excerpt)

The Liability Convention builds on the Rescue Treaty by addressing responsibility for any damage inflicted by debris that reaches Earth. This convention was invoked in 1978 when the Soviet Cosmos 954 reconnaissance satellite malfunctioned and de-orbited, dropping radioactive debris over Canada.

CONVENTION ON INTERNATIONAL LIABILITY FOR DAMAGE CAUSED BY SPACE OBJECTS

ARTICLE I

For the purposes of this Convention:

(a) The term "damage" means loss of life, personal injury or other impairment of health; or loss of or damage to property of States or of persons, natural or juridical, or property of international intergovernmental organizations;

(b) The term "launching" includes attempted launching;

(c) The term "launching State" means:
    (i) A State which launches or procures the launching of a space object;
    (ii) A State from whose territory or facility a space object is launched;

(d) The term "space object" includes component parts of a space object as well as its launch vehicle and parts thereof.

ARTICLE II

A launching State shall be absolutely liable to pay compensation for damage caused by its space object on the surface of the earth or to aircraft flight.

ARTICLE III

In the event of damage being caused elsewhere than on the surface of the earth to a space object of one launching State or to persons or property on board such a space object by a space object of another launching State, the latter shall be liable only if the damage is due to its fault or the fault of persons for whom it is responsible.

ARTICLE IV

1. In the event of damage being caused elsewhere than on the surface of the earth to a space object of one launching State or to persons or property on board such a space object by a space object of another launching State, and of damage thereby being caused to a third State or to its natural or juridical persons, the first two States shall be jointly and severally liable to the third State, to the extent indicated by the following:

    (a) If the damage has been caused to the third State on the surface of the earth or to aircraft in flight, their liability to the third State shall be absolute;

(b) If the damage has been caused to a space object of the third State or to persons or property on board that space object elsewhere than on the surface of the earth, their liability to the third State shall be based on the fault of either of the first two States or on the fault of persons for whom either is responsible.

2. In all cases of joint and several liability referred to in paragraph 1 of this article, the burden of compensation for the damage shall be apportioned between the first two States in accordance with the extent to which they were at fault; if the extent of the fault of each of these States cannot be established, the burden of compensation shall be apportioned equally between them. Such apportionment shall be without prejudice to the right of the third State to seek the entire compensation due under this Convention from any or all of the launching States which are jointly and severally liable.

ARTICLE V

1. Whenever two or more States jointly launch a space object, they shall be jointly and severally liable for any damage caused.
2. A launching State which has paid compensation for damage shall have the right to present a claim for indemnification to other participants in the joint launching. The participants in a joint launching may conclude agreements regarding the apportioning among themselves of the financial obligation in respect of which they are jointly and severally liable. Such agreements shall be without prejudice to the right of a State sustaining damage to seek the entire compensation due under this Convention from any or all of the launching States which are jointly and severally liable.
3. A State from whose territory or facility a space object is launched shall be regarded as a participant in a joint launching.

ARTICLE VI

1. Subject to the provisions of paragraph 2 of this Article, exoneration from absolute liability shall be granted to the extent that a launching State establishes that the damage has resulted either wholly or partially from gross negligence or from an act or omission done with intent to cause damage on the part of a claimant State or of natural or juridical persons it represents.
2. No exoneration whatever shall be granted in cases where the damage has resulted from activities conducted by a launching State which are not in conformity with international law including, in particular, the

MILITARIZATION OF SPACE

Charter of the United Nations and the Treaty on Principles Governing the Activities of States in the Exploration and Use of Outer Space,
including the Moon and Other Celestial Bodies.

ARTICLE VII

The provisions of this Convention shall not apply to damage caused by a
space object of a launching State to:

(a) nationals of that launching State;
(b) foreign nationals during such time as they are participating in the
operation of that space object from the time of its launching or at
any stage thereafter until its descent, or during such time as they
are in the immediate vicinity of a planned launching or recovery
area as the result of an invitation by that launching State.

ARTICLE VIII

1. A State which suffers damage, or whose natural or juridical persons suffer damage, may present to a launching State a claim for compensation
for such damage.
2. If the State of nationality has not presented a claim, another State may,
in respect of damage sustained in its territory by any natural or juridical
person, present a claim to a launching State.
3. If neither the State of nationality nor the State in whose territory the
damage was sustained has presented a claim or notified its intention of
presenting a claim, another State may, in respect of damage sustained
by its permanent residents, present a claim to a launching State.

ARTICLE IX

A claim for compensation for damage shall be presented to a launching
State through diplomatic channels. If a State does not maintain diplomatic
relations with the launching State concerned, it may request another State
to present its claim to that launching State or otherwise represent its
interests under this Convention. It may also present its claim through the
Secretary-General of the United Nations, provided the claimant State and
the launching State are both Members of the United Nations.

ARTICLE X

1. A claim for compensation for damage may be presented to a launching
State not later than one year following the date of the occurrence of the
damage or the identification of the launching State which is liable.

2. If, however, a State does not know of the occurrence of the damage or has not been able to identify the launching State which is liable, it may present a claim within one year following the date on which it learned of the aforementioned facts; however, this period shall in no event exceed one year following the date on which the State could reasonably be expected to have learned of the facts through the exercise of due diligence.

3. The time-limits specified in paragraphs 1 and 2 of this Article shall apply even if the full extent of the damage may not be known. In this event, however, the claimant State shall be entitled to revise the claim and submit additional documentation after the expiration of such time-limits until one year after the full extent of the damage is known.

ARTICLE XI

1. Presentation of a claim to a launching State for compensation for damage under this Convention shall not require the prior exhaustion of any local remedies which may be available to a claimant State or to natural or juridical persons it represents.

2. Nothing in this Convention shall prevent a State, or natural or juridical persons it might represent, from pursuing a claim in the courts or administrative tribunals or agencies of a launching State. A State shall not, however, be entitled to present a claim under this Convention in respect of the same damage for which a claim is being pursued in the courts or administrative tribunals or agencies of a launching State or under another international agreement which is binding on the States concerned.

[. . .]

ARTICLE XXI

If the damage caused by a space object presents a large-scale danger to human life or seriously interferes with the living conditions of the population or the functioning of vital centres, the States Parties, and in particular the launching State, shall examine the possibility of rendering appropriate and rapid assistance to the State which has suffered the damage, when it so requests. However, nothing in this article shall affect the rights or obligations of the States Parties under this Convention.

[. . .]

Source: UN Office for Outer Space Affairs. "Convention on International Liability for Damage Caused by Space Objects." Available online. URL: http://www.oosa.unvienna.org/oosa/en/SpaceLaw/gares/html/gares_26_2777. html. Accessed April 6, 2010.

Treaty between the United States of America and the Union of Soviet Socialist Republics on the Limitation of Anti-Ballistic Missile Systems (1972) (excerpt)

In conjunction with the US-USSR Strategic Arms Limitation Treaty (SALT), the two countries agreed to limit production of antiballistic missile systems to two sites in each country. Eventually technology, especially SDI, made the agreement obsolete. The United States withdrew from the ABM Treaty in 2002.

ARTICLE I

1. Each Party undertakes to limit anti-ballistic missile (ABM) systems and to adopt other measures in accordance with the provisions of this Treaty.
2. Each Party undertakes not to deploy ABM systems for a defense of the territory of its country and not to provide a base for such a defense, and not to deploy ABM systems for defense of an individual region except as provided for in Article III of this Treaty.

ARTICLE II

1. For the purpose of this Treaty an ABM system is a system to counter strategic ballistic missiles or their elements in flight trajectory, currently consisting of:

    (a) ABM interceptor missiles, which are interceptor missiles constructed and deployed for an ABM role, or of a type tested in an ABM mode;
    (b) ABM launchers, which are launchers constructed and deployed for launching ABM interceptor missiles; and
    (c) ABM radars, which are radars constructed and deployed for an ABM role, or of a type tested in an ABM mode.

2. The ABM system components listed in paragraph 1 of this Article include those which are:

    (a) operational; (b) under construction; (c) undergoing testing; (d) undergoing overhaul, repair or conversion; or (e) mothballed.

ARTICLE III

Each Party undertakes not to deploy ABM systems or their components except that:

    (a) within one ABM system deployment area having a radius of one hundred and fifty kilometers and centered on the Party's national capital, a Party may deploy:

206

    (1)   no more than one hundred ABM launchers and no more than one hundred ABM interceptor missiles at launch sites, and

    (2)   ABM radars within no more than six ABM radar complexes, the area of each complex being circular and having a diameter of no more than three kilometers; and

(b)   within one ABM system deployment area having a radius of one hundred and fifty kilometers and containing ICBM silo launchers, a Party may deploy:

    (1)   no more than one hundred ABM launchers and no more than one hundred ABM interceptor missiles at launch sites,

    (2)   two large phased-array ABM radars comparable in potential to corresponding ABM radars operational or under construction on the date of signature of the Treaty in an ABM system deployment area containing ICBM silo launchers, and

    (3)   no more than eighteen ABM radars each having a potential less than the potential of the smaller of the above-mentioned two large phased-array ABM radars.

ARTICLE IV

The limitations provided for in Article III shall not apply to ABM systems or their components used for development or testing, and located within current or additionally agreed test ranges. Each Party may have no more than a total of fifteen ABM launchers at test ranges.

ARTICLE V

1.   Each Party undertakes not to develop, test, or deploy ABM systems or components which are sea-based, air-based, space-based, or mobile land-based.

2.   Each Party undertakes not to develop, test or deploy ABM launchers for launching more than one ABM interceptor missile at a time from each launcher, not to modify deployed launchers to (provide them with such a capacity, not to develop, test, or deploy automatic or semi-automatic or other similar systems for rapid reload of ABM launchers.

ARTICLE VI

To enhance assurance of the effectiveness of the limitations on ABM systems and their components provided by the Treaty, each Party undertakes:

(a)   not to give missiles, launchers, or radars, other than ABM interceptor missiles, ABM launchers, or ABM radars, capabilities to counter strategic ballistic missiles or their elements in flight trajectory, and not to test them in an ABM mode; and

(b) not to deploy in the future radars for early warning of strategic ballistic missile attack except at locations along the periphery of its national territory and oriented outward.

[. . .]

ARTICLE XII

1. For the purpose of providing assurance or compliance with the provisions of this Treaty, each Party shall use national technical means of verification at its disposal in a manner consistent with generally recognized principles of international law.

2. Each Party undertakes not to interfere with the national technical means of verification of the other Party operating in accordance with paragraph 1 of this Article.

3. Each Party undertakes not to use deliberate concealment measures which impede verification by national technical means of compliance with the provisions of this Treaty. This obligation shall not require changes in current construction, assembly, conversion, or overhaul practices.

[. . .]

Source: U.S. Department of State. "Treaty Between the United States of America and the Union of Soviet Socialist Republics on the Limitation of Anti-Ballistic Missile systems." Available online. URL: http://www.state.gov/www/global/arms/treaties/abm/abm2.html. Accessed April 6, 2010.

Registration Convention (1976) (excerpt)

To facilitate the assignment of liability, the United Nations began requiring all launching authorities to maintain a registry of launched objects and to report the same information to a similar registry maintained by the United Nations. The UN Registry is available online.

CONVENTION ON REGISTRATION OF OBJECTS LAUNCHED INTO OUTER SPACE

ARTICLE I

For the purposes of this Convention:

(a) The term "launching State" means:

    (i)   A State which launches or procures the launching of a space object;

    (ii) A State from whose territory or facility a space object is launched;

  (b) The term "space object" includes component parts of a space object as well as its launch vehicle and parts thereof;

  (c) The term "State of registry" means a launching State on whose registry a space object is carried in accordance with article II.

ARTICLE II

1.   When a space object is launched into earth orbit or beyond, the launching State shall register the space object by means of an entry in an appropriate registry which it shall maintain. Each launching State shall inform the Secretary-General of the United Nations of the establishment of such a registry.

2.   Where there are two or more launching States in respect of any such space object, they shall jointly determine which one of them shall register the object in accordance with paragraph 1 of this article, bearing in mind the provisions of article VIII of the Treaty on Principles Governing the Activities of States in the Exploration and Use of Outer Space, including the Moon and Other Celestial Bodies, and without prejudice to appropriate agreements concluded or to be concluded among the launching States on jurisdiction and control over the space object and over any personnel thereof.

3.   The contents of each registry and the conditions under which it is maintained shall be determined by the State of registry concerned.

ARTICLE III

1.   The Secretary-General of the United Nations shall maintain a Register in which the information furnished in accordance with article IV shall be recorded.

2. There shall be full and open access to the information in this Register.

ARTICLE IV

1.   Each State of registry shall furnish to the Secretary-General of the United Nations, as soon as practicable, the following information concerning each space object carried on its registry:

  (a) name of launching State or States;

  (b) an appropriate designator of the space object or its registration number;

  (c) date and territory or location of launch;

(d) basic orbital parameters, including:
(i) nodal period;
(ii) inclination;
(iii) apogee;
(iv) perigee;
(e) general function of the space object.

2. Each State of registry may, from time to time, provide the Secretary-General of the United Nations with additional information concerning a space object carried on its registry.
3. Each State of registry shall notify the Secretary-General of the United Nations, to the greatest extent feasible and as soon as practicable, of space objects concerning which it has previously transmitted information, and which have been but no longer are in earth orbit.

ARTICLE V

Whenever a space object launched into earth orbit or beyond is marked with the designator or registration number referred to in article IV, paragraph 1 (b), or both, the State of registry shall notify the Secretary-General of this fact when submitting the information regarding the space object in accordance with article IV. In such case, the Secretary-General of the United Nations shall record this notification in the Register.

ARTICLE VI

Where the application of the provisions of this Convention has not enabled a State Party to identify a space object which has caused damage to it or to any of its natural or juridical persons, or which may be of a hazardous or deleterious nature, other States Parties, including in particular States possessing space monitoring and tracking facilities, shall respond to the greatest extent feasible to a request by that State Party, or transmitted through the Secretary-General on its behalf, for assistance under equitable and reasonable conditions in the identification of the object. A State Party making such a request shall, to the greatest extent feasible, submit information as to the time, nature and circumstances of the events giving rise to the request. Arrangements under which such assistance shall be rendered shall be the subject of agreement between the parties concerned.

ARTICLE VII

1. In this Convention, with the exception of articles VIII to XII inclusive, references to States shall be deemed to apply to any international

intergovernmental organization which conducts space activities if the organization declares its acceptance of the rights and obligations provided for in this Convention and if a majority of the States members of the organization are States Parties to this Convention and to the Treaty on Principles Governing the Activities of States in the Exploration and Use of Outer Space, including the Moon and Other Celestial Bodies.

2. States members of any such organization which are States Parties to this Convention shall take all appropriate steps to ensure that the organization makes a declaration in accordance with paragraph 1 of this article.

[. . .]

Source: UN Office for Outer Space Affairs. "Convention on the Registration of Objects Launched into Outer Space." Available online. URL: http://www.oosa.unvienna.org/oosa/en/SpaceLaw/gares/html/gares_29_3235 .html. Accessed April 6, 2010.

Bogotá Declaration (1976) (excerpt)

In 1976, Brazil, Colombia, Congo, Ecuador, Indonesia, Kenya, Uganda, and Zaire announced that their sovereignty extended not only across their geographic territory, but also upward to geostationary altitude. Thus, in contradiction to the Outer Space Treaty, any object flying across their claimed territory or in their claimed airspace would be subject to their national law. Their assertion has not been widely accepted.

DECLARATION OF THE FIRST MEETING OF EQUATORIAL COUNTRIES

1. THE GEOSTATIONARY ORBIT AS A NATURAL RESOURCE

The geostationary orbit is a circular orbit on the Equatorial plane in which the period of sideral revolution of the satellite is equal to the period of sideral rotation of the Earth and the satellite moves in the same direction of the Earth's rotation. When a satellite describes this particular orbit, it is said to be geostationary; such a satellite appears to be stationary in the sky, when viewed from the earth, and is fixed on the zenith of a given point of the Equator, whose longitude is by definition that of the satellite.

This orbit is located at an approximate distance of 35,871 Kmts. over the Earth's Equator.

Equatorial countries declare that the geostationary synchronous orbit is a physical fact linked to the reality of our planet because its existence depends exclusively on its relation to gravitational phenomena generated by the earth, and that is why it must not be considered part of the outer space. Therefore, the segments of geostationary synchronous orbit are part of the territory over which Equatorial states exercise their national sovereignty. The geostationary orbit is a scarce natural resource, whose importance and value increase rapidly together with the development of space technology and with the growing need for communication; therefore, the Equatorial countries meeting in Bogota have decided to proclaim and defend on behalf of their peoples, the existence of their sovereignty over this natural resource. The geostationary orbit represents a unique facility that it alone can offer for telecommunication services and other uses which require geostationary satellites.

The frequencies and orbit of geostationary satellites are limited natural resources, fully accepted as such by current standards of the International Telecommunications Union. Technological advancement has caused a continuous increase in the number of satellites that use this orbit, which could result in a saturation in the near future.

The solutions proposed by the International Telecommunications Union and the relevant documents that attempt to achieve a better use of the geostationary orbit that shall prevent its imminent saturation, are at present impracticable and unfair and would considerably increase the exploitation costs of this resource especially for developing countries that do not have equal technological and financial resources as compared to industrialized countries, who enjoy an apparent monopoly in the exploitation and use of its geostationary synchronous orbit. In spite of the principle established by Article 33, sub-paragraph 2 of the International Telecommunications Convention, of 1973, that in the use of frequency bands for space radiocommunications, the members shall take into account that the frequencies and the orbit for geostationary satellites are limited natural resources that must be used efficiently and economically to allow the equitable access to this orbit and to its frequencies, we can see that both the geostationary orbit and the frequencies have been used in a way that does not allow the equitable access of the developing countries that do not have the technical and financial means that the great powers have. Therefore, it is imperative for the equatorial countries to exercise their sovereignty over the corresponding segments of the geostationary orbit.

2. SOVEREIGNTY OF EQUATORIAL STATES OVER THE CORRESPONDING SEGMENTS OF THE GEOSTATIONARY ORBIT

In qualifying this orbit as a natural resource, equatorial states reaffirm "the right of the peoples and of nations to permanent sovereignty over their wealth and natural resources that must be exercised in the interest of their national development and of the welfare of the people of the nation concerned," as it is set forth in Resolution 2692 (XXV) of the United Nations General Assembly entitled "permanent sovereignty over the natural resources of developing countries and expansion of internal accumulation sources for economic developments".

Furthermore, the charter on economic rights and duties of states solemnly adopted by the United Nations General Assembly through Resolution 3281 (XXIV), once more confirms the existence of a sovereign right of nations over their natural resources, in Article 2 subparagraph i, which reads:

"All states have and freely exercise full and permanent sovereignty, including possession, use and disposal of all their wealth, natural resources and economic activities".

Consequently, the above-mentioned provisions lead the equatorial states to affirm that the synchronous geostationary orbit, being a natural resource, is under the sovereignty of the equatorial states.

3. LEGAL STATE OF THE GEOSTATIONARY ORBIT

Bearing in mind the existence of sovereign rights over segments of geostationary orbit, the equatorial countries consider that the applicable legal consultations in this area must take into account the following:

(a) The sovereign rights put forward by the equatorial countries are directed towards rendering tangible benefits to their respective people and for the universal community, which is completely different from the present reality when the orbit is used to the greater benefit of the most developed countries.

(b) The segments of the orbit corresponding to the open sea are beyond the national jurisdiction of states will be considered as common heritage of mankind. Consequently, the competent international agencies should regulate its use and exploitation for the benefit of mankind.

(c) The equatorial states do not object to the free orbital transit of satellites approved and authorized by the International Telecommunications Convention, when these satellites pass through their outer space in their gravitational flight outside their geostationary orbit.

(d) The devices to be placed permanently on the segment of a geostationary orbit of an equatorial state shall require previous and expressed authorization on the part of the concerned state, and the operation of the device should conform with the national law of that territorial country over which it is placed. It must be understood that the said authorization is different from the co-ordination requested in cases of interference among satellite systems, which are specified in the regulations for radiocommunications. The said authorization refers in very clear terms to the countries' right to allow the operation of fixed radiocommunications stations within their territory.

(e) Equatorial states do not condone the existing satellites or the position they occupy on their segments of the Geostationary Orbit nor does the existence of said satellites confer any rights of placement of satellites or use of the segment unless expressly authorized by the state exercising sovereignty over this segment.

4. TREATY OF 1967

The Treaty of 1967 on "The Principles Governing the Activities of States in the Exploration and Use of Outer Space, Including the Moon and Other Celestial Bodies," signed on 27 January, 1967, cannot be considered as a final answer to the problem of the exploration and use of outer space, even less when the international community is questioning all the terms of international law which were elaborated when the developing countries could not count on adequate scientific advice and were thus not able to observe and evaluate the omissions, contradictions and consequences of the proposals which were prepared with great ability by the industrialized powers for their own benefit.

There is no valid or satisfactory definition of outer space which may be advanced to support the argument that the geostationary orbit is included in the outer space. The legal affairs sub-commission which is dependent on the United Nations Commission on the Use of Outer Space for Peaceful Purposes, has been working for a long time on a definition of outer space, however, to date, there has been no agreement in this respect.

Therefore, it is imperative to elaborate a juridical definition of outer space, without which the implementation of the Treaty of 1967 is only a way to give recognition to the presence of the states that are already using the geostationary orbit. Under the name of a so-called non-national appropriation, what was actually developed was technological partition of the orbit, which is simply a national appropriation, and this must be denounced by the equatorial countries. The experiences observed up to the present and the development foreseeable for the coming years bring to light the obvious omissions of the Treaty of 1967 which force the equatorial states to claim the exclusion of the geostationary orbit.

The lack of definition of outer space in the Treaty of 1967, which has already been referred to, implies that Article II should not apply to geostationary orbit and therefore does not affect the right of the equatorial states that have already ratified the Treaty.

5. DIPLOMATIC AND POLITICAL ACTION

While Article 2 of the aforementioned Treaty does not establish an express exception regarding the synchronous geostationary orbit, as an integral element of the territory of equatorial states, the countries that have not ratified the Treaty should refrain from undertaking any procedure that allows the enforcement of provisions whose juridical omission has already been denounced.

The representatives of the equatorial countries attending the meeting in Bogota, wish to clearly state their position regarding the declarations of Colombia and Ecuador in the United Nations, which affirm that they consider the geostationary orbit to be an integral part of their sovereign territory; this declaration is a historical background for the defense of the sovereign rights of the equatorial countries. These countries will endeavour to make similar declarations in international agencies dealing with the same subject and to align their international policy in accordance with the principles elaborated in this document.

Signed in Bogotá 3 December 1976 by the Heads of Delegations.
Brasil, Colombia, Congo, Ecuador, Indonesia, Kenya, Uganda, Zaire

Source: Japanese Space Agency. "Declaration of the First Meeting of Equatorial Countries." Available online. URL: http://www.jaxa.jp/library/space_law/chapter_2/2-2-1-2_e.html. Accessed April 6, 2010.

Convention on the Prohibition of Military or Other Hostile Use of Environmental Modification Techniques (1977) (excerpt)

In addition to banning weapons of mass destruction in space, countries moved to ban hostile use of environmental modification, such as manipulating the weather.

ARTICLE I

1. Each State Party to this Convention undertakes not to engage in military or any other hostile use of environmental modification techniques having widespread, long-lasting or severe effects as the means of destruction, damage or injury to any other State Party.

2. Each State Party to this Convention undertakes not to assist, encourage or induce any State, group of States or international organization to engage in activities contrary to the provisions of paragraph 1 of this article.

ARTICLE II

As used in Article I, the term "environmental modification techniques" refers to any technique for changing—through the deliberate manipulation of natural processes—the dynamics, composition or structure of the Earth, including its biota, lithosphere, hydrosphere and atmosphere, or of outer space.

ARTICLE III

1. The provisions of this Convention shall not hinder the use of environmental modification techniques for peaceful purposes and shall be without prejudice to the generally recognized principles and applicable rules of international law concerning such use.

2. The States Parties to this Convention undertake to facilitate, and have the right to participate in, the fullest possible exchange of scientific and technological information on the use of environmental modification techniques for peaceful purposes. States Parties in a position to do so shall contribute, alone or together with other States or international organizations, to international economic and scientific co-operation in the preservation, improvement, and peaceful utilization of the environment, with due consideration for the needs of the developing areas of the world.

ARTICLE IV

Each State Party to this Convention undertakes to take any measures it considers necessary in accordance with its constitutional processes to prohibit and prevent any activity in violation of the provisions of the Convention anywhere under its jurisdiction or control.

Source: Federation of American Scientists. Available online. URL: http://www.fas.org/nuke/control/enmod/text/environ2.htm. Accessed April 6, 2010.

Moon Agreement (1984) (excerpt)

The Moon Agreement adapts the general coverage of the cosmos included in the 1967 Outer Space Treaty to the specific case of the Moon. While the Outer Space Treaty declares that all countries should have equal access to the Moon, regardless of their level of economic development, the Moon Agreement mentions that exploration "shall be carried out for the benefit and in the interests of all countries, irrespective of their degree of economic or scientific development." The agreement also calls for equitable sharing of the Moon's natural resources among all countries involved in their discovery.

AGREEMENT GOVERNING THE ACTIVITIES OF STATES ON THE MOON AND OTHER CELESTIAL BODIES

ARTICLE 1

1. The provisions of this Agreement relating to the moon shall also apply to other celestial bodies within the solar system, other than the earth, except in so far as specific legal norms enter into force with respect to any of these celestial bodies.
2. For the purposes of this Agreement reference to the moon shall include orbits around or other trajectories to or around it.
3. This Agreement does not apply to extraterrestrial materials which reach the surface of the earth by natural means.

ARTICLE 2

All activities on the moon, including its exploration and use, shall be carried out in accordance with international law, in particular the Charter of the United Nations, and taking into account the Declaration on Principles of International Law concerning Friendly Relations and Co-operation among States in accordance with the Charter of the United Nations, adopted by the General Assembly on 24 October 1970, in the interest of maintaining international peace and security and promoting international co-operation and mutual understanding, and with due regard to the corresponding interests of all other States Parties.

ARTICLE 3

1. The moon shall be used by all States Parties exclusively for peaceful purposes.
2. Any threat or use of force or any other hostile act or threat of hostile act on the moon is prohibited. It is likewise prohibited to use the moon in order to commit any such act or to engage in any such threat in relation to the earth, the moon, spacecraft, the personnel of spacecraft or man-made space objects.
3. States Parties shall not place in orbit around or other trajectory to or around the moon objects carrying nuclear weapons or any other kinds of weapons of mass destruction or place or use such weapons on or in the moon.
4. The establishment of military bases, installations and fortifications, the testing of any type of weapons and the conduct of military manuvres on the moon shall be forbidden. The use of military personnel for scientific

research or for any other peaceful purposes shall not be prohibited. The use of any equipment or facility necessary for peaceful exploration and use of the moon shall also not be prohibited.

ARTICLE 4

1. The exploration and use of the moon shall be the province of all mankind and shall be carried out for the benefit and in the interests of all countries, irrespective of their degree of economic or scientific development. Due regard shall be paid to the interests of present and future generations as well as to the need to promote higher standards of living and conditions of economic and social progress and development in accordance with the Charter of the United Nations.

2. States Parties shall be guided by the principle of co-operation and mutual assistance in all their activities concerning the exploration and use of the moon. International co-operation in pursuance of this Agreement should be as wide as possible and may take place on a multilateral basis, on a bilateral basis or through international intergovernmental organizations.

ARTICLE 5

1. States Parties shall inform the Secretary-General of the United Nations as well as the public and the international scientific community, to the greatest extent feasible and practicable, of their activities concerned with the exploration and use of the moon. Information on the time, purposes, locations, orbital parameters and duration shall be given in respect of each mission to the moon as soon as possible after launching, while information on the results of each mission, including scientific results, shall be furnished upon completion of the mission. In the case of a mission lasting more than sixty days, information on conduct of the mission, including any scientific results, shall be given periodically, at thirty-day intervals. For missions lasting more than six months, only significant additions to such information need be reported thereafter.

2. If a State Party becomes aware that another State Party plans to operate simultaneously in the same area of or in the same orbit around or trajectory to or around the moon, it shall promptly inform the other State of the timing of and plans for its own operations.

3. In carrying out activities under this Agreement, States Parties shall promptly inform the Secretary-General, as well as the public and the international scientific community, of any phenomena they discover in

outer space, including the moon, which could endanger human life or health, as well as of any indication of organic life.

ARTICLE 6

1. There shall be freedom of scientific investigation on the moon by all States Parties without discrimination of any kind, on the basis of equality and in accordance with international law.

2. In carrying out scientific investigations and in furtherance of the provisions of this Agreement, the States Parties shall have the right to collect on and remove from the moon samples of its mineral and other substances. Such samples shall remain at the disposal of those States Parties which caused them to be collected and may be used by them for scientific purposes. States Parties shall have regard to the desirability of making a portion of such samples available to other interested States Parties and the international scientific community for scientific investigation. States Parties may in the course of scientific investigations also use mineral and other substances of the moon in quantities appropriate for the support of their missions.

3. States Parties agree on the desirability of exchanging scientific and other personnel on expeditions to or installations on the moon to the greatest extent feasible and practicable.

ARTICLE 7

1. In exploring and using the moon, States Parties shall take measures to prevent the disruption of the existing balance of its environment, whether by introducing adverse changes in that environment, by its harmful contamination through the introduction of extra-environmental matter or otherwise. States Parties shall also take measures to avoid harmfully affecting the environment of the earth through the introduction of extraterrestrial matter or otherwise.

2. States Parties shall inform the Secretary-General of the United Nations of the measures being adopted by them in accordance with paragraph 1 of this article and shall also, to the maximum extent feasible, notify him in advance of all placements by them of radio-active materials on the moon and of the purposes of such placements.

3. States Parties shall report to other States Parties and to the Secretary-General concerning areas of the moon having special scientific interest in order that, without prejudice to the rights of other States Parties, consideration may be given to the designation of such areas as international scientific preserves for which special protective arrangements

are to be agreed upon in consultation with the competent bodies of the United Nations.

ARTICLE 8

1. States Parties may pursue their activities in the exploration and use of the moon anywhere on or below its surface, subject to the provisions of this Agreement.
2. For these purposes States Parties may, in particular:
    (a) Land their space objects on the moon and launch them from the moon;
    (b) Place their personnel, space vehicles, equipment, facilities, stations and installations anywhere on or below the surface of the moon.
    (c) Personnel, space vehicles, equipment, facilities, stations and installations may move or be moved freely over or below the surface of the moon.
3. Activities of States Parties in accordance with paragraphs 1 and 2 of this article shall not interfere with the activities of other States Parties on the moon. Where such interference may occur, the States Parties concerned shall undertake consultations in accordance with article 15, paragraphs 2 and 3, of this Agreement.

ARTICLE 9

1. States Parties may establish manned and unmanned stations on the moon. A State Party establishing a station shall use only that area which is required for the needs of the station and shall immediately inform the Secretary-General of the United Nations of the location and purposes of that station. Subsequently, at annual intervals that State shall likewise inform the Secretary-General whether the station continues in use and whether its purposes have changed.
2. Stations shall be installed in such a manner that they do not impede the free access to all areas of the moon of personnel, vehicles and equipment of other States Parties conducting activities on the moon in accordance with the provisions of this Agreement or of article I of the Treaty on Principles Governing the Activities of States in the Exploration and Use of Outer Space, including the Moon and Other Celestial Bodies.

ARTICLE 10

1. States Parties shall adopt all practicable measures to safeguard the life and health of persons on the moon. For this purpose they shall regard any person on the moon as an astronaut within the meaning of article

V of the Treaty on Principles Governing the Activities of States in the Exploration and Use of Outer Space, including the Moon and Other Celestial Bodies and as part of the personnel of a spacecraft within the meaning of the Agreement on the Rescue of Astronauts, the Return of Astronauts and the Return of Objects Launched into Outer Space.

2. States Parties shall offer shelter in their stations, installations, vehicles and other facilities to persons in distress on the moon.

ARTICLE 11

1. The moon and its natural resources are the common heritage of mankind, which finds its expression in the provisions of this Agreement, in particular in paragraph 5 of this article.

2. The moon is not subject to national appropriation by any claim of sovereignty, by means of use or occupation, or by any other means.

3. Neither the surface nor the subsurface of the moon, nor any part thereof or natural resources in place, shall become property of any State, international intergovernmental or non- governmental organization, national organization or non-governmental entity or of any natural person. The placement of personnel, space vehicles, equipment, facilities, stations and installations on or below the surface of the moon, including structures connected with its surface or subsurface, shall not create a right of ownership over the surface or the subsurface of the moon or any areas thereof. The foregoing provisions are without prejudice to the international regime referred to in paragraph 5 of this article.

4. States Parties have the right to exploration and use of the moon without discrimination of any kind, on the basis of equality and in accordance with international law and the terms of this Agreement.

5. States Parties to this Agreement hereby undertake to establish an international regime, including appropriate procedures, to govern the exploitation of the natural resources of the moon as such exploitation is about to become feasible. This provision shall be implemented in accordance with article 18 of this Agreement.

6. In order to facilitate the establishment of the international regime referred to in paragraph 5 of this article, States Parties shall inform the Secretary-General of the United Nations as well as the public and the international scientific community, to the greatest extent feasible and practicable, of any natural resources they may discover on the moon.

7. The main purposes of the international regime to be established shall include:

(a) The orderly and safe development of the natural resources of the moon;

(b) The rational management of those resources;

(c) The expansion of opportunities in the use of those resources;

(d) An equitable sharing by all States Parties in the benefits derived from those resources, whereby the interests and needs of the developing countries, as well as the efforts of those countries which have contributed either directly or indirectly to the exploration of the moon, shall be given special consideration.

8. All the activities with respect to the natural resources of the moon shall be carried out in a manner compatible with the purposes specified in paragraph 7 of this article and the provisions of article 6, paragraph 2, of this Agreement.

ARTICLE 12

1. States Parties shall retain jurisdiction and control over their personnel, vehicles, equipment, facilities, stations and installations on the moon. The ownership of space vehicles, equipment, facilities, stations and installations shall not be affected by their presence on the moon.

2. Vehicles, installations and equipment or their component parts found in places other than their intended location shall be dealt with in accordance with article 5 of the Agreement on the Rescue of Astronauts, the Return of Astronauts and the Return of Objects Launched into Outer Space.

3. In the event of an emergency involving a threat to human life, States Parties may use the equipment, vehicles, installations, facilities or supplies of other States Parties on the moon. Prompt notification of such use shall be made to the Secretary-General of the United Nations or the State Party concerned.

ARTICLE 13

A State Party which learns of the crash landing, forced landing or other unintended landing on the moon of a space object, or its component parts, that were not launched by it, shall promptly inform the launching State Party and the Secretary-General of the United Nations.

ARTICLE 14

1. States Parties to this Agreement shall bear international responsibility for national activities on the moon, whether such activities are carried on by governmental agencies or by non- governmental entities, and for assuring that national activities are carried out in conformity with

the provisions set forth in this Agreement. States Parties shall ensure that non-governmental entities under their jurisdiction shall engage in activities on the moon only under the authority and continuing supervision of the appropriate State Party.

2. States Parties recognize that detailed arrangements concerning liability for damage caused on the moon, in addition to the provisions of the Treaty on Principles Governing the Activities of States in the Exploration and Use of Outer Space, including the Moon and Other Celestial Bodies and the Convention on International Liability for Damage Caused by Space Objects, may become necessary as a result of more extensive activities on the moon. Any such arrangements shall be elaborated in accordance with the procedure provided for in article 18 of this Agreement.

ARTICLE 15

1. Each State Party may assure itself that the activities of other States Parties in the exploration and use of the moon are compatible with the provisions of this Agreement. To this end, all space vehicles, equipment, facilities, stations and installations on the moon shall be open to other States Parties. Such States Parties shall give reasonable advance notice of a projected visit, in order that appropriate consultations may be held and that maximum precautions may be taken to assure safety and to avoid interference with normal operations in the facility to be visited. In pursuance of this article, any State Party may act on its own behalf or with the full or partial assistance of any other State Party or through appropriate international procedures within the framework of the United Nations and in accordance with the Charter.

2. A State Party which has reason to believe that another State Party is not fulfilling the obligations incumbent upon it pursuant to this Agreement or that another State Party is interfering with the rights which the former State has under this Agreement may request consultations with that State Party. A State Party receiving such a request shall enter into such consultations without delay. Any other State Party which requests to do so shall be entitled to take part in the consultations. Each State Party participating in such consultations shall seek a mutually acceptable resolution of any controversy and shall bear in mind the rights and interests of all States Parties. The Secretary-General of the United Nations shall be informed of the results of the consultations and shall transmit the information received to all States Parties concerned.

3. If the consultations do not lead to a mutually acceptable settlement which has due regard for the rights and interests of all States Parties,

the parties concerned shall take all measures to settle the dispute by other peaceful means of their choice appropriate to the circumstances and the nature of the dispute. If difficulties arise in connection with the opening of consultations or if consultations do not lead to a mutually acceptable settlement, any State Party may seek the assistance of the Secretary-General, without seeking the consent of any other State Party concerned, in order to resolve the controversy. A State Party which does not maintain diplomatic relations with another State Party concerned shall participate in such consultations, at its choice, either itself or through another State Party or the Secretary-General as intermediary.

ARTICLE 16

With the exception of articles 17 to 21, references in this Agreement to States shall be deemed to apply to any international intergovernmental organization which conducts space activities if the organization declares its acceptance of the rights and obligations provided for in this Agreement and if a majority of the States members of the organization are States Parties to this Agreement and to the Treaty on Principles Governing the Activities of States in the Exploration and Use of Outer Space, including the Moon and Other Celestial Bodies. States members of any such organization which are States Parties to this Agreement shall take all appropriate steps to ensure that the organization makes a declaration in accordance with the foregoing.

[. . .]

Source: UN Office for Outer Space Affairs. "Agreement Governing the Activities of States on the Moon and Other Celestial Bodies." Available online. URL: http://www.oosa.unvienna.org/oosa/en/SpaceLaw/gares/html/gares_34_0068.html. Accessed April 6, 2010.

UN Resolution against an Arms Race in Outer Space (1997) (excerpt)

Because the Outer Space Treaty does not ban all types of weapons in outer space, members of the United Nations have called upon the standing Conference on Disarmament to negotiate more comprehensive agreements to ban non-peaceful activities. Resolutions have been periodically made since 1981, often with strong backing from the USSR and China. The United States, however, always abstained from voting, insisting the resolution was pointless as there was no arms race in space.

52/37 PREVENTION OF AN ARMS RACE IN OUTER SPACE

Date: 9 December 1997
Vote: 128-0-39

The General Assembly,

Recognizing the common interest of all mankind in the exploration and use of outer space for peaceful purposes,

Reaffirming the will of all States that the exploration and use of outer space, including the Moon and other celestial bodies, shall be for peaceful purposes and shall be carried out for the benefit and in the interest of all countries, irrespective of their degree of economic or scientific development.

Reaffirming also provisions of articles III and IV of the Treaty on Principles Governing the Activities of States in the Exploration and Use of Outer Space, including the Moon and Other Celestial Bodies,

Recalling the obligation of all States to observe the provisions of the Charter of the United Nations regarding the use or threat of use of force in their international relations, including in their space activities,

Reaffirming paragraph 80 of the Final Document of the Tenth Special Session of the General Assembly, in which it is stated that in order to prevent an arms race in outer space further measures should be taken and appropriate international negotiations held in accordance with the spirit of the Treaty,

Recalling its previous resolutions on this issue and taking note of the proposals submitted to the General Assembly at its tenth special session and at its regular sessions, and of the recommendations made to the competent organs of the United Nations and to the Conference on Disarmament,

Recognizing that prevention of an arms race in outer space would avert a grave danger for international peace and security,

Emphasizing the paramount importance of strict compliance with existing arms limitation and disarmament agreements relevant to outer space, including bilateral agreements, and with the existing legal regime concerning the use of outer space,

Considering that wide participation in the legal regime applicable to outer space could contribute to enhancing its effectiveness,

Noting that the Ad Hoc Committee on the Prevention of an Arms Race in Outer Space, taking into account its previous efforts since its establishment in 1985 and seeking to enhance its functioning in qualitative terms, continued the examination and identification of various issues, existing agreements and existing proposals, as well as future initiatives relevant to the prevention of an arms race in outer space, and that this contributed to a better understanding of a number of problems and to a clearer perception of the various positions,

Noting also that there were no objections in principle in the Conference on Disarmament during its 1997 session to the re-establishment of the Ad Hoc Committee, subject to re-examination of the mandate contained in the decision of the Conference on Disarmament of 13 February 1992,

Emphasizing the mutually complementary nature of bilateral and multilateral efforts in the field of preventing an arms race in outer space, and hoping that concrete results will emerge from those efforts as soon as possible,

Convinced that further measures should be examined in the search for effective and verifiable bilateral and multilateral agreements in order to prevent an arms race in outer space, including the weaponization of outer space,

Stressing that the growing use of outer space increases the need for greater transparency and better information on the part of the international community,

Recalling in this context its previous resolutions, in particular resolutions 45/55 B of 4 December 1990, 47/51 of 9 December 1992 and 48/74 A of 16 December 1993, in which, *inter alia*, it reaffirmed the importance of confidence-building measures as means conducive to ensuring the attainment of the objective of the prevention of an arms race in outer space,

Conscious of the benefits of confidence- and security-building measures in the military field,

Recognizing that negotiations for the conclusion of an international agreement or agreements to prevent an arms race in outer space remain a priority task of the Ad Hoc Committee and that the concrete proposals on confidence-building measures could form an integral part of such agreements,

1. *Reaffirms* the importance and urgency of preventing an arms race in outer space, and the readiness of all States to contribute to that common objective, in conformity with the provisions of the Treaty on Principles Governing the Activities of States in the Exploration and Use of Outer Space, including the Moon and Other Celestial Bodies;
2. *Reaffirms its recognition,* as stated in the report of the Ad Hoc Committee on the Prevention of an Arms Race in Outer Space, that the legal regime applicable to outer space by itself does not guarantee the prevention of an arms race in outer space, that this legal regime plays a significant role in the prevention of an arms race in that environment, that there is a need to consolidate and reinforce that

regime and enhance its effectiveness, and that it is important strictly to comply with existing agreements, both bilateral and multilateral;

3. *Emphasizes* the necessity of further measures with appropriate and effective provisions for verification to prevent an arms race in outer space;

4. *Calls upon* all States, in particular those with major space capabilities, to contribute actively to the objective of the peaceful use of outer space and of the prevention of an arms race in outer space and to refrain from actions contrary to that objective and to the relevant existing treaties in the interest of maintaining international peace and security and promoting international cooperation;

5. *Reiterates* that the Conference on Disarmament, as the single multilateral disarmament negotiating forum, has the primary role in the negotiation of a multilateral agreement or agreements, as appropriate, on the prevention of an arms race in outer space in all its aspects;

6. *Invites* the Conference on Disarmament to re-examine the mandate contained in its decision of 13 February 1992, with a view to updating it as appropriate, thus providing for the re-establishment of the Ad Hoc Committee during the 1998 session of the Conference on Disarmament;

7. *Recognizes,* in this respect, the growing convergence of views on the elaboration of measures designed to strengthen transparency, confidence and security in the peaceful uses of outer space;

8. *Urges* States conducting activities in outer space, as well as States interested in conducting such activities, to keep the Conference on Disarmament informed of the progress of bilateral or multilateral negotiations on the matter, if any, so as to facilitate its work;

9. *Decides* to include in the provisional agenda of its fifty-third session the item entitled "Prevention of an arms race in outer space."

Source: United Nations Office for Outer Space Affairs. "2222 (XXI). Treaty on Principles Governing the Activities of States in the Exploration and Use of Outer Space, including the Moon and Other Celestial Bodies." Available online. URL:http://www.oosa.unvienna.org/oosa/SpaceLaw/gares/html/gares_21_2222.html. Accessed May 28, 2010.

U.N. Resolution on Prevention of an Arms Race in Outer Space (2003) (excerpts)

This incarnation of the ongoing UN Resolutions on the Prevention of an Arms Race in Outer Space (PAROS) coincided with the George W. Bush administration's National Policy on Ballistic Missile Defense in 2003. This new U.S.

policy asserted that the changing international security situation made it imperative for the United States to prepare to defend its territory against emerging threats. While previous U.S. administrations had abstained from PAROS votes, Washington voted against the resolution in 2005.

RESOLUTION 58/36

Prevention of an Arms Race in Outer Space

The General Assembly,

Recognizing the common interest of all mankind in the exploration and use of outer space for peaceful purposes,

Reaffirming the will of all States that the exploration and use of outer space, including the Moon and other celestial bodies, shall be for peaceful purposes and shall be carried out for the benefit and in the interest of all countries, irrespective of their degree of economic or scientific development,

Reaffirming also the provisions of articles III and IV of the Treaty on Principles Governing the Activities of States in the Exploration and Use of Outer Space, including the Moon and Other Celestial Bodies,

Recalling the obligation of all States to observe the provisions of the Charter of the United Nations regarding the use or threat of use of force in their international relations, including in their space activities,

Reaffirming paragraph 80 of the Final Document of the Tenth Special Session of the General Assembly, in which it is stated that in order to prevent an arms race in outer space, further measures should be taken and appropriate international negotiations held in accordance with the spirit of the Treaty,

Recalling its previous resolutions on this issue, and taking note of the proposals submitted to the General Assembly at its tenth special session and at its regular sessions, and of the recommendations made to the competent organs of the United Nations and to the Conference on Disarmament,

Recognizing that prevention of an arms race in outer space would avert a grave danger for international peace and security,

Emphasizing the paramount importance of strict compliance with existing arms limitation and disarmament agreements relevant to outer space, including bilateral agreements, and with the existing legal regime concerning the use of outer space,

Considering that wide participation in the legal regime applicable to outer space could contribute to enhancing its effectiveness,

Noting that the Ad Hoc Committee on the Prevention of an Arms Race in Outer Space, taking into account its previous efforts since its establishment in 1985 and seeking to enhance its functioning in qualitative terms, continued the examination and identification of various issues, existing agreements and existing proposals, as well as future initiatives relevant to the prevention of an arms race in outer space, and that this contributed to a better understanding of a number of problems and to a clearer perception of the various positions,

Noting also that there were no objections in principle in the Conference on Disarmament to the re-establishment of the Ad Hoc Committee, subject to re-examination of the mandate contained in the decision of the.Conference on Disarmament of 13 February 1992,

Emphasizing the mutually complementary nature of bilateral and multilateral efforts in the field of preventing an arms race in outer space, and hoping that concrete results will emerge from those efforts as soon as possible,

Convinced that further measures should be examined in the search for effective and verifiable bilateral and multilateral agreements in order to prevent an arms race in outer space, including the weaponization of outer space,

Stressing that the growing use of outer space increases the need for greater transparency and better information on the part of the international community,

Recalling, in this context, its previous resolutions, in particular resolutions 45/55 B of 4 December 1990, 47/51 of 9 December 1992 and 48/74 A of 16 December 1993, in which, inter alia, it reaffirmed the importance of confidence-building measures as a means conducive to ensuring the attainment of the objective of the prevention of an arms race in outer space,

Conscious of the benefits of confidence- and security-building measures in the military field,

Recognizing that negotiations for the conclusion of an international agreement or agreements to prevent an arms race in outer space remain a priority task of the Ad Hoc Committee and that the concrete proposals on confidence-building measures could form an integral part of such agreements,

1. *Reaffirms* the importance and urgency of preventing an arms race in outer space and the readiness of all States to contribute to that common objective, in conformity with the provisions of the Treaty on Principles Governing the Activities of States in the Exploration and Use of Outer Space, including the Moon and Other Celestial Bodies;
2. *Reaffirms its recognition,* as stated in the report of the Ad Hoc Committee on the Prevention of an Arms Race in Outer Space, that the legal regime applicable to outer space does not in and of itself guarantee the prevention of an arms race in outer space, that the regime plays a significant role in the prevention of an arms race in that

environment, that there is a need to consolidate and reinforce that regime and enhance its effectiveness and that it is important to comply strictly with existing agreements, both bilateral and multilateral;

3. *Emphasizes* the necessity of further measures with appropriate and effective provisions for verification to prevent an arms race in outer space;

4. *Calls upon* all States, in particular those with major space capabilities, to contribute actively to the objective of the peaceful use of outer space and of the prevention of an arms race in outer space and to refrain from actions contrary to that objective and to the relevant existing treaties in the interest of maintaining international peace and security and promoting international cooperation;

5. *Reiterates* that the Conference on Disarmament, as the single multilateral disarmament negotiating forum, has the primary role in the negotiation of a multilateral agreement or agreements, as appropriate, on the prevention of an arms race in outer space in all its aspects;

6. *Invites* the Conference on Disarmament to complete the examination and updating of the mandate contained in its decision of 13 February 19924 and to establish an ad hoc committee as early as possible during its 2004 session;

7. *Recognizes,* in this respect, the growing convergence of views on the elaboration of measures designed to strengthen transparency, confidence and security in the peaceful uses of outer space;

8. *Urges* States conducting activities in outer space, as well as States interested in conducting such activities, to keep the Conference on Disarmament informed of the progress of bilateral and multilateral negotiations on the matter, if any, so as to facilitate its work;

9. *Decides* to include in the provisional agenda of its fifty-ninth session the item entitled "Prevention of an arms race in outer space."

Source: Resolution 58/36. Available online. URL: http://disarmament.un.org/yearbook-2003/ResDec9.html. Accessed April 26, 2010.

Cheng Jingye, Statement on PAROS (2006) (excerpts)

Addressing the UN Committee on Prevention of an Arms Race in Outer Space (PAROS) in 2006, China's ambassador for disarmament affairs, H. E. Mr. Cheng Jingye, stated his country's support for a ban on weapons deployment in outer space.

International Documents

STATEMENT ON PAROS BY H.E. MR. CHENG JINGYE, AMBASSADOR FOR DISARMAMENT AFFAIRS OF CHINA, AT THE PLENARY OF THE CONFERENCE ON DISARMAMENT

The Chinese Delegation welcomes the focused debate on PAROS. It is the first time over recent years that we have an opportunity to conduct in-depth discussions on this important issue which has been on the agenda of the CD since 1982. As a matter of fact, as early as in the late 1950s when the exploration of outer space by mankind just started, how to secure the peaceful use of outer space was under the deliberation of the UN General Assembly. Several decades later, this issue has proved more important and urgent today and safeguarding outer space security has increasingly become the consensus of the international community.

Over the past five decades and more, the exploration and utilization of outer space has given a giant impetus to the development of human society. The outer space, like the land, the ocean and the sky, has become an integral part of our life on which we increasingly depend. Peaceful use of outer space is the common aspiration of peoples of all countries.

However, outer space technologies, a two-edged sword like nuclear and coloning [sic] technologies, can either contribute to the well-being of mankind or cause severe detriment to the world if applied improperly or without control. During the Cold War, we witnessed an arms race in outer space, which, fortunately, did not lead to weaponization of outer space. Yet the end of the Cold War has not dispersed this shadow. The development of outer space weapons keeps progressing quietly, and relevant military doctrine is taking shape. The deployment of weapons in outer space would bring unimaginable consequences. The outer space assets of all countries would be endangered, mankind's peaceful use of outer space threatened, and international peace and security undermined. It is in the interest of all countries to protect the humanity from the threat of outer space weapons.

It is true that so far there are still no weapons in outer space, but this should not become our excuse for sitting idly by. Drawing lessons from the past, we have in recent years underscored the necessity of preventive diplomacy in the UN and in other multilateral arenas. The outer space is just one such field that requires our preventive efforts. Prevention is far better than facing the consequences. The history of the development of nuclear weapons constantly reminds us that once outer space weapons become full-fledged, how difficult it would be to control them and to prevent their proliferation, let alone to eliminate them. We simply cannot afford to delay actions and wait until the deployment of outer space weapons and an arms race in outer space become a reality. The price would be too high. As s

231

result, we should spare no efforts to avoid repeating the history of nuclear weapon development, and the effective way to achieve this goal is to conclude a new international legal instrument.

It is true that we already have some international legal instruments in this field, such as, inter alia, the 1967 Outer Space Treaty, the 1979 Moon Agreement and the 1972 ABM Treaty, which contributed significantly to preventing arms race in outer space. However, all of them have apparent lacunae. Some focus on the WMD only, some are limited to a certain celestial body or area in outer space and lack universality, and some have even been scrapped. In order to remedy the lacunae and close the loopholes of existing legal framework and fundamentally prevent the weaponization of and an arms race in outer space, a new international legal instrument is obviously needed.

Mr. President,

The Chinese Delegation is of the view that there is a sound basis and the conditions are ripe for negotiating such a legal instrument. It is time for us to carry out the substantive work.

Firstly, we enjoy broad political support. Over the recent 20 odd years, the UNGA has annually adopted, by overwhelming majority, the resolution on PAROS, in which the GA calls for the negotiation of an international agreement on PAROS. Last year, the number of Member States voting in favor of the resolution amounted to as many as 180. An overwhelming majority of the CD members agree to establish an Ad Hoc Committee dedicated to PAROS. It is fair to say that initiating the substantive work on PAROS at an early date conforms to the common will of the international community.

Secondly, the CD has had the experience of establishing an Ad Hoc Committee dealing with PAROS. For 10 consecutive years from 1985 to 1994, an Ad Hoc Committee was created on definition, principles, existing legal instruments and confidence-building measures, etc.. Though the Committee did not achieve tangible results given the circumstances in those years, it surely built a good basis for our work today.

Thirdly, there has been growing awareness and broader common ground on the importance of PAROS in the international community. In recent years, various seminars on this subject have been held. The UNIDIR, in collaboration with parties concerned, has convened in Geneva five international conferences on outer space in succession, during which many valuable ideas and proposals were expressed. Though their views on the ways and methods to deal with the outer space issue might vary, all participants shared a common understanding that preventing weaponization of outer space and maintaining outer space security serves the interests of all countries.

Last but not least, the framework of a new legal instrument on outer space is taking shape. In 2002, seven countries, namely, Russia, China, Indonesia, Belarus, Vietnam, Zimbabwe and Syria, jointly presented to the CD the working paper entitled "Possible Elements for a Future International Legal Agreement on the Prevention of the Deployment of Weapons in Outer Space, the Threat or Use of Force Against Outer Space Objects," contained in CD/1679. The document, in a form of treaty, puts forward detailed proposals on all composing elements of a new legal instrument on outer space, which provides a clear and feasible blueprint for our future work. In addition, China and the Russian Federation have jointly submitted to the CD four thematic papers regarding definition, verification, transparency and confidence-building measures, etc.

Mr. President,

It is our view that PAROS, together with other main CD agenda items, has a bearing on global security, which is closely linked to the maintenance of world peace and stability and therefore they all deserve serious consideration by the CD. A world free of outer space weapons is no less important than a world free of the WMD.

From time to time, we have heard the assertion of the so-called "linkage." Isn't it a form of "linkage" when some people insist on negotiating one issue, while refusing to conduct any substantive work on others? Each and every country has its one's own priority. To focus exclusively on one's own priority, while disregarding the priorities of others, would only lead to an unbreakable deadlock of CD. As it is known to all, China is in favor of negotiation on PAROS and this position remains unchanged. However, in order to help bring this body back to substantive work, we have demonstrated repeated flexibility. In August 2003, we agreed to the mandate for Ad Hoc Committee on PAROS contained in "Five Ambassador's Proposal" (CD/1693/Rev. 1) and indicated our readiness to join in the consensus on this Proposal.

China shares the concerns of all sides over the protracted deadlock of the CD, and likewise hopes for a positive turnaround in the CD as soon as possible. The "Five Ambassador's Proposal" accepted by the vast majority of members offers a practical and feasible way out. I would like to stress that, any idea that aims at circumventing the "program of work" and initiating negotiation solely on one issue while refraining from substantive work on other issues will not fly.

The Chinese experts are ready in the coming days to have in-depth exchanges of views with all sides on the definition, scope, transparency, confidence-building measures and assets security in outer space, with a view to further enriching our discussion on outer space. It is our belief that this debate will help create the conditions for the CD to agree on a "program of work" and to begin substantive work on PAROS in the near future.

Thank you, Mr. President.

Source: Chinese Permanent Mission to the United Nations. "Statement on PAROS by H. E. Mr. Cheng Jingye, Ambassador for Disarmament Affairs of China, at the Plenary of the Conference on Disarmament." Available online. URL: http://www.china-un.ch/eng/xwdt/t257105.htm. Accessed April 6, 2010.

Draft Treaty on Prevention of the Placement of Weapons in Outer Space and of the Threat or Use of Force Against Outer Space Objects (February 29, 2008) (Excerpt)

In early 2008, China and Russia submitted a draft of a new treaty banning the use of weapons or force in outer space to the UN Conference on Disarmament. The draft emerged from discussions about a 2002 working paper. While it bans the deployment of weapons in orbit or on celestial bodies, it does not ban research and development of antisatellite, missile defense, or ground-based antisatellite systems. The U.S. government dismissed the draft treaty as a "diplomatic ploy by the two nations to gain a military advantage."

The States Parties to this Treaty,

Reaffirming that outer space is playing an ever-increasing role in the future development of mankind,

Emphasizing the right to explore and use outer space freely for peaceful purposes,

Interested in preventing outer space from becoming an arena for military confrontation and ensuring security in outer space and the undisturbed functioning of space objects,

Recognizing that prevention of the placement of weapons in outer space and of an arms race in outer space would avert a grave danger for international peace and security,

Desiring to keep outer space as a sphere where no weapon of any kind is placed,

Noting that the existing agreements on arms control and disarmament relevant to outer space, including bilateral agreements, and the existing legal regimes concerning the use of outer space play a positive role in exploration of outer space and in regulating outer space activities, and should be strictly complied with, although they are unable to effectively prevent the placement of weapons in outer space and an arms race in outer space,

Recalling the United Nations General Assembly resolution on "Prevention of an arms race in outer space," in which, inter alia, the Assembly expressed conviction that further measures should be examined in the search for effective and verifiable bilateral and multilateral agreements in order to prevent an arms race in outer space,

Have agreed on the following:

ARTICLE I

For the purposes of this Treaty:

(a) The term "outer space" means the space above the Earth in excess of 100 km above sea level;

(b) The term "outer space object" means any device designed to function in outer space which is launched into an orbit around any celestial body, or located in orbit around any celestial body, or on any celestial body, except the Earth, or leaving orbit around any celestial body towards this celestial body, or moving from any celestial body towards another celestial body, or placed in outer space by any other means;

(c) The term "weapon in outer space" means any device placed in outer space, based on any physical principle, which has been specially produced or converted to destroy, damage or disrupt the normal functioning of objects in outer space, on the Earth or in the Earth's atmosphere, or to eliminate a population or components of the biosphere which are important to human existence or inflict damage on them;

(d) A weapon shall be considered to have been "placed" in outer space if it orbits the Earth at least once, or follows a section of such an orbit before leaving this orbit, or is permanently located somewhere in outer space;

(e) The "use of force" or the "threat of force" mean any hostile actions against outer space objects including, inter alia, actions aimed at destroying them, damaging them, temporarily or permanently disrupting their normal functioning or deliberately changing their orbit parameters, or the threat of such actions.

ARTICLE II

The States Parties undertake not to place in orbit around the Earth any objects carrying any kinds of weapons, not to install such weapons on celestial bodies and not to place such weapons in outer space in any other manner; not to resort to the threat or use of force against outer space objects; and not to assist or induce other States, groups of States or international organizations to participate in activities prohibited by this Treaty.

ARTICLE III

Each State Party shall take all necessary measures to prevent any activity prohibited by this Treaty on its territory or in any other place under its jurisdiction or control.

ARTICLE IV

Nothing in this Treaty may be interpreted as impeding the exercise by the States Parties of their right to explore and use outer space for peaceful purposes in accordance with international law, including the Charter of the United Nations and the Outer Space Treaty.

ARTICLE V

Nothing in this Treaty may be interpreted as impeding the exercise by the States Parties of their right of self-defence in accordance with Article 51 of the Charter of the United Nations.

ARTICLE VI

With a view to promoting confidence in compliance with the provisions of the Treaty and ensuring transparency and confidence-building in outer space activities, the States Parties shall implement agreed confidence-building measures on a voluntary basis, unless agreed otherwise.

Measures to verify compliance with the Treaty may form the subject of an additional protocol.

ARTICLE VII

If a dispute arises between States Parties concerning the application or the interpretation of the provisions of this Treaty, the parties concerned shall first consult together with a view to settling the dispute by negotiation and cooperation.

If the parties concerned do not reach agreement after consultation, an interested State Party may refer the situation at issue to the executive organization of the Treaty, providing the relevant argumentation.

Each State Party shall undertake to cooperate in the settlement of the situation at issue with the executive organization of the Treaty.

ARTICLE VIII

To promote the implementation of the objectives and provisions of this Treaty, the States Parties shall establish the executive organization of the Treaty, which shall:

(a) Accept for consideration communications from any State Party or group of States Parties relating to cases where there is reason to believe that a violation of this Treaty by any State Party is taking place;

(b) Consider matters concerning compliance with the obligations entered into by States Parties;

(c) Organize and conduct consultations with the States Parties with a view to resolving any situation that has arisen in connection with the violation of this Treaty by a State Party;

(d) Take steps to put an end to the violation of this Treaty by any State Party. The title, status, specific functions and forms of work of the executive organization of the Treaty shall be the subject of an additional protocol to this Treaty.

Source: United Nations Office at Geneva. "Disarmament: CD Documents related to Prevention of an Arms Race in Outer Space." Available online. URL: http://www.unog.ch/80256EE600585943/%28httpPages%29/D4C4FE00A7302FB2C12575E4002DED85?OpenDocument. Accessed May 28, 2010.

Council of the European Union, Draft Code of Conduct for Outer Space Activities (December 8–9, 2008) (excerpt)

Following on the UN General Assembly's 2006 call for individual countries to propose confidence building and transparency measures, the European Union issued a draft Code of Conduct. The EU document applies to both military and civilian uses, urging parties to voluntarily refrain from harming space objects, to better control debris, and to create channels for joint communication and consultation. After the United States shot down a damaged satellite in February 2008, the draft included an exception allowing space objects to be destroyed due to safety considerations.

I. CORE PRINCIPLES AND OBJECTIVES

1. Purpose and scope

1.1. The purpose of the present code is to enhance the safety, security and predictability of outer space activities for all.

1.2. The present Code is applicable to all outer space activities conducted by a Subscribing State or jointly with other State(s) or by non-governmental entities under the jurisdiction of a Subscribing State, including those activities within the framework of international intergovernmental organisations.

1.3. This Code, in codifying new best practices, contributes to transparency and confidence-building measures and is complementary to the existing framework regulating outer space activities.

1.4. Adherence to this Code and to the measures contained in it is voluntary and open to all States.

2. General principles

The Subscribing States resolve to abide by the following principles:

- the freedom of access to, exploration and use of outer space and exploitation of space objects for peaceful purposes without interference, fully respecting the security, safety and integrity of space objects in orbit;
- the inherent right of individual or collective self-defence in accordance with the United Nations Charter;
- the responsibility of States to take all the appropriate measures and cooperate in good faith to prevent harmful interference in outer space activities;
- the responsibility of States, in the conduct of scientific, commercial and military activities, to promote the peaceful exploration and use of outer space and take all the adequate measures to prevent outer space from becoming an area of conflict;

3. Compliance with and promotion of treaties, conventions and other commitments relating to outer space activities

3.1. The Subscribing States reaffirm their commitment to:

- the existing legal framework relating to outer space activities;
- making progress towards adherence to, and implementation of:

(a) the existing framework regulating outer space activities, inter alia:

- the Treaty on Principles Governing the Activities of States in the Exploration and Use of Outer Space, including the Moon and Other Celestial Bodies (1967);
- the Agreement on the Rescue of Astronauts, the Return of Astronauts and the Return of Objects Launched into Outer Space (1968);
- the Convention on International Liability for Damage Caused by Space Objects (1972);

- the Convention on Registration of Objects Launched into Outer Space (1975);
- the Constitution and Convention of the International Telecommunications Union and its Radio Regulations (2002);
- the Treaty banning Nuclear Weapon Tests in the Atmosphere, in Outer Space and under Water (1963) and the Comprehensive Nuclear Test Ban Treaty (1996);
- the International Code of Conduct against Ballistic Missile Proliferation (2002).

(b) declarations and Principles, inter alia:

- the Declaration of Legal Principles Governing the Activities of States in the Exploration and Use of Outer Space as stated in UNGA Resolution 1962 (XVIII);
- the Principles Relevant to the Use of Nuclear Power Sources in Outer Space as stated in UNGA Resolution 47/68;
- the Declaration on International Cooperation in the Exploration and Use of Outer Space for the Benefit and in the Interest of All States, Taking into Particular Account the Needs of Developing Countries as stated in UNGA Resolution 51/122;
- the Recommendations on the Practice of States and International Organisations in Registering Space Objects as stated in UNGA Resolution 62/101;
- the Space Debris Mitigation Guidelines of the United Nations Committee for the Peaceful Uses of Outer Space as stated in UNGA Resolution 62/217.

3.2. The Subscribing States also reiterate their support to encourage coordinated efforts in order to promote universal adherence to the above mentioned instruments.

II. GENERAL MEASURES
4. Measures on space operations

4.1. The Subscribing States will establish and implement national policies and procedures to minimise the possibility of accidents in space, collisions between space objects or any form of harmful interference with other States' right to the peaceful exploration and use of outer space.

4.2. The Subscribing States will, in conducting outer space activities:

- refrain from any intentional action which will or might bring about, directly or indirectly, the damage or destruction of outer space objects unless such action is conducted to reduce the creation of outer space debris and/or justified by imperative safety considerations;

- take appropriate steps to minimise the risk of collision;
- abide by and implement all International Telecommunications Union recommendations and regulations on allocation of radio spectra and orbital assignments.

4.3. When executing manoeuvres of space objects in outer space, for example to supply space stations, repair space objects, mitigate debris, or reposition space objects, the Subscribing States agree to take all reasonable measures to minimise the risks of collision.

4.4. The Subscribing States resolve to promote the development of guidelines for space operations within the appropriate fora for the purpose of protecting the safety of space operations and long term sustainability of outer space activities.

5. Measures on space debris control and mitigation

In order to limit the creation of space debris and reduce its impact in outer space, the Subscribing States will:

- refrain from intentional destruction of any on-orbit space object or other harmful activities which may generate long-lived space debris;
- adopt, in accordance with their national legislative processes, the appropriate policies and procedures in order to implement the Space Debris Mitigation Guidelines of the United Nations Committee for the Peaceful Uses of Outer Space as endorsed by UNGA Resolution 62/217.

III. COOPERATION MECHANISMS
6. Notification of outer space activities

6.1. The Subscribing States commit to notify, in a timely manner, to the greatest extent feasible and practicable, all potentially affected Subscribing States on the outer space activities conducted which are relevant for the purposes of this Code, inter alia:

- the scheduled manoeuvres which may result in dangerous proximity to space objects;
- orbital changes and re-entries, as well as other relevant orbital parameters;
- collisions or accidents which have taken place;

- the malfunctioning of orbiting space objects with significant risk of re-entry into the atmosphere or of orbital collision.

6.2. The Subscribing States reaffirm their commitment to the Principles Relevant to the Use of Nuclear Power Sources in Outer Space as stated in UNGA Resolution 47/68.

7. Registration of space objects

The Subscribing States undertake to register space objects in accordance with the Convention on Registration of Objects launched in Outer Space and to provide the United Nations Secretary-General with the relevant data as set forth in this Convention and in the Recommendations on the Practice of States and International Organisations in Registering Space Objects as stated in UNGA Resolution 62/101.

8. Information on outer space activities

8.1. The Subscribing States resolve to share, on an annual basis, and, where available, information on:

- national space policies and strategies, including basic objectives for security and defence related activities;
- national space policies and procedures to prevent and minimise the possibility of accidents, collisions or other forms of harmful interference;
- national space policies and procedures to minimise the creation of space debris;
- efforts taken in order to promote universal adherence to legal and political regulatory instruments concerning outer space activities.

8.2. The Subscribing States may also consider providing timely information on space environmental conditions and forecasts to other Subscribing States or private entities through their national space situational awareness capabilities.

9. Consultation mechanism

9.1. Without prejudice to existing consultation mechanisms provided for in Article IX of the Outer Space Treaty of 1967 and in Article 56 of the ITU Constitution, the Subscribing States have decided on the creation of the following consultation mechanism:

- A Subscribing State with reason to believe that certain outer space activities conducted by one or more Subscribing State(s) are, or may be, contrary to the purposes of the Code may request consultations

241

with a view to achieving acceptable solutions regarding measures to be adopted in order to prevent or minimise the inherent risks.

- The Subscribing States involved in a consultation process will decide on a timeframe consistent with the timescale of the identified risk triggering the consultations.
- Any other Subscribing State which may be affected by the risk and requests to take part in the consultations will be entitled to take part.
- The Subscribing States participating in the consultations shall seek solutions based on an equitable balance of interests.

9.2. In addition, the Subscribing States may propose to create a mechanism to investigate proven incidents affecting space objects. The mechanism, to be agreed upon at a later stage, could be based on national information and/or national means of investigation provided on a voluntary basis by the Subscribing States and on a roster of internationally recognised experts to undertake an investigation.

Source: Council of the European Union, "Council Conclusions on the draft Code of Conduct for outer space activities." December 8–9, 2008. Available online. URL: http://www.docstoc.com/docs/.33977074/PUBLIC. Accessed July 7, 2010.

Statement by North Korea on Nuclear Tests (April 29, 2009) (excerpt)

When North Korea launched a satellite on April 5, 2009, the international community immediately denounced the test as being hostile and provocative. While China and Russia argued that countries have the right to test equipment, the United States, Japan, and South Korea called for the United Nations to impose sanctions. When the UN Security Council announced sanctions, North Korea issued the following statement.

In accordance with its "presidential statement" which has no binding force, on April 24 the UNSC officially designated three companies of the DPRK [North Korea] as targets of sanctions and many kinds of military supplies and materials as embargo items over the DPRK's peaceful satellite launch, a DPRK's exercise of sovereignty, and thus committed such illegal provocations as setting in motion its sanctions on the DPRK, the statement notes, and says:

Such sanctions can never work on the DPRK which has been subject to all sorts of sanctions and blockade by the hostile forces for the past scores of years.

What is serious is the fact that the UNSC has set out in directly jeopardizing the security of the country and the nation, the supreme interests of the DPRK, though it had already wantonly infringed the sovereignty of a sovereign state, pursuant to the U.S. moves.

The hostile forces are foolishly scheming to suffocate the DPRK's defense industry by physical methods as they failed to attain their aims for disarming the DPRK through the six-way talks.

In the 1990s the DPRK already declared that any anti-DPRK sanctions to be put by the United Nations, a legal party to the Korean Armistice Agreement, would be regarded as a termination of the agreement, that is, a declaration of war.

The desire for denuclearization of the Korean peninsula has gone forever with the six-way talks and the situation is inching to the brink of war by the hostile forces. The DPRK Ministry of Foreign Affairs solemnly gives the following warnings to cope with such grave situation:

The UNSC should promptly make an apology for having infringed the sovereignty of the DPRK and withdraw all its unreasonable and discriminative 'resolutions' and decisions adopted against the DPRK.

This is the only way for it to regain confidence of the UN member nations and fulfill its responsibility for maintaining international peace and security, not serving as a tool for the U.S. highhanded and arbitrary practices any longer.

In case the UNSC does not make an immediate apology, such actions will be taken as:

Firstly, the DPRK will be compelled to take additional self-defensive measures in order to defend its supreme interests.

The measures will include nuclear tests and test-firings of intercontinental ballistic missiles.

Secondly, the DPRK will make a decision to build a light water reactor power plant and start the technological development for ensuring self-production of nuclear fuel as its first process without delay.

Source: Council on Foreign Relations Essential Documents. "Statement by North Korea on Nuclear Tests, April 2009." Available online. URL: http://www.cfr.org/publication/19254/statement_by_north_korea_on_nuclear_tests_april_2009. html?breadcrumb=%2Fregion%2F276%2Fnorth_korea. Accessed April 6, 2010.

PART III

Research Tools

6

How to Research the Militarization of Space

The military uses of space can be explored from multiple angles. Historians might focus on the use of hot-air balloons during the U.S. Civil War, the Nazi V-2 rocket program, or the early days of the Apollo program. Policy makers need to know costs, development times, and implications. Military analysts would look at the strategic and tactical uses of space-based assets and the offensive and defensive needs that space might affect. Lawyers would need to understand the operating specifications of new space applications to see if they adhere to—or are even covered by—existing treaties and agreements. Business analysts and economists would look at the cost effectiveness of a program, while scientists might focus on astronomy or environmental impacts.

HOW TO START

Researchers should first identify what they want to know. Consider the subject using the basic journalism questions: Why put weapons in space? How can countries get weapons in space and prevent being attacked by weapons other countries have deployed in space? Who owns space? What are the advantages and drawbacks of military space assets? Another possibility is to explore a counterfactual assumption. In other words, what if history had been different? What if the USSR had beaten the United States to the Moon? What if the United States had launched a satellite before *Sputnik*? What would happen if countries were not required to announce their launches? What if *Skylab* or other expiring satellites had actually crashed to Earth? What if there were no curbs on antiballistic missile technology? What if SDI had been deployed? Once the question (or questions) is identified, it is time to start researching the topic.

The first goal is to establish the facts; one must know what happened before one can explain why it happened. Determine which country, which time period, which weapon, and which laws are involved. To determine why the Soviet Union was first to launch a satellite, establish the date, the type of rocket used, the development and testing history, and how long did it take for the United States to launch a similar satellite.

The next step is to head to a library or log on to a research portal for space-related military topics and gather the information to back up your argument. Articles in daily newspapers are likely too numerous to be a good starting point for research and an Internet search for "space weapon" will return more than 36 million hits. More productive searches come from key phrases such as "military weapon" or "military space," as they weed out some of the science fiction materials. As early as possible, determine whether a particular weapon was actually deployed or whether it was merely researched. Was it tested in a laboratory or in the real world?

Some books are likely to provide more information and technical detail than needed during this phase, while others will be outdated. Instead, try searching for the specific country, weapon, treaty, or summit meeting in weekly or monthly news magazines such as *Time, Newsweek,* or the *Economist.* Also look in an encyclopedia or specialized almanacs, such as *Air Force Magazine*'s "2009 Space Almanac: The U.S. Military Space Operation in Facts and Figures," which is downloadable from http://www.airforce-magazine.com. The Air Force publication has budget data, chronologies, command structure, rocket descriptions, and space junk inventories. The Space Security Index also has excellent downloadable yearbooks (http://www.spacesecurity.org/publications.htm).

As evidence is gathered, put together a time line of key events. This will help researchers find gaps in their research and provide a quick reference if details become overwhelming. Many agencies and military branches publish their own chronologies that can be very helpful as well, including the Federation of American Scientists (http://www.fas.org/spp/starwars/program/milestone.htm) and the U.S. Missile Defense Agency (http://www.mda.mil/careers/timeline.html). A time line will also help keep data in chronological order, a critical detail because so many weapons or scientific missions are cumulative, building on and refining previous incarnations. The time line can also be included as an appendix to a final research paper.

Try to locate and use primary sources whenever possible. Primary sources are statements that come from the actual people involved; secondary sources talk about what people said and did. A research project needs to stay as close to actual events as possible to guarantee accuracy. In addition, it is often more interesting to read an original statement rather than read a report about a statement. Similarly, rather than just quoting a press release, take the time to find the actual text under discussion. While it is unlikely that an undergraduate researcher will secure an interview, other people have met the relevant people. Numerous quotations and primary documents are available both online and in print. However, an exception can be made when the original sources use advanced astrophysics, space engineering, or another highly technical field. Then using a secondary source to translate the formulas into strategy is perfectly reasonable.

Finally, verify the legitimacy of your source. In the library or, especially, on the Internet, it is very easy to wander into highly one-sided blogs and science fiction or commercial groups trying to sell you a rocket or a ride to the *International Space Station.* If your document suddenly begins discussing warp drives, alien abductions, or Jedi knights, it's time to find another resource.

RESEARCHING THE TOPIC

Researchers must make sure to use reliable sources to support their argument. Fortunately, many publishers of reference materials already have procedures in place to verify their information. Research tools can be divided into two categories: traditional and nontraditional.

Traditional Research Tools

"Traditional" research tools are printed-paper sources, many of which predate the Internet. While they may not be as convenient as computer sources, many have undergone critical evaluations by editorial boards and review panels before being published and are trustworthy sources. The line between traditional and electronic research sources is becoming blurred, as many long-established publishers have embraced electronic publishing and produce online versions of their products as well. However, online editions tend to be fee-based, while the print versions available at public and school libraries are free. For a comprehensive and annotated selection of relevant books on the topic, please refer to chapter 10 in this volume.

ENCYCLOPEDIAS AND ALMANACS

Encyclopedia sets are a staple of public and school libraries, and the two best-known are the *Encyclopedia Britannica* and the *World Book Encyclopedia.* Almanacs are also valuable for basic country information and time lines. More specialized publications include the annual reports from

Space Security Index (www.spacesecurity.org), which uses eight indicators to compile a "comprehensive and integrated assessment of space security." *Encyclopedia Astronautica*, which focuses on astronomy, is also useful (http://www.astronautix.com/).

MAGAZINES

Weekly newsmagazines are a great reference tool as they provide more detailed and considered coverage of events. U.S. newsmagazines include *Newsweek, Time,* and the British *Economist* is highly regarded as well. All three have search engines available on the Internet, although the *Economist* charges non-subscribers for full-text articles.

There are also a variety of specialized magazines. One of the most highly respected monthly magazines is *Air and Space* from the Smithsonian Institution. The topics covered in *Air and Space* range from space exploration to NASA missions and military aviation. International space issues are also covered, such as the Soviet and Russian space programs and special coverage of the 50th anniversary of *Sputnik.*

Air Force Magazine (http://www.airforce-magazine.com) is even more useful for military space applications. *Astronomy* magazine's Web site also includes teaching materials and "Astro" which is aimed at kids (http://www.astronomy.com), while *Mercury Magazine* is the official publication of the Astronomy Society of the Pacific (http://www.astrosociety.org/pubs/mercury/mercury.html), which also has a more technical journal, *Publications of the Astronomical Society of the Pacific.* Archives of both can be found on their Web site. The British Interplanetary Society publishes *Spaceflight,* a magazine older than *Sputnik. Sky and Telescope* magazine (http://www.skyandtelescope.com/) focuses more on astronomy, while *Scientific American* has the occasional article on astrophysics, cosmology, and extraterrestrial life. *Popular Science* and *Popular Mechanics* are other possibilities. Many online searches will turn up references to *Jane's Space Systems and Industry, Jane's Defense Weekly,* or other publications from this UK–based consultant. Such articles tend to cater to the defense contracting world and tend to be high quality, and access is limited without a subscription. Few public libraries can afford subscriptions, which cost more than $1,700 per year.

JOURNALS

Journals offer more detailed studies of topics by leading scholars. They typically assume some prior knowledge or even expertise on a topic, and articles usually go through a rigorous peer-review process. Scholarly journals are not always available in public or high school libraries, but online research tools for these publications are likely available, and a librarian can often order a copy of the article. *Astropolitics: The International Journal of Space Politics and Policy* is published three times per year. *Air and Space Power Journal* (http://www.airpower.maxwell.af.mil/airchronicles/apje.html) is published by the U.S. Air Force. *Aerospace Power Journal* is the same publication, although some sources confuse the two names. The British Interplanetary Society publishes *Journal of the BIS*, which debuted in 1934 and is now complemented by the less technical *Space Chronicle. High Frontier* (http://www. afspc.af.mil/library/highfrontierjournal.asp), the Air Force Space Command's premier space professional journal, is published quarterly. *World Politics*, the *Brown Journal of World Affairs*, and *International Security* occasionally carry space security studies.

NEWSPAPERS

National newspapers are preferable to local newspapers, unless it is a local paper from a town near a relevant military or launch facility. The *Houston Chronicle*, for example, has a space site (http://www.chron.com/content/ chronicle/space/), which is especially good for business-related topics. Often articles from the national papers are reprinted in local papers with at least one day's delay. Look for the *New York Times, Wall Street Journal, Washington Post*, and the *Los Angeles Times*. Paper indexes to the newspapers may be available at local libraries and most newspapers' Web sites also allow extensive searching of their archives—but they may charge to print the article. The *Washington Post*, for example, charges for any full-text article more than 60 days old.

The *New York Times* compiled an engaging collection of articles, photos, and U.S. reactions from the first week following Sputnik's launch (http:// www.nytimes.com/partners/aol/special/sputnik/). The *Wall Street Journal* also has a large, illustrated space-topic area (http://topics.wsj.com/subject/s/ Space/3926).

TRANSCRIPTS

Networks have long made transcripts of their news reports available, but it often took days, if not weeks, to receive them. Now reports from the British Broadcasting Company (BBC: www.bbc.co.uk) and Cable News Network (CNN: www.cnn.com) are posted online immediately. Both BBC and CNN online also provide links to additional articles on a topic. Other cable networks are also making their reports available online, often with additional background materials and teacher guides. Transcripts of White House press briefings, press releases, government and agency statements, and congressional testimony can prove useful as well.

Electronic Research Tools

The Internet has brought thousands of references sources into homes, libraries, and coffee houses, any place with a modem or WiFi access. As mentioned above, many long-standing publications have chosen to make their products available on the Web on a limited basis. The Internet is strongest covering current events. It can provide weapons schematics, outer space diagrams, photos from the Hubble telescope or a scientific voyage within hours, in contrast to the next day for newspapers, next week for newsmagazines, and next year for scholarly journals, and next couple of years for books and print encyclopedia entries. At the same time, because information can be posted online without consideration or evaluation, errors naturally occur.

False information is a particular problem with electronic sources. Anyone with Internet access can upload his or her own research studies, blogs, rumors, abductee reports, and gossip. The Internet is filled with treasures and trash—the challenge is to differentiate between the two.

WIKIPEDIA: A WORD OF CAUTION

Internet searches on almost any topic tend to return hits to Wikipedia articles. Wikipedia (www.en.wikipedia.org) is an online encyclopedia that has become a first stop for research on almost any topic. It is an open-source, user-run project that features more than 1 million articles in English and hundreds of thousands of articles in more than 200 other languages. While it is a good tool to orient researchers and provide a broad overview of a topic, it should not be considered a launchpad, not a definitive source. Librarians dislike the site for its lack of quality control, and increasingly teachers and college professors are refusing to accept it as a valid source in research projects.

The problem stems from its open-source structure. Because so many people can create and alter entries, intellectual vandalism has begun to occur. Wikipedia vandalism has become so widespread that the topic has its own Web sites (http://wikipediavandalism.tumblr.com). Put simply, many Wikipedia entries are written by experts in their field, but the entries can be rewritten by pranksters or people with their own motives or axes to grind. In addition, some authors are not as expert as they believe, and factual errors may appear; it is quite possible inadvertently to replace facts with mistakes. Wikipedia has no editorial review board in place, but there is a team of 800 or so volunteers who check the accuracy of posts *after* they appear online; there is no screening before posting. This makes the process somewhat like shooting first and asking questions later.

The most infamous case of Wikipedia vandalism to date concerned John Seigenthaler, Sr., a retired newspaper editor who had once been an assistant to Robert F. Kennedy. A prankster anonymously altered Seigenthaler's entry to say erroneously that he had been a suspect in the assassinations of both Robert Kennedy and his brother, President John F. Kennedy. Seigenthaler blasted Wikipedia in *USA Today*, provoking obscenity-laced attacks on his Wikipedia entry.

The Seigenthaler case illustrates academia's basic problem with Wikipedia: The information cannot be fully trusted. What if a student was researching a topic, such as the Kennedy assassinations, and happened to find and quote the erroneous Seigenthaler allegation before it had been corrected? The disinformation was online for nearly four months before it was discovered. At minimum, the student might fail the assignment; if the allegation was repeated, Seigenthaler's reputation might be ruined and lawsuits might ensue. Most in-print publications and reputable online sources will have procedures in place to catch most errors of this type before they ever reach term papers.

Wikipedia can provide links to articles, webpages, and other sources of information. But perhaps the main reason to read through an entire Wikipedia entry on a particular topic is to be prepared to recognize it in other places on the web. Many, many sites claiming to offer new information have simply copied Wikipedia entries in full. If the methodology was substandard on Wikipedia, it will still be substandard on another page.

WEB PORTALS

Since popular search engines such as Google produce millions of hits on a term such as "space weapon" they are not a good place for someone to begin research. Instead of trying to wade through those results, use a Web portal to filter and organize the information.

There are a variety of high-quality Internet portals for studies of the militarization of space. The **Union of Concerned Scientists** (http://www.ucsusa.org/nuclear_weapons_and_global_security/space_weapons/technical_issues/) Nuclear Weapons and Global Security page offers a special category on space weapons that includes technical information and policy analysis. UCS also has a database of satellites, containing information about the 900+ satellites currently orbiting the earth.

The **Federation of American Scientists** (FAS) was founded in 1945 by scientists who had worked on the Manhattan Project to develop the first atomic bombs. These scientists believed that scientists had a unique responsibility to both warn the public and policy leaders of potential dangers from scientific and technical advances and to show how good policy could increase the benefits of new scientific knowledge. The FAS Weapons in Space program (www.fas.org/programs/ssp/man/wpnsinspace/index.html) addresses threats from high-altitude nuclear explosions, space mines, ASAT weapons, orbital debris, and others.

Global Security (http://www.globalsecurity.org/space/index.html) has a space page with streaming and archived articles related to space defense topics as well as analytical articles. It is particularly good for locating government information, congressional debates, weapons systems, space facilities, and budgets.

Space Today (http://www.spacetoday.org/) began as a print magazine in 1986 but has completely migrated to the Internet and tries to offer information useful to experts and amateurs alike. It provides links to "authoritative and highly reliable open sources," such as space agencies around the world, nonprofit organizations, educational institutions, research centers, embassies and consulates, corporate press relations offices, museum archives, library collections, and international media reports.

The *Encyclopedia Astronautica* (http://www.astronautix.com/) is now online, with over 25,000 pages of information and 10,000 images. Subsections include spacecraft, biographies, program by countries, rocket and missile families. There is also a convenient index feature.

Space Facts (http://spacefacts.de/) focuses more on space exploration than military uses, but it has valuable information on specific missions, astronauts, and spacecraft, including space stations.

Former NASA aerospace engineer **James Oberg** is a prolific author and commentator on space issues for MSNBC and other news outlets. His personal Web site (www.jamesoberg.com) digests his numerous publications on a variety of space and space weapons topics. Themes covered include China's space program, tourism, the space shuttle, even space folklore, humor, and urban legends. He also provides links to favorite Internet sites.

Finally, **Space Debate** (http://www.spacedebate.org/) offers a mixture of portal and opinion. The site declares it is "expanding the debate on space weaponization" by using a content system similar to Wikipedia's. As such, it is a mixture of information gathered from other sources with reader comments added. Many discussions are organized as debates, with arguments clearly grouped as for or against.

BLOGS

While blogs often are the assorting ramblings and rantings of individuals, several experts in the field of space militarization have their own sites that offer good-quality instant analysis of breaking news. *Washington Post* science reporter Joel Achenbach regularly covers NASA and defense issues in **Achenblog** (http://voices.washingtonpost.com/achenblog). Michael Krepon, cofounder of the Stimson Center think tank, has published over

350 articles on space militarization, and he offers frequent commentary on **Michael Krepon's Shoebox,** part of Arms Control Wonk (http://www. armscontrolwonk.com/2264/michael-krepons-shoebox). **Arms Control Wonk** (http://www.armscontrolwonk.com) is compiled by Jeffrey Lewis, director of the Nuclear Strategy and Nonproliferation Initiative at the New America Foundation, and James Acton of the Center for Science and Security Studies at King's College London. Finally, Smithsonian Air and Space historian Roger Launius has his own blog (http://launiusr.wordpress.com/) that discusses spaceflight.

GENERAL NEWS

Many of the news stories gathered in the Web portals listed above come from mainstream news agencies. Both the **BBC** (http://www.bbc.co.uk/science/ space/) and **CNN** (http://www.cnn.com/TECH/space/) have sections devoted to technology and space. **MSNBC** (http://www.msnbc.com) has a smaller technology and space site.

Original news stories can also be found at **Space.com,** which has narrowly focused, free daily newsletters, such as *Space War, Space Daily, Space Mart, Space Travel, Russo Daily,* and *Sino Daily.* **Spaceref.com** offers 21 news and reference Web sites that "are designed to allow both the novice and specialist alike to explore outer space and Earth observation." The **Space Review** (www.thespacereview.com) describes itself as a "publication whose focus is on publishing in-depth articles, essays, editorials, and reviews on a wide range of space-related topics." The Review has a companion site (http:// www.spacetoday.net/) that "links to dozens of space news articles published online each day." **Space News** (http://www.spacenews.com/) bills itself as the "publication of record for commercial space, military space and satellite communication businesses."

HISTORICAL INFORMATION

NASA, especially its History Division, has made a tremendous amount of information available (http://www.hq.nasa.gov/office/pao/History/ or http://history.nasa.gov). NASA also has an extensive listing of publications related to international cooperation in space (http://www.hq.nasa. gov/office/hqlibrary/ppm/ppm34.htm). The **Marshall Spaceflight Facility** in Huntsville, Alabama (http://www.nasa.gov/centers/marshall/home/index. html), has information on the U.S. space and rocketry programs. For the Soviet space program, see **Russian Space Web** (http://www.russianspaceweb. com/I), which has subsections on rockets, spacecraft, space centers, people, and chronology of the Soviet space program and its successors. **Sputnikbook. net** is tied to Paul Dickson's 2001 book, *Sputnik: The Shock of the Century.*

The site predictably focuses on the 1957 launch of the first man-made satellite, including photographs, voice recordings from the media and key players, and an excellent collection of links (http://www.sputnikbook.net/home.php).

LAWS/TREATIES

The United Nations has full texts of space law treaties and international agreements. For the treaties, see the **United Nations Office for Outer Space Affairs** (http://www.oosa.unvienna.org/). The **UN Space Law Bibliography,** a 230+ page PDF file (http://www.unoosa.org/pdf/reports/ac105/AC105_636. pdf) contains a wealth of sources on space law in theory and practice. The **Federation of American Scientists Intelligence Program** (http://fas.org/ irp/offdocs/) has full text of almost all U.S. presidential National Security Directives since the Truman administration, as well as congressional reports, debates, and laws related to intelligence gathering.

INFORMATION ON WEAPONS

Researchers may need help keeping straight all of the military hardware mentioned in studies of the militarization of space, especially as one rocket may have a name, a series designation, and possibly a name in a different language. The **Federation of American Scientists' Military Analysis Network** (http://www.fas.org/man/dod-101/sys/) has specifications for multiple weapons systems, for both the United States and the rest of the world. **Nuclear ABMs of the United States** (http://www.nuclearabms.info/) is organized chronologically, with not only weapons specifications but the context of their development, regulation, and deployment. Researchers should scrutinize any Internet site purporting to show weapons, to make sure if the system is deployed, in development, merely proposed, or actually from science fiction.

7

Facts and Figures

UNITED STATES MILITARIZATION OF SPACE
1.1 U.S. Space Funding, 1959–2008

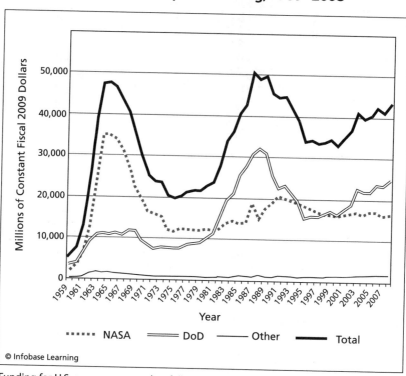

Funding for U.S. space programs has followed trends in presdential science and security policy. Dwight Eisenhower, John F. Kennedy, and Lyndon Johnson emphasized the scientific aspects of space exploration, while Ronald Reagan's Strategic Defense Initiative in 1983 triggered a shift toward an emphasis on defense applications.

Source: "2009 Space Almanac." *Air Force Magazine* (August 2009), p. 54.

1.2 U.S. Space Shuttle Missions

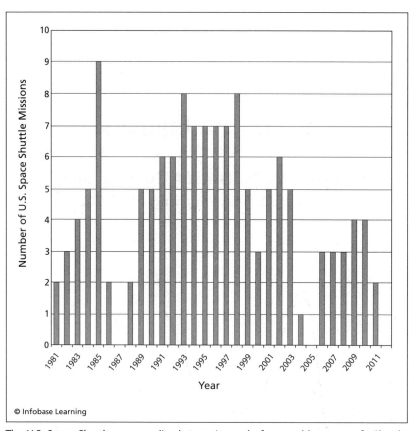

© Infobase Learning

The U.S. Space Shuttle program lived up to its goal of a reusable spacecraft. Shuttle missions became regular events, steadily averaging five missions per year. Routine did not mean space travel became less dangerous, as the loss of the *Challenger* (1986) and *Columbia* (2003) shuttles reminded a shocked public. Shuttle operations were halted while each disaster was fully investigated. (Data for 2010 and 2011 is based on NASA schedules.)

Source: NASA. "Shuttle Missions." Available online. URL: http://www.nasa.gov/mission_pages/shuttle/shuttlemissions/list_main.html. Accessed October 29, 2010.

1.3 U.S. Military Payloads by Mission, 1958–2008 (Orbital Only)

Applications	409
Communications	127
Weather	48
Navigation	100
Launch vehicle/spacecraft tests	6
Other military	128
Weapons-Related Activities	47
SDI tests	11
Antisatellite targets	2
Antisatellite interceptors	34
Reconnaissance	445
Photographic/radar imaging	256
Electronic intelligence	56
Ocean surveillance	48
Nuclear detection	12
Radar calibration	37
Early warning	36
Total	**901**

The U.S. military already makes extensive use of space-based assets, such as communications and imaging satellites, but to date no country has been discovered deploying weapons in outer space.

Source: "2009 Space Almanac." *Air Force Magazine* (August 2009), p. 58.

1.4 U.S. Satellites Placed in Orbit or Deep Space

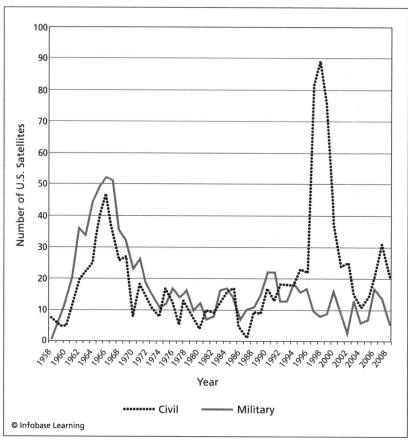

After the cold war ended in 1991, the civilian satellite market grew exponentially, and it now outpaces military applications. GPS, the Internet, and high definition television are among the popular applications driving the civilian market.

Source: "2009 Space Almanac." *Air Force Magazine* (August 2009), p. 59.

1.5 U.S. Military and Civil Launches

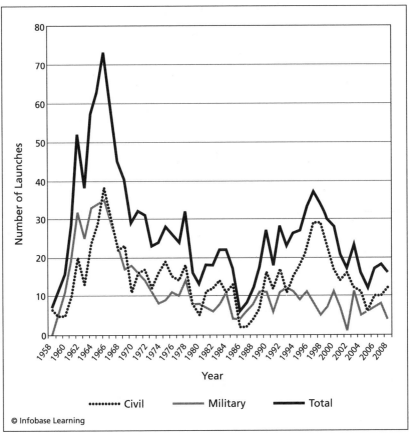

Nonmilitary launches have increased in the United States, as a growing number of private firms and international consortia enter the lucrative space market.

Source: "2009 Space Almanac." *Air Force Magazine* (August 2009), p. 60.

INTERNATIONAL MILITARIZATION OF SPACE
2.1 What's Up There?
(AS OF DECEMBER 31, 2008)

	Satellites	Space probes	Debris	Total
United States	1,003	61	2,590	3,654
Russia*	1,379	35	2,104	3,518
People's Republic of China	78	1	2,650	2,729
France	49	0	219	268
Japan	105	10	32	147
India	36	1	98	135
European Space Agency	38	6	30	74
Int'l. Telecom. Sat. Org.	65	0	0	65
CHBZ	3	0	59	62
Globalstar	60	0	0	60
Germany	33	2	0	35
Orbcomm	35	0	0	35
Canada	29	0	2	31
European Telecom. Sat. Org.	28	0	0	28
United Kingdom	26	0	0	26
Luxembourg	16	0	0	16
Italy	15	0	0	15
Saudi Arabia	12	0	0	12
Brazil	12	0	0	12
Int'l. Maritime Sat. Org.	12	0	0	12
Australia	11	0	0	11
Sweden	11	0	0	11
Argentina	10	0	0	10
Indonesia	10	0	0	10
South Korea	10	0	0	10
Arab Sat. Comm. Org.	9	0	0	9
Israel	9	0	0	9
Spain	9	0	0	9
NATO	8	0	0	8
Taiwan	8	0	0	8
Mexico	7	0	0	7
Thailand	7	0	0	7

	Satellites	Space probes	Debris	Total
Netherlands	6	0	0	6
Turkey	6	0	0	6
Czech Republic	5	0	0	5
International Space Station	1	3	2	6
Other**	47	0	3	50
Total	**3,208**	**119**	**7,789**	**11,116**

Notes:
* Russia includes Commonwealth of Independent States (CIS) and former Soviet Union.
** Other refers to countries or organizations that have placed fewer than five objects in space.

Outer space is a surprisingly crowded realm, cluttered with working and defunct satellites, space stations, probes exploring the universe, and the junk left behind by these activities. The United States and Russia were the first countries to launch satellites and lead the way in the amount of debris left behind in space.

Source: "2009 Space Almanac." *Air Force Magazine* (August 2009), p. 60.

2.2 States' First Dedicated Military Satellites and Their Function

Year	State/Actor	Satellite	Description
1958	United States	Project SCORE	Communications and experimental satellite
1962	USSR	Cosmos 4	Remote sensing (optical)
1969	United Kingdom	Skynet 1A	Communications
1970	NATO	NATO 1	Communications
1975	China	FSW-0 No. 1	Remote sensing (optical)
1988	Israel	Ofeq 1	Remote sensing (optical)
1995	France	Helios 1A	Remote sensing (optical)
1995	Chile	Fasat Alfa	Communications and remote sensing (optical)
1998	Thailand	TMSAT	Communications
2001	Italy	Sicral	Communications
2003	Australia	Optus and Defense 1	Communications
2003	Japan	ISG-1A and IGS-1B	Remote sensing (optical)
2006	Spain	Spainsat	Communications
2006	Germany	SARLupe-1	Remote sensing (radar)

Note: Other satellites may have some military usage.

The United States and Russia bore the cost of launching the earliest satellites, but improved technology is allowing a growing number of countries to develop and deploy their own dedicated military satellites. Governments can also use civilian satellites for military needs, such as communications and imagery.

Source: Space Security 2010. Waterloo, Ontario, Canada: Project Ploughshares, 2009, p. 139. Available online. URL: http://www.spacesecurity.org. Accessed October 22, 2009.

2.3 Space Surveillance Capabilities

Country	Optical Sensors	Radar Sensors	Orbital Sensors	Global Coverage	Centralized Tasking	Catalog	Public Data
Amateur observers	■		□	□	□	■	
Bolivia*	■						
Canada	■		(□)				
China	■	■					
European Union	■	■			(□)	(□)	
France	■	■					
Georgia*	■						
Germany		■					
Great Britain	■	■					
Japan	■	■					
India	■						
Norway		■					
Russia	■	■			■	□	
South Africa	■						
Spain*	■						
Switzer-land	■						
Tajikistan*	■						
Ukraine	■						
United States	■	■	(□)	□	■	■	□
Uzbeki-stan*	■						

Notes:
■ = Full capability.
□ = Some capability.

264

(□) = Under development.
*Part of the International Scientific Optical Network (ISON).

Space debris is tracked using a combination of transmitters and receivers operated by amateur astronomers and governments. No dedicated international surveillance system is in place, although policy makers have mooted the idea since the 1980s because such a coordinated system would provide greater information to avoid collisions with space debris.

Source: Space Security 2010. Waterloo, Ontario, Canada: Project Ploughshares, 2010, p. 52. Available online. URL: http://www.spacesecurity.org. Accessed October 22, 2009.

8

Key Players A to Z

NEIL A. ARMSTRONG (1930–) Trained as a naval aviator, Armstrong flew dozens of missions during the Korean War before becoming a test pilot for the X-15 and other projects. He transferred to the astronaut corps in 1962, joining the Gemini project. As commander of the Apollo 11 mission, Armstrong became the first human to walk on the moon in 1969.

ANATOLI A. BLAGONRAVOV (1895–1975) Soviet aerospace engineer. Blagonravov taught at the Artillery Academy in Moscow before becoming director of the Institute of Machine Studies of the Academy of Sciences. He supervised the early years of the Soviet space program, including *Sputnik-1*, *Sputnik-2* with Laika the dog, and Yuri Gagarin's pioneering flight in 1961. He later was one of the driving factors behind U.S.-USSR space cooperative efforts in the early 1970s, culminating in the Apollo-Soyuz Test Project in July 1975. He died five months before the Test Project took place.

RICHARD BRANSON (1950–) Founder of Virgin Galactic, the company that plans to offer suborbital spaceflights and, in the future, orbital spaceflights to the paying public using the *SpaceShipOne* spacecraft from Scaled Composites. Branson's other ventures include Virgin Music, Virgin Atlantic Airways, and a large group of affiliated commercial ventures and charitable organizations. Branson is a billionaire and was knighted in 1999 for his entrepreneurial and philanthropic endeavors.

WERNHER VON BRAUN (1912–1977) A charismatic engineer who led the Nazi and, later, U.S. rocket programs. Born in Germany, von Braun developed the V-2 rocket for the Nazis. As World War II drew to a close, U.S. and Soviet officials raced to capture the V-2 program. Moscow got the laboratory buildings, but Washington got von Braun, offering the German and his team asylum in the United States. Von Braun's team set up at the Redstone Arsenal in Huntsville, Alabama, in 1950. Von Braun had a talent for publicity but was also criticized as being arrogant and too independent. His main interest was not weapons but space travel, and he eventually joined NASA.

KALPANA CHAWLA (1962–2003) First Indian-born female astronaut. Chawla was born in India but moved to the United States in 1982 for graduate school. She earned two master's degrees and was awarded a doctorate in aerospace engineering in 1988. She married an American flight instructor in 1983 and eventually earned her own certified flight instructor certificate as well as a commercial pilot rating. After becoming a naturalized U.S. citizen in 1990, Chawla joined the U.S. astronaut corps in 1995. She flew aboard the *Columbia* space shuttle in November 1997 and again in January 2003. During the second mission she directed the crew in performing over 80 experiments on different aspects of microgravity. All seven members of the *Columbia*'s crew perished when their shuttle disintegrated upon reentry on February 1, 2003.

VLADIMIR CHELOMEI (1914–1984) Chelomei joined the USSR Ministry of Aviation in 1942 and spearheaded development of the Soviet jet engine program. Two years later, Great Britain sent a recovered Nazi V-1 rocket to Moscow and Chelomei was assigned to reverse-engineer the weapon. He eventually developed his own version of a cruise missile, but technical glitches made the USSR air force reluctant to embrace the new technology. But in 1955, the Soviet navy called upon Chelomei to adapt his missiles for use on submarines.

JAMES DOOLITTLE (1896–1993) Legendary pioneer of military aviation in the United States. "Jimmy" Doolittle enlisted in the Army Signal Corps in World War I and eventually rose to the rank of major general in the U.S. Air Force. Doolittle was awarded the U.S. Medal of Honor for leading the legendary "30-seconds over Tokyo" raid in 1942, which was the first bomber attack on mainland Japan. His theories of airpower established standard operating procedures for decades.

HUGH L. DRYDEN (1898–1965) Director of the National Advisory Committee for Aeronautics (NACA) and selected by President Dwight D. Eisenhower to be deputy director of the new National Aeronautics and Space Administration in 1958. Trained as a physicist, Dryden supervised the development of the X-15 rocket plane. He participated in the discussion of cooperative projects that led to the Apollo-Soyuz Test Project of 1975.

YURI GAGARIN (1934–1968) Soviet cosmonaut who became first human to orbit the Earth. On April 12, 1961, Gagarin's *Vostok 1* spacecraft made one orbit around the Earth in a flight lasting 89 minutes. Celebrated as a hero across the world, Gagarin was named to the Supreme Soviet legislature and became deputy director of training at the Star City cosmonaut facility. Gagarin died at age 38 when his MiG fighter jet crashed on a routine training flight.

TREVOR GARDNER (1915–1963) Civilian special assistant to Secretary of the Air Force Harold E. Talbott and responsible for U.S. ICBM program in the 1950s. Gardner successfully lobbied to adapt and accelerate the existing Atlas rocket project rather than starting from scratch.

JOHN GLENN (1921–   ) A former Marine Corps fighter pilot, Glenn was selected to be a member of the *Mercury 7* team and became the first American to orbit the Earth, making three trips around the planet on February 20, 1962, aboard *Friendship 7*. At 77, Glenn became the oldest American in space by participating in a *Discovery* space shuttle mission in 1998. After retiring from military service, Glenn was a U.S. senator from Ohio from 1974 to 1999.

T. KEITH GLENNAN (1905–1995) First administrator of the National Aeronautics and Space Administration, founded in 1958. Previously Glennan had been president of the Case Institute of Technology in Cleveland.

ROBERT HUTCHINGS GODDARD (1882–1945) Considered the father of modern rocketry, Goddard developed the first liquid-fueled rocket and eventually received more than 200 patents for rocket technology. Eccentric and secretive, Goddard preferred to work alone, but his work laid the foundation for all of the major U.S. rocket systems. In 1959, NASA established the Goddard Space Flight Center in his honor.

MICHAEL GRIFFIN (1949–   ) NASA administrator from 2005 to 2009. Griffin frequently clashed with the George W. Bush administration over NASA funding. Specifically, he accused the Bush White House of neglecting programs such as the space shuttle and *International Space Station* in favor of a military buildup. Prior to becoming NASA's top administrator, he worked at a variety of NASA facilities on projects including the *Hubble Space Telescope* and the Mars rover, as well as at several universities and commercial firms. He returned to NASA in 1993 as chief engineer. Griffin earned six graduate degrees, in fields such as aerospace engineering, electrical engineering, and applied physics.

HELMUT GRÖTTRUP (1916–1981) He was the only member of the Nazi V-2 rocket team who chose to go to the USSR rather than the United

States. Only a mid-level designer in Germany, in the Soviet Union Gröttrup supervised a team of nearly 5,000.

NIKOLAI KAMANIN (1908–1982) Served as head of Soviet cosmonaut training from 1960 to 1971. Prior to joining the space program, he had been named a Hero of the Soviet Union for aiding in the rescue of passengers from a ship trapped in the Arctic Ocean. He advocated training women and recruiting civilians.

THEODORE VON KARMAN (1881–1963) A Hungarian engineer who later became a U.S. citizen. Von Karman formulated the "Karman primary jurisdiction line," which is the lowest point at which an object can maintain its orbit. This line offers one possible answer to the question of where air ends and space begins.

KERIM KERIMOV (1917–2003) One of the leaders of the Soviet rocket and space program. When SERGEI KOROLEV died, Kerimov succeeded him as head of the Soviet space program, but his identity was unknown by the West for decades.

JAMES R. KILLIAN, JR. (1904–1988) President of the Massachusetts Institute of Technology, Killian became the first White House science adviser, being appointed one month after *Sputnik* was launched. He helped create NASA and called for a revision of national science and technology education to improve U.S. standing in the world scientific community.

YURI N. KOPTEV (1940–) Appointed by Russian president Boris Yeltsin in 1992 to organize a new space program for the Russian Federation in the wake of the USSR's demise. Koptev had spent years working in the Soviet program as a rocket builder and now had to address the bleak economic outlook for the Russian program.

SERGEI P. KOROLEV (1907–1966) Korolev led the Soviet space program from the end of World War II until his premature death two decades later. Despite being arrested for "sabotaging" the Soviet jet propulsion program, Korolev was released in 1944 and sent to Germany in 1945 to try and recover Nazi V-2 rocket technology. Following World War II Korolev worked in the Soviet rocketry program and successfully lobbied to create a manned spaceflight program.

KO SAN (1976–) Selected in 2007 to be South Korea's first astronaut. He was later accused of removing classified materials from the Russian training facility and fired. Trained as a mathematician, Ko subsequently left South Korea to study science and technology at Harvard University.

SERGEI K. KRIKALEV (1958–)   Soviet cosmonaut left stranded aboard *Mir* space station when USSR collapsed at the end of 1991. Krikalev spent an additional five months at *Mir* while Russia tried to find funds to send a spacecraft to relieve him. Krikalev logged multiple spaceflights, including the first joint U.S.-Russian space shuttle mission in 1994. In 1998, he again took part in an international mission, taking the space shuttle *Endeavour* for the first stage of assembling the *International Space Station.* Krikalev has spent over 800 days in space.

IGOR KURCHATOV (1903–1960)   Considered the "Father of the Soviet Atomic Bomb." During World War II, the USSR became interested in atomic weapons, and in 1941 Kurchatov was named to direct the project. He worked on both atomic weapons and atomic energy, directing the design of the USSR's first cyclotron in 1949 and the world's first nuclear power plant in 1954. Later in life, Kurchatov focused on peaceful uses of nuclear energy and opposed further testing of nuclear weapons.

LAIKA (1955?–1957)   She was a two-year-old dog launched into space. On November 3, 1957, the USSR launched a spacecraft carrying Laika to analyze how the launch, orbit, and weightlessness would affect a living organism. But because technology did not yet exist to allow a spacecraft to safely reenter the Earth's atmosphere, Laika was doomed. Although her space capsule, *Sputnik-2*, was outfitted with food and cooling fans, as well as an automatic lethal injection, she died of stress shortly after the launch.

SAMUEL P. LANGLEY (1834–1906)   In 1898, the U.S. Signal Corps commissioned an airplane from Langley. He delivered his invention in 1903, only to have it crash into the Potomac River.

ALEXEI LEONOV (1934–)   Soviet cosmonaut and first man to walk in space during the 1965 *Voskhod 2* mission. He also was part of the *Soyuz 19* crew that docked with *Apollo 18* in 1975. After the mission he was named head of the cosmonaut team.

ROBERT S. McNAMARA (1916–2009)   U.S. secretary of defense under Presidents John F. Kennedy and Lyndon B. Johnson. Applied systems management techniques to the armed forces and encouraged use of air power as a deterrent.

JOHN B. MEDARIS (1902–1990)   Army major-general Medaris was the first commander of the Army Ballistic Missile Agency, overseeing production of the Jupiter-C rocket, the Redstone ballistic missile, and the *Explorer* satellite. He also supervised the German rocket team led by WERNHER VON BRAUN.

271

ELON MUSK (1971–) Philanthropist entrepreneur, and innovator. Born in South Africa, Musk immigrated to his mother's native Canada in the 1980s. In 1999, he founded X.com to provide financial services to Internet users. The company eventually came to be known as PayPal. Elon founded SpaceX—Space Exploration Technologies—in 2002 with plans to explore Mars. Due to the problems he encountered trying to purchase rockets, he decided to build his own. In June 2010, SpaceX successfully launched its Falcon 9 rocket, a significant milestone in Musk's ambition to provide cargo and crew services to the *International Space Station.*

SIR ISAAC NEWTON (1643–1727) An English scientist, mathematician, and philosopher, Newton laid the theoretical groundwork that would ultimately lead to rocketry and artificial satellites. He codified the laws of gravity, formulated three laws of motion, and built an early optical telescope.

HERMANN OBERTH (1894–1989) Born in Transylvania, Oberth was a mathematician based in Germany who popularized rocketry and a mentor to WERNHER VON BRAUN. As a child Oberth was fascinated by Jules Verne's book, *From Earth to the Moon,* and spent his career trying to realize Verne's vision. Author of the 1923 work *By Rocket to Space,* Oberth wrote in accessible language that encouraged nonscientists to study rocketry. As president of the Society for Space Travel, his enthusiasm led to the creation of many rocketry clubs. Oberth was a special guest at the launch of *Apollo 11,* the craft that finally put a man on the moon.

VIRGILIU POP (1974–) Member of the Romanian Academy of Sciences, Pop is considered the leading international expert in space property rights. Pop is the author of *Who Owns the Moon? Extraterrestrial Aspects of Land and Mineral Resources Ownership.*

QIAN XUESEN (1911–2009) Qian pioneered the Chinese space program. He came to the United States in the 1930s to study engineering at California Institute of Technology. He worked on advanced rocket projects for the U.S. military and was one of the founders of the Jet Propulsion Laboratory at Caltech. He was part of the team sent to Germany late in World War II to recover Nazi rocket technology. Qian became swept up in the anticommunist hysteria of McCarthyism and was stripped of his security clearance and accused of Communist sympathies. Although no evidence was ever produced against him, he and his family lived for five years under house arrest while undergoing government surveillance. Finally, in 1955, he was permitted to return to China. Chinese communist leaders immediately put him to work, and he helped China launch its own ballistic missile program in 1956.

SALLY RIDE (1951–)   Holding a doctorate in physics, Dr. Ride joined the astronaut corps in 1978 and qualified for the new space shuttle program. On June 18, 1983, aboard the second flight of the space shuttle *Challenger*, Ride became the first U.S. female astronaut, flying nearly 20 years after Soviet cosmonaut Valentina Tereshkova. She participated in a second Challenger mission in 1984. Upon retirement from NASA, Dr. Ride established her own foundation, Sally Ride Science, to encourage girls to pursue careers in science, math, and technology.

BURT RUTAN (1943–)   Aerospace engineer and entrepreneur. Rutan's company, Scaled Composites, designed and built the first privately financed aircraft that could one day make space tourism available to ordinary citizens. His *SpaceShipOne* made its inaugural flight in June 2004, after being launched from *White Knight* airplane, which he had also designed. Virgin Galactic's passenger service will use Rutan's spacecraft. After working for the Air Force and Bede Aircraft Company, Rutan opened the Rutan Aircraft Factory in 1974. The company specialized in designs for home-built aircraft and other small planes. In 1986, his *Voyager-1* plane became the first aircraft to circle the Earth without needing to refuel. Rutan founded Scaled Composites in 1982 to develop aerospace prototypes.

VIKRAM AMBALAL SARABHAI (1919–1971)   Considered the "father" of India's space program. Trained as a physicist, Sarabhai spent time studying in the United Kingdom. Upon his return to India in 1947, he established the Physical Research Laboratory that focused on a variety of space sciences, including astronomy, astrophysics, and theoretical physics. Inspired by the Soviet Sputnik program 10 years later, Sarabhai lobbied the government to establish the Indian Space Research Organization, and he was named the ISRO's first director, a post he held until his death.

SVETLANA SAVITSKAYA (1948–)   Soviet cosmonaut who became the second woman in space and the first female to walk in space. Although cosmonaut VALENTINA TERESHKOVA became the first woman in space in 1963, another two decades passed before another female orbited Earth. An experienced parachutist and test pilot, Savitskaya became a cosmonaut in 1980. She participated in two separate space missions; in 1982 she traveled to the *Salyut-7* spaceship during the Soyuz T-7 mission. She returned to *Salyut-7* in 1984 and performed a space walk on July 25, 1984. Savitskaya was scheduled to lead an all-female crew that was to launch *Salyut-7* on March 1987 for International Women's Day, but that mission was later canceled. She retired from the cosmonaut corps in 1993.

BERNARD SCHRIEVER (1910–2005)   U.S. Air Force general considered the architect of the Air Force's ballistic missile and military space program. He pushed for a military manned spaceflight program separate from NASA's civilian program, especially the short-lived Blue Gemini program. Schriever was born in Germany, but his family immigrated to the United States in 1917. He joined the Army Air Corps in 1933, eventually rising to the rank of four-star general. Schriever believed rocketry could be a major asset to the air force arsenal and often clashed with more traditional officers who favored long-distance bombers. His arguments proved convincing, and in 1961 he was put in charge of the air force's new systems command division, the primary missile procurement office. In 1998 Falcon Air Force Base in Colorado was renamed in his honor.

LEONID I. SEDOV (1907–1999)   Sedov was a Soviet academician and chair of the prestigious Commission for Interplanetary Communications. The USSR Academy of Sciences created the committee to take part in the International Geophysical Year and its efforts to place an artificial satellite in Earth's orbit.

RAKESH SHARMA (1949–   )   India's first astronaut. Sharma spent his military career in the Indian Air Force, where he served as a test pilot and retired at the rank of wing commander. Sharma began training with Soviet cosmonauts in 1984 as part of an international program between the Indian Space Research Organization and the USSR Intercosmos organization and spent eight days in space aboard the *Salyut-7* space station. His mission involved taking a series of multispectral photographs over India to be used to build a hydropower station in the Himalayan mountains, as well as practicing a series of yoga asanas modified for zero gravity. Upon his return, Sharma was named a Hero of the Soviet Union and received India's top award for bravery, the Ashoka Chakra.

ALAN B. SHEPARD, JR. (1923–1998)   A naval test pilot, Shepard joined the first class of *Mercury* astronauts. In 1961, he became the first American in space, conducting a 15-minute suborbital flight aboard *Freedom 7*, making a key step in the U.S. manned space program. Grounded for a medical problem, Shepard headed NASA's Astronaut Office from 1963 to 1969. Cleared to fly again in 1969, he commanded the *Apollo 14* Moon mission in early 1971. He returned to the Astronaut Office in June 1971, remaining there until he retired as a rear admiral in 1974.

VALENTINA TERESHKOVA (1937–   )   Tereshkova was the first female Soviet cosmonaut, becoming the first woman in space on June 16, 1963, when she orbited the Earth 48 times over the next three days. Many observers believe she was more a passenger in a publicity stunt than a trained pilot, citing her background as a textile worker and amateur parachutist. As a result, Tereshkova is frequently omitted from Western cosmonaut rosters.

GERMAN TITOV (1935–2000) Soviet cosmonaut Titov became the second man in space, spending more than 24 hours in orbit in August 1961 aboard the *Vostok 2*. While in orbit, Titov reported his experiences with motion sickness and learning to sleep in space.

GRIGORI TOKATY (1909–2003) Tokaty (also spelled Tokady) taught at the City University of London's Department of Aeronautics and Space Technology from 1967 to 1975. Born in Russia as Grigori Tokaev, he trained as an aeronautical engineer, eventually becoming head of aeronautics at the prestigious Zhukovsky Academy from 1938 to 1941. He concentrated on developing high-speed flight and rocketry and was dispatched to Germany in 1945 to recover as much information as possible from the Nazi V-2 missile program. Instead, aware of the dark side of Stalinism and fearful of a new round of purges, Tokaty crossed into the British sector of Berlin in 1947 and defected. His inside knowledge of the Soviet missile program was invaluable to Western intelligence agencies.

FRIDRIKH ARTUROVICH TSANDER (1887–1933) Tsander was a student of Konstantin Tsiolkovsky's work. In 1920 Tsander met with Lenin to discuss the possible uses of rocketry in the new Soviet state. An ethnic Lithuanian, he was a graduate of the Riga Polytechnic Institute and was so devoted to space topics that he named his children Astra and Mercury.

KONSTANTIN TSIOLKOVSKY (1857–1935) Tsiolkovsky was the Russian scientist considered to be the father of the Soviet space program. His pioneering work laid the theoretical foundation for liquid-fueled rockets and multistage booster rockets. His 1895 book *Dream of Earth and Sky* was standard reading for 20th-century rocket scientists, and in 1898 he devised a formula for propelling an object into Earth's orbit despite the pull of gravity.

JAMES A. VAN ALLEN (1914–2006) Van Allen was a physicist, rocket scientist, and discoverer of the two Van Allen Radiation Belts that ring the Earth. Trained at the University of Iowa, Van Allen began his career at the Johns Hopkins University Applied Physics Laboratory outside Washington D.C., later returning to Iowa. He became part of the team analyzing Nazi V-2 rockets, led the early U.S. satellite program, and chaired the National Academy of Sciences Rocket and Satellite Research Panel.

YANG LIWEI (1965–) China's first human in space, Yang piloted the *Shenzhou 5* in 2003. He remained in space for 21 hours, completing 14 orbits before a hard landing in Inner Mongolia. Yang became a national hero and received the title Space Hero from President Jiang Zemin.

YI SO-YEON (1978–) Selected in 2007 to be South Korea's second astronaut and first female astronaut. When KO SAN was fired from the space program in 2008, Yi took his place aboard a Russian *Soyuz* flight to the *International Space Station,* becoming the first South Korean in space.

ZHAI ZHIGANG (1966–) Zhai was the first Chinese citizen to undertake a space walk, leaving the *Shenzhou 7* spacecraft on September 27, 2008. A fighter pilot, holding the rank of colonel in the People's Liberation Army, Zhai's official biography reports a suitably proletarian background. He was born in Heilongjiang Province near the border with Russia to a mother who sold sunflower seeds to pay for his education.

9

Organizations and Agencies

Air Force Space Command
URL: http://www.afspc.af.mil
150 Vandenberg Street, suite 1105
Peterson AFB, CO 80914-4500
Phone: (719) 554-3731

Created in 1982, the Air Force Space Command is a major command head-quartered at Peterson Air Force Base, Colorado. The 40,000 people working under AFSPC defend North America with its space and intercontinental ballistic missile operations. Reports, fact sheets, biographies, photos, and more are available at the AFSPC Web site.

Arianespace
URL: http://www.arianespace.com
Boulevard de l'Europe
BP 177 91006 Evry-Courcouronnes CEDEX
France
Phone: (011-33-1) 6087-6000

Founded in 1980, Arianespace provides commercial launch services for satellites and other payloads, and for both government and commercial clients. The company's shareholders are located in 10 European countries. The largest shareholders are the French CNES agency (34%) and EADS Astrium (30%). Arianespace launches its *Ariane 5* and *Vega* rockets at its spaceport in Kourou, French Guiana. In 2003, Russia contracted with Arianespace to build a facility in Guiana that would accommodate its *Soyuz 2* launcher, which is capable of launching manned spacecraft.

Bigelow Aerospace
URL: http://www.bigelowaerospace.com
4640 S. Eastern Avenue
Las Vegas, NV 89119
Email: contact_BA@bigelowaerospace.com
Phone: (702) 688-6600

Bigelow Aerospace is developing next-generation crewed space complexes to revolutionize space commerce and open up the final frontier to all of humanity. Currently, Bigelow is operating unmanned prototype spacecraft, *Genesis I* and *Genesis II*. Founded by Las Vegas hotel entrepreneur Robert Bigelow, the company hopes to have space tourism opportunities ready by 2012 at an estimated price of $15 million per ticket.

Canadian Space Agency
URL: http://www.space.gc.ca
John H. Chapman Space Centre
6767 Route de l'Aéroport
Saint-Hubert, Quebec J3Y 8Y9
Canada
Phone: (450) 926-4800

CSA was established by an act of parliament in March 1989 in order to meet the needs of Canadians for scientific knowledge, space technology, and information. The agency focuses on scientific exploration, concentrating in four areas: Earth observation, space science and exploration, satellite communications, and space awareness and learning. In 1962, NASA launched the Alouette satellite, making Canada the third country to have a satellite in space. Today, it has Earth-observation satellites, as well as communications and scientific satellites.

Center for Defense Information (CDI)
URL: http://www.cdi.org
1779 Massachusetts Avenue NW
Washington, DC 20036-2109
Phone: (202) 332-0600
Fax: (202) 462-4559

Founded in 1972, the Center for Defense Information provides expert, nonpartisan analysis on various components of U.S. national security, international security, and defense policy. CDI promotes wide-ranging discussion and debate on security issues such as nuclear weapons, space security, missile defense, and military transformation.

China National Space Administration
URL: http://www.cnsa.gov.cn
A8, Fucheng Road
Haidian District
Beijing, China

In 1993, Beijing split China's Ministry of Aerospace Industry into two units: CNSA and the China Aerospace Science and Technology Corporation. CNSA serves as Beijing's agent for international cooperation and joint ventures. CNSA offers commercial launch services, while its military counterpart, China Aerospace Science and Technology Corporation, is responsible for research and development. CNSA itself has four divisions: the departments of general planning; system engineering; science, technology and quality control; and foreign affairs.

European Aeronautic Defense and Space Company (EADS)
URL: http://www.eads.eu
N.V. Le Carré
Beechavenue 130-132
1119 PR Schiphol Rijk
The Netherlands

EADS develops innovative technology to foster economic growth, security, and environmental protection. The company manufactures the Airbus fleet, the Eurocopter, and Eurofighter Typhoon. The EADS Astrium division is a commercial vendor of satellites, launchers, and other space services. In 2010, the company bid for the contract to develop a new U.S. Air Force tanker airplane.

European Space Agency
URL: http://www.esa.int
8–10, rue Mario Nikis
75738 Paris Cedex 15
France
Phone: (011-33-1) 5369-7654

Washington, D.C. Office
955 L'Enfant Plaza SW
Suite 7800
Washington, DC 20024
Phone: (202) 488-4158

The European Space Agency (ESA) claims to be Europe's gateway to space. Its many activities are intended to influence the development of Europe's space capability and to ensure that investment in space continues to deliver benefits to the citizens of Europe and the world. As an intergovernmental organization of 18 countries, ESA is not a component of the European Union but rather a frequent partner. The company has facilities across Europe and a launching facility in Kourou, Guiana.

European Space Policy Institute
URL: http://www.espi.or.at
Schwarzenbergplatz 6
A-1030 Vienna, Austria
Phone: (011-43-1) 718 11 18 35

The European Space Policy Institute was established to provide decision-makers with an independent view and analysis on mid- to long-term issues relevant to the use of space. It maintains a network of experts that work with similar centers across the globe to develop books, articles, conferences, and more to further its mission. ESPI was created in 2003 by the Council of the European Space Agency.

Excalibur Almaz
URL: http://www.excaliburalmaz.com
15-19 Athol Street
Douglas
Isle of Man IM1 1LB
British Isles

The USSR developed and deployed the *Almaz* space station alongside the civilian *Salyut* stations. Today, Excalibur Almaz offers its own launch services to the *International Space Station*, using time-tested hardware. Thanks to its existing spacecraft, Excalibur Almaz will pave the way for reliable, affordable, and routine access to space for exploration, experimentation, and enjoyment by customers around the world.

Federal Aviation Administration (FAA)
URL: http://www.faa.gov/about/office_org/headquarters_offices/ast/
Office of Commercial Space Transportation
800 Independence Avenue SW
Washington, DC 20591
Phone: (866) TELL-FAA, (866) 835-5322

The Federal Aviation Administration is tasked with ensuring the protection of the public, property, and the national security and foreign policy interests of the United States during commercial launch or reentry activities, and to encourage, facilitate, and promote U.S. commercial space transportation.

Federation of American Scientists
URL: http://www.fas.org
1725 DeSales Street, NW, 6th Floor
Washington, DC 20036
Phone: (202) 546-3300

The Federation of American Scientists (FAS) was founded in 1945 by scientists who had worked on the Manhattan Project to develop the first atomic bombs. They believed that scientists had a unique responsibility to both warn the public and policy leaders of potential dangers from scientific and technical advances and to show how good policy could increase the benefits of new scientific knowledge. FAS publishes backgrounders and policy papers on a variety of subjects, including space security.

Global Network Against Weapons & Nuclear Power in Space
URL: http://www.space4peace.org
P.O. Box 652
Brunswick, ME 04011

The Global Network Against Weapons and Nuclear Power in Space is urgently trying to convince leaders of the United States and other countries to not take the arms race into outer space. The grassroots movement has established a fun, collaborative strategy to keep outer space a peaceful sanctuary.

Group of Earth Observations
URL: http://www.earthobservations.org/contact.shtml
GEO Secretariat
7 bis, avenue de la Paix
Case postale 2300
CH-1211 Geneva 2
Switzerland
Phone: (011-41-22) 730-8505

GEO is a nongovernmental organization coordinating efforts to build a Global Earth Observation System of Systems (GEOSS). The idea grew out of the

2002 World Summit on Sustainable Development and various Group of Eight conferences. Participants in this meeting realized that international cooperation on satellite imagery would facilitate and improve the decision-making process on critical issues.

Henry L. Stimson Center
URL: http://www.stimson.org
1111 19th Street NW, 12th Floor
Washington, DC 20036
Phone: (202) 233-5956

Founded in 1989, the Henry L. Stimson Center is a nonprofit, nonpartisan institution devoted to enhancing international peace and security through a unique combination of rigorous analysis and outreach. The Stimson Center's work is focused on three priorities that are essential to global security: strengthening institutions for international peace and security, building regional security, and reducing weapons of mass destruction and transnational threats. The agency hosts regular seminars on security issues, full-text and often video versions of the presentations are available on their Web site. A sizable library contains reports from related research projects and background briefs, many of which are downloadable.

Indian Space Research Organization (ISRO)
URL: http://www.isro.org
Antariksh Bhavan, New BEL Road
Bangalore-560 231
India
Phone: (011-91-80) 23 41 52 75 or (011-91-80) 22 17 22 96

In 1969, the government of India established the Indian Space Research Organization to develop space technology and applications to benefit the country and its people. Since then, ISRO has established a network of communication, television broadcasting, and meteorological services (INSAT) and a remote sensing satellite program (IRS). The agency is now directing the establishment of a manned spaceflight program.

Institute for Cooperation in Space
URL: http://www.peaceinspace.com
3339 West 41 Avenue
Vancouver, B.C. V6N 3E5
Canada
Phone: (604) 733-8134

ICIS is a nonprofit organization seeking to change the way people think about space. Instead of focusing on the space-based weapons industry, ICIS envisions a world where countries cooperate to explore space but not to weaponize space. It seeks to establish a verifiable ban on space-based weapons in the United States and the world and enshrine it in a World Space Preservation Treaty.

Inter-Agency Space Debris Coordination Committee
URL: http://www.iadc-online.org

The Inter-Agency Space Debris Coordination Committee consists of 11 national space agencies: Italy, UK, France, China, Germany, European Space Agency, India, Japan, United States, Ukraine, and Russia. Since 1993 the agency has coordinated debris-related research projects by the members, reported on activities, and identified debris-mitigation options. IADC does not have a permanent office, instead its meetings rotate among its member agencies.

International Institute of Space Law
URL: http://www.iislweb.org
94bis, Avenue de Suffren
75015 Paris, France

Founded in 1960, the International Institute of Space Law (IISL) seeks to secure cooperation with appropriate international organizations and national institutions in the field of space law and conducts activities to foster the development of space law.

International Launch Services
URL: http://www.ilslaunch.com
1875 Explorer Street
Suite 700
Reston, VA 20190
Phone: (571) 633-7400
Fax: (571) 633-7500

International Launch Services is the U.S.-based company with exclusive rights for worldwide commercial sales and mission management of satellite launches on Russia's premier vehicle, the *Proton*.

International Telecommunications Satellite Consortium (INTELSAT)
URL: http://www.intelsat.com/
Intelsat Global S.A.

23, Avenue Monterey
L-2086 Luxembourg
Grand-Duchy of Luxembourg
Phone: (011-35-2) 24 87 99 20
Washington, DC, Office
3400 International Drive, NW
Washington, DC 20008
Phone: (202) 944-6800

In 1961, the United Nations General Assembly passed Resolution 1721, calling for unfettered access to global communications. One year later, U.S. president John F. Kennedy signed the Communications Satellite Act as the first step in creating an international organization. Other countries followed suit, and in 1964 the International Telecommunications Satellite Consortium (INTELSAT) was born. INTELSAT launched its first satellite in 1965, creating the first commercial global satellite communications agency. INTELSAT broadcast the Moon landing in 1969. It remains the largest commercial provider of fixed satellite communications services.

International Telecommunications Union
URL: http://www.itu.int
Place des Nations
1211 Geneva 20
Switzerland
Phone: (011-41-22) 730 5111

Founded in 1865, the International Telecommunication Union is the leading UN agency for information and communication technology issues. Today ITU works to standardize communication protocols and to bridge the digital divide so that all people can participate in and benefit from the emerging information society and global economy.

James Martin Center For Nonproliferation Studies
URL: http://cns.miis.edu
460 Pierce Street
Monterey, CA 93940
Phone: (831) 647-4154

CNS combats the spread of weapons of mass destruction (WMD) by training the next generation of nonproliferation specialists and disseminating timely information and analysis. Affiliated with the Monterey Institute of

International Studies, CNS is home to a team of researchers and visiting scholars who produce a variety of studies on nonproliferation issues.

Japan Aerospace Exploration Agency (JAXA)
URL: http://www.jaxa.jp/index_e.html
7-44-1 Jindaiji Higashi-machi
Chofu-shi
Tokyo 182-8522
Japan
Phone: (011-81-3) 6266-6400

The Japan Aerospace Exploration Agency (JAXA) is the official government agency for satellite and space industries. Its projects range from space transportation to human space activities, satellites, and space science research. While Japan has decades of experience with space-related industries, JAXA only dates to October 2003, when the Institute of Space and Aeronautical Science (ISAS), the National Aerospace Laboratory of Japan (NAL), and the National Space Development Agency of Japan (NASDA) were combined into one organization.

Korea Aerospace Research Institute (KARI)
URL: http://new.kari.re.kr/english/
115 Gwahangno (45 Eoeun-dong)
Yuseong Daejeon
Korea (305+333)
Phone: (011-82-42) 860-2164

Founded in 1989, KARI is South Korea's space agency. It focuses on aerospace research and development, and it has guided the country into the space age. KARI began to develop South Korea's first rockets and satellites, including the Korea Sounding Rocket and the Korean Space Launch Vehicle programs. It is charged with developing LEO and GEO satellites, reliable space launch vehicles, and launching a lunar satellite. It also supervises the national aviation program. It oversees the South Korean manned space program.

NASA Ames Research Center
URL: http://www.nasa.gov/centers/ames/about/overview.html
Moffett Field, California

Founded in 1939, the Ames Research Center in California began as an aircraft research lab under the National Advisory Committee for Aeronautics. It became part of NASA upon the agency's creation in 1958. Ames continues to research aircraft issues and to work with the Silicon Valley industry on information technology applications. In particular, Ames focuses on issues of how gravity affects living things and how stars, planets, and life are distributed across the universe.

NASA Dryden Flight Research Center
URL: http://www.nasa.gov/center/dryden/
home.index.html

Dryden's mission is "advancing technology and science through flight." It conducts flight research programs to revolutionize aviation and aerospace technology. Dryden also supports space shuttle and *International Space Station* operations.

NASA Glenn Research Center
URL: http://www.nasa.gov/center/glenn/home/index.html

Named after pioneering U.S. astronaut John Glenn, the Glenn Research Center develops aeropropulsion and communications technologies. The center works in conjunction with U.S. industry and higher education on projects involving spaceflight and aeronautics. It was designated to support the Constellation, Orion, and Ares programs.

NASA Goddard Space Flight Center
URL: http://www.nasa.gov/centers/goddard/
visitor/home/index.html
Greenbelt Road
Greenbelt, MD 20771-0001
Phone: (301) 286-8981

The Goddard Space Flight Center is named for Dr. Robert Hutchings Goddard, who is considered the father of modern rocket propulsion. The facility focuses on astronomy, space observation, and space physics. Unlike other centers that focus on research, Goddard is open for public tours and has interactive exhibits suitable for adults and children.

NASA Jet Propulsion Laboratory
URL: http://www.jpl.nasa.gov
California Institute of Technology
4800 Oak Grove Drive
Pasadena, California 91109
Phone: (818) 354-4321

The Jet Propulsion Laboratory is older than NASA and has been involved in the U.S. satellite program since its inception, helping build *Explorer 1*. JPL manages the Deep Space Network of antennas, which are deployed across the Earth and facilitate communications between spacecraft and land-based mission staff. Recent projects include the Kepler spaceborne telescope and robotic Moon explorers.

NASA Johnson Space Center
URL: http://www.nasa.gov/centers/johnson/home/index.html
Houston, TX 77058
Phone: (281) 244-2105

Johnson Space Center is the training and operation facility for all U.S. astronauts and operates as mission control center for missions in progress. The space shuttle program was based at Johnson, and JSC is the lead international agency for operating the *International Space Station.*

NASA Kennedy Space Center
URL: http://www.nasa.gov/centers/kennedy/about/visit/index.html
Spaceport U.S.A.
Kennedy Space Center, FL 32899-0001
Phone: (321) 452-2121

While major space missions are planned and rehearsed at the Johnson Space Center in Houston, Texas, almost all launch from the Kennedy Space Center in Florida. Consequently, Kennedy is home to rocket and launch vehicle development and operation. The center describes itself as "our nation's gateway to exploring, discovering and understanding our universe."

NASA Marshall Space Flight Center
URL: http://www.spacecamp.com/museum/
U.S. Space & Rocket Center
One Tranquility Base
Huntsville, AL 35805
Phone: (256) 837-3400

The Marshall Space Flight Center is home to NASA's famed space camps. The museum boasts incredible artifacts from the U.S. space program, hands-on interactive exhibits and space travel simulators. The Saturn V Restoration Project is based at Marshall and is not only restoring a Saturn V, but also collecting related artifacts and compiling a database of people who worked on the Saturn and Apollo projects.

National Air and Space Museum
URL: http://www.nasm.si.edu
Independence Avenue at 6th Street, SW
Washington, DC 20560
Phone: (202) 663-1000

Steven F. Udvar Hazy Center
URL: http://www.nasm.si.edu/museum/udvarhazy
14390 Air & Space Museum Parkway
Chantilly, VA 20151
Phone: (202) 663-1000

The National Air and Space Museum is a division of the Smithsonian Institution and the most popular museum in the world. Its collections of exhibits range from the Wright flyer to the *Spirit of St. Louis* and moon rocks. The museum has an extensive program of research, restoration, and education spread among its Aeronautics, Archives, Center for Earth and Planetary Studies, and Space History Divisions. Much of this information is available online. The Udvar Hazy Center opened in 2003 and is located outside of Washington, D.C. With no space to expand on the National Mall, the Smithsonian built a large, auxiliary space to exhibit more of its collections.

National Aviation and Space Agency
URL: http://www.nasa.gov
Public Communications Office
NASA Headquarters
300 E Street SW
Suite 5K39
Washington, DC 20546-0001
Phone: (202) 358-0001

Founded in 1958, the National Aviation and Space Agency directs the U.S. civilian space program and works in cooperation with national security agencies on military space matters. It has a variety of subsidiaries and resource centers.

National Reconnaissance Office
URL: http://www.nro.gov
14675 Lee Road
Chantilly, VA 20151-1715
Phone: (703) 808-1198

The National Reconnaissance Office designs, builds, and operates the nation's reconnaissance satellites. Agencies such as the Central Intelligence Agency and the Department of Defense use NRO information to identify potential trouble spots around the world, plan military operations, and monitor the environment. The NRO Web site contains a section devoted to the CORONA cold war reconnaissance operations, including many previously classified photos.

Nuclear Threat Initiative (NTI)
URL: http://www.ndi.org
1747 Pennsylvania Avenue NW, 7th Floor
Washington, DC 20006
Phone: (202) 296-4810

The Nuclear Threat Initiative (NTI) is a nonprofit organization with a mission to strengthen global security by reducing the risk of use and preventing the spread of nuclear, biological and chemical weapons, and to work to build the trust, transparency and security which are preconditions to the ultimate fulfillment of the Non-Proliferation Treaty's goals and ambitions. NTI is cochaired by former Senator Sam Nunn and Ted Turner, founder of Cable News Network. NTI works by creating leverage to change prevailing ideas about disarmament. NTI's documentary *Last Best Chance* is a 45-minute film intended to raise awareness of the global nuclear terrorism threat linked to unsecured nuclear weapons and materials around the world.

Pakistan Space and Upper Atmosphere Research
Commission (SUPARCO)
URL: http://www.suparco.gov.pk
SUPARCO Hqs
Sector 28
Gulzar-e-Hijri
Off University Road
Karachi 75270
P. O. Box 8402
Pakistan

Pakistan's Space and Upper Atmosphere Research Commission was established in 1961. The agency develops and leads projects in space science and space technology, focusing on peaceful applications to benefit the entire country. It works toward developing indigenous capabilities in space technology and promoting space applications for socio-economic uplift of the country. Its satellite programs focus on remote sensing, meteorology, environmental observation, and various scientific applications. SUPARCO oversees Pakistan's satellite manufacturing industry.

Russian Federal Space Agency (ROSKOSMOS)
URL: http://www.roscosmos.ru/index.asp?Lang=ENG
Schepkina st. 42
Moscow, Russia 107996

The Russian Federal Space Agency is responsible for communications, intelligence, and remote-sensing satellites, navigation, including development of the GLONASS system. It also supports television broadcasting and fixed communications services for the entire country. Roskosmos also supervises the Russian satellite launch Baikonur space port in Kazakhstan and manned spaceflight programs including the Star City training facility. Roskosmos is a descendant of the Soviet space program and was established in 1992 following the collapse of the USSR.

Scaled Composites
URL: http://www.scaled.com
Hangar 78 Airport
1624 Flight Line
Mojave, CA 93501
Phone: (661) 824-4541

Founded in 1982 by noted aviation engineer Burt Rutan, Scaled Composites focuses on air vehicle design, tooling and manufacturing, specialty composite structure design, analysis and fabrication, and developmental flight tests of air and space vehicles. The company's SpaceShipOne project won the $10 million Ansari X prize in 2004 as the world's first privately developed manned spacecraft.

School of Advanced Air and Space Studies
URL: http://www.maxwell.af.mil
URL: http://space.au.af.mil
600 Chennault Circle
Maxwell AFB AL 36112-6424
Phone: (334) 953-2014

The School of Advanced Air and Space Studies is the primary advanced training facility for members of the U.S. Air Force. Established in 1946, the Air University provides the full spectrum of Air Force education, from pre-commissioning to the highest levels of professional military education, including degree granting and professional continuing education. AU programs educate Airmen on the capabilities of air and space power and its role in national security. AU's publishing division produces a wide variety of resources, and many of their reports and publications are available online or at the school's library.

Sea Launch
URL: http://www.sea-launch.com/
2700 Nimitz Road
Long Beach, CA 90802
Phone: (562) 499-4729

Sea Launch was established in 1995 as a joint venture among U.S. (Boeing), Russian (Energia), Norwegian (Aker), and Ukrainian (Yuzhnoye) companies. The company differed from other commercial providers in that it used an offshore platform as its launch and control facility. Sea Launch launched its first rocket in 1999 and 30 more followed. But the company was not financially successful and filed for bankruptcy in 2009. It is looking for new investment partners and hopes to come out of bankruptcy in 2011 or 2012.

Secure World Foundation
URL: http://www.secureworldfoundation.org
314 W. Charles St.
Superior, CO 80027
Phone: (303) 554-1560
1779 Massachusetts Avenue, NW
Washington, DC 20036
Phone: (202) 462-1842

The Secure World Foundation is a private foundation working to keep space a sustainable environment open to all of the peoples of Earth. It works with a variety of academics, policy makers, scientists and more to explain the importance of efficient and effective global systems of governance. The foundation compiles a number of briefing materials about the status of space use around the world. This information, as well as links to publications by other organizations, is available on the SWF Web site.

Space Adventures
URL: http://www.spaceadventures.com
8000 Towers Crescent Drive
Suite 1000
Vienna, VA 22182
Phone: (888) 85-SPACE or (703) 524-7172

Founded in 1998, Space Adventures is the only private company currently providing opportunities for actual private spaceflight and space tourism today. Using proven equipment and working side-by-side with professional astronauts and cosmonauts, Space Adventures has sent five clients to the International Space Station. Space Adventures is advertising the first private trip to the moon, with a $100 million price tag.

Space Policy Institute
URL: http://www.gwu.edu/~spi/
George Washington University
1957 E Street, NW, Suite 403
Washington, DC 20052
Phone: (202) 994-7292

Part of the George Washington University Elliott School of International Affairs, the Space Policy Institute focuses its activities on policy issues related to the space efforts of the United States and cooperative and competitive interactions in space between the United States and other countries. SPI takes advantage of its convenient location in the middle of Washington, D.C., to bring together scholars, policy analysts, practitioners, and students. SPI faculty and visiting scholars produce a variety of research reports, articles, and books, many of which are available on their Web site.

SpaceX (Space Exploration Technologies Corp.)
URL: http://www.spacex.com
1 Rocket Road
Hawthorne, CA 90250
Phone: (310) 363-6000

Established in 2002 by PayPal founder Elon Musk, SpaceX has developed its Falcon family of rockets and the Dragon crew and cargo capsule. The company has a lengthy launch manifest, and it has received seed money from NASA to provide cargo delivery and return services to the *International Space Station.*

Starsem
URL: http://www.starsem.com
2, rue François Truffaut
91042 Evry Cedex
France
Phone: (011-33-1) 69 87 01 10

Created in 1996 as a joint venture between Arianespace and Russia to offer commercial launch services and the Soyuz spacecraft for a broad range of mission needs, including satellite telecommunications systems, scientific spacecraft, and Earth observation/meteorological platforms. Starsem declares that the Soyuz is "the world's most versatile launch vehicle." Starsem offers a comprehensive program, starting from the manufacture of the launch vehicle to mission preparations at the Baikonur Cosmodrome and successful in-orbit delivery of payloads.

United Nations Conference on Disarmament
URL: http://www.unog.ch/
Palais des Nations
1211 Geneva 10, Switzerland
Phone: (011-41-22) 917 1234

In 1920, the League of Nations held its first assembly at the Palais Wilson in Geneva, Switzerland. When the United Nations superseded the League of Nations, officials turned over the facility, now known as the Palais of Nations, to the United Nations. The UN uses the facility to host diplomatic conferences and an operational base for a great number of activities in the economic and social fields, including an ongoing dialogue on disarmament. The Committee on the Prevention of an Arms Race in Outer Space (PAROS), as a subgroup of the conference, is also based in Geneva.

United Nations Office for Outer Space Affairs
URL: http://www.oosa.unvienna.org
United Nations Office at Vienna
Vienna International Centre,
Wagramerstrasse 5,
A-1220 Vienna
AUSTRIA
Phone: (011-43-1) 260 60 4950

The United Nations Office for Outer Space Affairs is the United Nations office responsible for promoting international cooperation in the peaceful uses of outer space. UNOOSA serves as the secretariat for the United Nations Committee on the Peaceful Uses of Outer Space (COPUOS).

U.S. Air Force National Air and Space Intelligence Center
URL: http://www.afisr.af.mil
Lackland Air Force Base
San Antonio, TX 78236
Phone: (210) 671-1110

The National Air and Space Intelligence Center provides the training, intelligence, and equipment needed for the Air Force to conduct intelligence, surveillance, and reconnaissance activities. NASIC's Web site includes a library containing relevant documents, fact sheets, and materials declassified through the Freedom of Information Act.

U.S. Department of Defense
URL: http://www.defense.gov
1400 Defense Pentagon
Washington, DC 20301-1400
Phone: (703) 571-3343

The Department of Defense coordinates and supervises all branches of the U.S. armed forces. The Pentagon Web site is a portal that leads to individual service branches, press releases, battle information, multimedia resources, and other publications.

U.S. Missile Defense Agency
URL: http://www.mda.mil
7100 Defense Pentagon
Washington, DC 20301-7100

A division of the Department of Defense, the Missile Defense Agency is a research, development, and acquisition agency to develop, test, and deploy a comprehensive missile defense system. MDA's Web site contains links to resources explaining the principles of missile defense including high quality graphics of proposed installations and technologies.

Virgin Galactic
URL: http://www.virgingalactic.com (company)
URL: http://www.virgingalactic.org (blog)

In 2004, the British entrepreneur Sir Richard Branson established Virgin Galactic to offer suborbital spaceflights and, in the future, orbital spaceflights to the paying public. The company will fly civilian passengers to an altitude slightly over 100 kilometers (62 mi) and allow them to experience weightlessness for up to six minutes. Tickets are being sold for $200,000. The Virgin Galactic service will use the *WhiteKnightTwo* spacecraft from Scaled Composites.

Xcor Aerospace
URL: http://www.xcor.com
PO Box 1163
Mojave, CA 93502
Phone: (661) 824-4714

XCOR Aerospace is a small, privately held corporation founded in 1999. The company is focused on the research, development, and production of safe, reliable, reusable launch vehicles (RLVs), rocket engines, and rocket propulsion systems. The company has produced two different spacecraft, the EZ-Rocket, completed in 2001, and the X-Racer. Between the two vehicles, XCOR has safely flown a piloted rocket operations demonstrator aircraft 66 times.

10

Annotated Bibliography

The annotated bibliography on the militarization of space is divided into the following broad subject areas:

General Works on the Militarization of Space
General Works on Spacecraft and Space Stations
General Works on Satellites
Antiballistic Missiles and Missile Defense Systems
Antisatellite (ASAT) Weapons
Debris and Environmental Management
International Space Law
Commercial Use of Outer Space
Case Studies

United States
Union of Soviet Socialist Republics/Russian Federation
China
Europe
Emerging Players

Each of these areas is subdivided into three subsections: Books and Book Chapters; Articles; and Web Documents.

GENERAL WORKS ON THE MILITARIZATION OF SPACE

Books and Book Chapters

Arbatov, Alexei G., and Vladimir Dvorkin. *Outer Space, Weapons, Diplomacy, Security.* Moscow: Carnegie Moscow Center, 2009. Russian scholars address the fundamental question of space security. "Will space become an arena for armed conflict, or will it remain a sphere of international cooperation?"

Brachet, Gerard. "Collective Security in Space: A Key Factor for Sustainable Long-Term Use of Space." In *Collective Security in Space: European Perspectives,* edited by John M. Logsdon, James Clay Moltz, and Emma S. Hinds. Washington, D.C.: George Washington University Space Policy Institute, 2007, pp. 1–16. The UN Committee on Peaceful Uses of Outer Space is the ideal forum for countries to exchange information about debris and other threats that could interfere with free access to space and its resources.

Bulkeley, Rip, and Graham Spinardi. *Space Weapons: Deterrence or Delusion?* New York: Polity Press, 1986. While SDI was introduced in 1983, space-weapons technology actually dates to at least the 1920s. The authors fit SDI into this historical time line and consider the technical, political, and strategic ramifications of the new program.

Cadbury, Deborah. *Space Race: The Epic Battle between America and the Soviet Union for Dominion of Space.* New York: Harper Collins, 2006. The U.S.-USSR space race is traced through the perspective of two leading figures: Wernher von Braun and Sergei Korolev.

Dolman, Everett C. *Astropolitik: Classical Geopolitics in the Space Age.* New York: Frank Cass, 2002. Following in the international relations tradition of realpolitik, Dolman argues that whichever state controls space will have not only military but also economic advantages. To find the best system for regulating space, Dolman evaluates earlier conventions, including use of the seas, air, and Antarctica. He believes that only a democratic, wealthy state has a genuine shot at controlling space, due to the high costs associated with developing appropriate technology and long-term commitment.

————. "Space Power and U.S. Hegemony: Maintaining a Liberal World Order in the 21st Century." In *Space Weapons: Are They Needed?,* edited by John M. Logsdon and Gordon Adams. Washington, D.C.: Space Policy Institute, George Washington University, 2003, pp. 51–114. Dolman suggests that the United States is in fact the "morally superior" choice in terms of which country should seize and control space, because its political values and position as a hegemonic power will allow it to ensure that space is used in ways that are "good for all humankind."

Garthoff, Raymond L. *Detente and Confrontation: American-Soviet Relations from Nixon to Reagan.* Washington, D.C.: Brookings Institution, 1994. Long-time scholar of superpower relations updates his classic 1985 work using newly declassified Russian and U.S. materials.

Handberg, Roger. *Seeking New World Vistas: The Militarization of Space*. Westport, Conn.: Greenwood, 2000. Space has becoming an increasingly important factor in military policy from *Sputnik* to the Gulf War. As more decision makers realize the value of space-based assets, competition for these assets and control of outer space could trigger a new space race.

Hitchens, Theresa. *Developments in Military Space: Movement Toward Space Weapons?* Washington, D.C.: Center for Defense Information, 2004. Surveys latest military applications of space assets and discusses the probability and impact of their deployment.

_____. *Future Security in Space: Charting a Cooperative Course*. Washington, D.C.: Center for Defense Information, 2004. There are three challenges involved in space security: degradation of the environment, misunderstanding due to the lack of transparency, and lack of accepted rules of behavior.

_____. "Weapons in Space: Silver Bullet or Russian Roulette? The Policy Implications of U.S. Pursuit of Space-Based Weapons." In *Space Weapons: Are They Needed?*, edited by John M. Logsdon and Gordon Adams. Washington, D.C.: Space Policy Institute, George Washington University, 2003, pp. 115–156. The United States has little to gain and much to lose in George W. Bush's campaign to make the country the first to deploy space-based weapons. Such a narrow-minded pursuit could ultimately endanger, not protect, national security. Includes data regarding the market for satellites and BMD systems.

Jasani, Bhupendra. "New Approaches to Achieving Space Security." In *Collective Security in Space: European Perspectives*, edited by John M. Logsdon, James Clay Moltz and Emma S. Hinds. Washington, D.C.: George Washington University Space Policy Institute, 2007, pp. 29–50. The international community has come to realize that threats to security, including space security, come from many non-state sources. Jasani calls for efforts to create an ASAT treaty, improve transparency in outer space, and mitigate debris.

_____, ed. *Space Weapons and International Security: An Overview*. New York: Oxford University Press, 1987. Experts from the United States, USSR, Japan, Europe, and the third world offer perspectives on the political, technical, and legal implications of developing ballistic missile defense systems.

Johnson-Freese, Joan. *Space as a Strategic Asset*. New York: Columbia University Press, 2007. How can the United States best protect its vital space-based military assets, especially its numerous satellites? Arguments for and against deploying weapons in space to protect these assets are explored.

Krepon, Michael, and Christopher Clary. *Space Assurance or Space Dominance: The Case against Weaponizing Space*. Washington, D.C.: Stimson Center, 2003. During the cold war, the Soviet Union and, especially, the United States began to make increasing use of space-based military assets. Most assets were satellites, but each side explored the possibility of space-based weapons. Krepon and Clary trace the evolution of the international political order from bipolar during the cold war to asymmetric today. They acknowledge that efforts at treaties banning the weaponization of outer space have been only moderately helpful and offers alternative cooperative strategies, such as a space code of conduct.

Annotated Bibliography

Lakoff, Sanford A., and Herbert F. York. *A Shield in Space? Technology, Politics, and the Strategic Defense Initiative.* Berkeley: University of California Press, 1989. Reagan's SDI lost support under President George H. W. Bush, in part because the required technology has not yet developed.

Lambeth, Benjamin S. *Mastering the Ultimate High Ground: Next Steps in the Military Uses of Space.* Santa Monica, Calif.: RAND, 2003. Analyzes conclusions of the Commission to Assess the Ballistic Missile Threat to the United States, chaired by Donald R. Rumsfeld.

Lochan, Rajeev. "Some Reflections on Collective Security in Space." In *Collective Security in Space: Asian Perspectives,* edited by John M. Logsdon and James Clay Moltz. Washington, D.C.: George Washington University Space Policy Institute, 2008, pp. 33–46. India has used its space program to promote economic development and insisted on home-grown technology for satellites and launchers. Lochan calls on other countries to follow India's example and eschew deploying weapons in outer space.

Logsdon, John M., and Gordon Adams. *Space Weapons: Are They Needed?* Washington, D.C.: Space Policy Institute, George Washington University, 2003. Publication of the Rumsfeld Commission's report has prompted a reevaluation of the appropriate uses of outer space for national and international security.

Logsdon, John M., and James Clay Moltz, eds. *Collective Security in Space: Asian Perspectives.* Washington, D.C.: George Washington University Space Policy Institute, 2008. Asia lacks a unified position on the relative advantages of space security. Drawing upon conference papers by influential policy makers across Asia, Logsdon and Moltz bring new perspectives to a persistent issue.

Logsdon, John M., James Clay Moltz, and Emma S. Hinds, eds. *Collective Security in Space: European Perspectives.* Washington, D.C.: George Washington University Space Policy Institute, 2007. Early approaches to space security involved international treaties. But as the United States began to resist more restrictive agreements, other spacefaring countries began to investigate possible steps toward collective space security. European space experts and policy makers offer their own suggestions in this collection of conference papers.

Lupton, David E. *On Space Warfare: A Spacepower Doctrine.* Maxwell AFB, Ala.: Air University Press, 1998. Space power is based on five factors: logistics, personnel, reconnaissance and surveillance, weapons, and organization of forces. The most important is logistics, as logistics determine which countries have access to space and which do not.

McDougall, Walter A. *The Heavens and the Earth.* New York: Basic Books, 1985. McDougall won the 1986 Pulitzer Prize in history for this work. The highly readable volume traces the origins of the U.S. and Soviet space programs, focusing on inter-agency rivalries, diplomacy, and transparency.

Mizin, Victor. "Russian Perspectives on Space Security." In *Collective Security in Space: European Perspectives,* edited by John M. Logsdon, James Clay Moltz, and Emma S. Hinds. Washington, D.C.: George Washington University Space Policy Institute, 2007, pp. 75–108. Russia is proud of its historic contributions to space exploration. Despite political and funding hardships, Moscow is committed to preserving open access to space for security and commercial purposes.

Moltz, James Clay. "Next Steps toward Space Security." In *Collective Security in Space: European Perspectives*, edited by John M. Logsdon, James Clay Moltz, and Emma S. Hinds. Washington, D.C.: George Washington University Space Policy Institute, 2007, pp. 109–130. Varying definitions of "space security" hamper the effort to create binding international agreements to preserve free access to outer space.

_____. *The Politics of Space Security: Strategic Restraint and the Pursuit of National Interests*. Stanford, Calif.: Stanford University Press, 2008. For much of the space age, the United States has enjoyed dominance in terms of space security, "the ability to place and operate assets outside the Earth's atmosphere without external interference, damage, or destruction." Moltz traces the history of space security from 1958 to 2008, detailing alternating periods of preference for military and diplomatic strategies and a growing international awareness of the environmental consequences of space-based conflict. But as the 21st century unfolds, China's 2007 ASAT test suggests a new potential rival to U.S. space hegemony. Moltz touches on the space security policies of the George W. Bush administration, noting that its bellicose statements have not been matched by dangerous actions.

Mowthorpe, Matthew. *The Militarization and Weaponization of Space*. Lanham, Md.: Rowman & Littlefield, 2004. Drawn from Mowthorpe's doctoral dissertation, this provides an exhaustive inventory of technical details about the space assets of the United States, Russia, China, and emerging powers. While potentially a helpful reference guide, the scant two-page index is almost useless. The book is written in a very dry tone, full of typos, and shows little sign of editing. It also assumes some technical background on the part of the reader, referring to weapons systems or acronyms without identifying them.

Mueller, Karl P. "Totem and Taboo: Depolarizing the Space Weaponization Debate." In *Space Weapons: Are They Needed?*, edited by John M. Logsdon and Gordon Adams. Washington, D.C.: Space Policy Institute, George Washington University, 2003, pp. 1–49. Should the United States place weapons in space? Mueller examines the advantages and disadvantages of space weaponization, while describing and categorizing the options available. He concludes that the cost and technological advances needed to develop and deploy space weapons are so vast that policy makers should not be frightened by any imminent threat from space.

O'Hanlon, Michael E. *Neither Star Wars Nor Sanctuary: Constraining the Military Use of Space*. Washington, D.C.: Brookings Institution, 2004. O'Hanlon provides an easy-to-understand overview of the physics behind space weapons, focusing on the types of weapons that could be deployed in space and the defensive measures—both Earth-based and space-based—that could defeat them. He particularly emphasizes the U.S. government's reliance on commercial satellites for military and intelligence purposes and calls on the government to find incentives for private firms to harden their satellites.

_____. "Preserving U.S. Dominance While Slowing the Weaponization of Space." In *Space Weapons: Are They Needed?*, edited by John M. Logsdon and Gordon Adams. Washington, D.C.: Space Policy Institute, George Washington University, 2003, pp. 239–270. The United States should delay the development of space weapons indefinitely in order to preserve its current military dominance.

Preston, Bob, Dana J. Johnson, Sean J. A. Edwards, Michael Miller, and Calvin Shipbaugh. *Space Weapons, Earth Wars.* Santa Monica, Calif.: RAND Project Air Force, 2002. This book provides standardized definitions of concepts and weapons systems involved in any military use of space to facilitate policy discussions and treaty negotiations. Published in cooperation with the U.S. Air Force.

Sheehan, Michael. *The International Politics of Space.* New York: Routledge, 2007. Space has become central to issues of war and peace, international law, justice, and international development, and cooperation between the world's leading states. Newcomers China and India are committed to becoming space powers themselves.

Spacy, William, II. "Assessing the Military Utility of Space-Based Weapons." In *Space Weapons: Are They Needed?*, edited by John M. Logsdon and Gordon Adams. Washington, D.C.: Space Policy Institute, George Washington University, 2003, pp. 157–238. Spacy discusses the technical and policy aspects of how well space weapons, especially ballistic missile defense systems, would solve a range of national security objectives. He concludes that existing land, sea, or airborne weapons can provide satisfactory and much cheaper solutions.

Sparrow, Giles. *Spaceflight: The Complete Story from Sputnik to Shuttle—and Beyond.* New York: DK, 2007. Lavishly illustrated coffee table book covering missions, spacecraft, astronauts, space suits, and more in great detail. Includes foreword by Buzz Aldrin.

Stares, Paul. *Space and U.S. National Security.* Washington, D.C.: Brookings Institution, 1987. As the U.S. military becomes increasingly reliant on space-based assets, those assets become highly vulnerable targets. The relative costs and benefits of deploying antisatellite systems are examined and compared for both the United States and the USSR. But in the long term, the United States might be safer if it avoids deploying such weapons.

Valasek, Tomas. "The Future of U.S.-European Space Security Cooperation." In *Collective Security in Space: European Perspectives*, edited by John M. Logsdon, James Clay Moltz, and Emma S. Hinds. Washington, D.C.: George Washington University Space Policy Institute, 2007, pp. 63–74. U.S. think tanks have offered a variety of proposals regarding space security cooperation between the United States and Europe but none have been received policy. The issues, relative interests, and overlapping interests are outlined.

Wolter, Detlev. "Common Security in Outer Space and International Law: A European Perspective." In *Collective Security in Space: European Perspectives*, edited by John M. Logsdon, James Clay Moltz, and Emma S. Hinds. Washington, D.C.: George Washington University Space Policy Institute, 2007, pp. 17–28. Europe has a long history of collaborative approaches to common problems and a shared commitment to the peaceful uses of space. The European Union, the Organization for Security and Cooperation in Europe, and NATO offer frameworks for developing codes of conduct, rules of the road, and other cooperative approaches to managing outer space.

Print Articles

Danielson, Dennis L. "Theater Missile Defense and the Anti-Ballistic Missile (ABM) Treaty: Either—Or?" Research Program in Arms Control. University of Illinois at Urbana–Champaign, February 1995. The 1972 ABM treaty succeeded in containing the nuclear arms race but did not reduce the number of U.S. and Soviet strategic warheads. Now the United States should push to create a binding international treaty eliminating nuclear ballistic missiles.

Deblois, Bruce M. "The Advent of Space Weapons." In *Astropolitics* (2003): 29–53. Analyzes the debate over a standard definition of space weapons in international law.

Dubey, Muchkund. "SDI from the Viewpoint of the Non-Aligned Nations." In *Medicine, Conflict, and Survival* (October 1987): 243–248. SDI was usually viewed as a bilateral issue between the United States and USSR, but the nonaligned states would also be affected.

Gray, Colin S. "The Influence of Spacepower upon History." *Comparative Strategy* (1996): 293–308. As space power develops and become more accessible, it will shift the relative balance of power in the international system.

Web Documents

Chin, Larry. "'Deep Impact' and the Militarization of Space: Official Policy, Not Science Fiction." Available online. URL: http://www.globalresearch.ca/index.php?context=va&aid=645. Accessed May 15, 2010. NASA's 2005 mission to crash its *Deep Impact* spacecraft into the Tempel-1 comet demonstrated one potential application of kinetic energy weapons in outer space combat.

David, Leonard. "E-Weapons: Directed Energy Warfare in the 21st Century." Available online. URL: http://www.space.com/businesstechnology/060111_e-weapons.html. Accessed June 15, 2010. New high-tech weapons, such as lasers, particle beams, and high-powered microwaves, may have more impact on warfare than the atomic bomb.

Garwin, Richard L. "Militarization of Space: Why Not Space Weapons?" Available online. URL: http://fas.org/rlg/092082mos.htm. Accessed June 16, 2010. Transcript of Garwin's testimony to the U.S. Senate Foreign Relations Committee on September 20, 1982, warning lawmakers that they cannot realistically make policy to regulate technologies that are not yet available. Presented six months before the announcement of SDI, many of Garwin's predictions are on the mark.

———. "Militarization of Space: Why Not Space Weapons? Space-based Missile Defense?" Available online. URL: www.fas.org/rlg/051605MOS.pdf. Accessed June 12, 2010. For over 30 years, Garwin has advocated for using outer space for civil and military purposes, so long as space is not weaponized. This presentation for a 2005 conference includes highly accessible graphics and schematics for leading projects.

Annotated Bibliography

Shah, Anup. "Militarization and Weaponization of Outer Space." Available online. URL: http://www.globalissues.org/article/69/militarization-and-weaponization-of-outer-space. Accessed May 10, 2010. The longstanding international consensus on the peaceful use of outer space was significantly challenged by the George W. Bush administration's 2006 space policy. This analysis of the policy implications features extensive citations from key documents and speeches with numerous links to original texts and related materials.

GENERAL WORKS ON SPACECRAFT AND SPACE STATIONS
Books and Book Chapters
Bonnet, Roger, and Vittorio Manno. *International Cooperation in Space: The Example of the European Space Agency.* Cambridge, Mass.: Harvard University Press, 1994. The formation of the European Space Agency has been an exemplar of international cooperation, registering multiple successes despite sometimes competing technological, economic, and political agendas. The ESA and NASA formed an equally impressive series of collaborative projects, especially the Ulysses solar probe and the *International Space Station.*

Butrica, Andrew J. *Single Stage to Orbit: Politics, Space Technology, and the Quest for Reusable Rocketry.* Baltimore, Md.: Johns Hopkins University Press, 2003. The race to space has had two main paths: one-time use rockets and reusable craft. The U.S. decision to use one path in the 1950s and 1960s and then change in the 1970s is the result of changing technologies and politics.

Catchpole, John E. *The International Space Station: Building for the Future.* New York: Springer, 2008. The International Space Station is a triumph in technology and international cooperation.

Ivanovich, Grujica S. *Salyut—The First Space Station: Triumph and Tragedy.* New York: Springer, 2008. Traces the development of the first Soviet space station, launched in 1971, the tragedy causing its redesign, the mysterious Almaz military version, and the replacement station, *Mir.*

Shayler, David J. *Skylab: America's Space Station.* New York: Springer, 2008. Documents, interviews, and other primary materials remind us that *Skylab* was a triumph of the U.S. space program before it returned to Earth.

Print Documents
Achenbach, Joel. "Where Will NASA's Next Giant Step Take Us?" *Washington Post* (27 October 2009), p. 1. The United States is behind schedule in replacing the space shuttle fleet. Until new systems come on line, Washington may need to send astronauts into space using commercial launch services.

Lemonick, Michael D. "Surging Ahead: The Soviets Overtake the U.S. as the No. 1 Spacefaring Nation." *Time* (5 October 1987), pp. 64–71. In the final years of the USSR, Soviet space science outpaced the United States in terms of astronomy, astrophysics, space medicine, and space transportation. Yet there are limits; one reason Moscow launched so many satellites is that Soviet-made satellites tend to rapidly disintegrate.

Sawyer, Kathy. "U.S. Proposes Space Merger with Russia." *Washington Post* (5 November 1993), p. A1. As the Russian space program struggles to survive the collapse of the USSR, President Bill Clinton proposes merging the two countries' space programs, allowing Russia to become a partner in the *International Space Station*.

Stone, Richard. "Russia's Last Shot at Space." *Science* (1997): 1,780–1,782. Instead of launching a new era of Russian space research, the 1996 probe to Mars fell back to Earth four hours after launch. With lost defense contracts and a plummeting budget, the Institute for Space Research in Moscow is struggling to survive.

Web Documents

Amos, Jonathan. "Berlin Unveils 'Crewed Spaceship.'" Available online. URL: http://news.bbc.co.uk/2/hi/science/nature/7419793.stm. Accessed June 1, 2010. The European Aeronautic Defense and Space Company has a new spaceship capable of carrying crew to the *International Space Station* and returning with cargo. EADSC bills the craft as an upgraded version of the non-manned *Jules Verne* transport used by the European Space Agency. Such a craft would allow Europeans to reach the station without U.S. or Russian help.

_____. "European Spaceport's Sky-High Ambition." Available online. URL: http://news.bbc.co.uk/2/hi/science/nature/7278409.stm. Accessed June 1, 2010. The European Space Agency opens its new space launch facility in French Guiana, paving the way for it to expand its already extensive manifest of satellite launches into cargo and perhaps crew shuttle services for the *International Space Station*.

_____. "Huge Space Truck Races into Orbit." Available online. URL: http://news.bbc.co.uk/2/hi/science/nature/7285796.stm. Accessed June 1, 2010. The *Jules Verne*, a European-made spacecraft, made its successful debut launching atop an Ariane rocket from the Kourou Spaceport in French Guiana. The spacecraft will allow the European Space Agency to ferry cargo to the *International Space Station* without help from the United States or Russia.

Phillips, Rich. "How Much Longer Will the Space Station Fly?" Available online. URL: http://edition.cnn.com/2009/TECH/space/10/12/space.station.future/index.html. Accessed June 16, 2010. Examines use and lifespan of the *International Space Station* and its future uses, ahead of the Obama administration's new space policy.

GENERAL WORKS ON SATELLITES
Books and Book Chapters

Brzezinski, Matthew. *Red Moon Rising: Sputnik and the Hidden Rivalries That Ignited the Space Age*. New York: Times Books, 2007. Traces the early years of the space age from both Soviet and U.S. perspectives, highlighting the contributions of key individuals.

Annotated Bibliography

DalBello, Richard. "Commercial Communications Satellites: Assessing Vulnerability in a Changing World." In *Space Weapons: Are They Needed?*, edited by John M. Logsdon and Gordon Adams. Washington, D.C.: Space Policy Institute, George Washington University, 2003, pp. 271–296. The United States should delay the development of space weapons indefinitely in order to preserve its current military dominance.

D'Antonio, Michael. *A Ball, a Dog, and a Monkey: 1957—The Space Race Begins.* New York: Simon and Schuster, 2008. While the early days of the space age drew on hard science, a surprising number of innovations were improvisation.

Dickson, Paul. *Sputnik: The Shock of the Century.* New York: Walker, 2001. Fifty years after the Soviet Union launched the first artificial Earth satellite, Paul Dickson examines the panic it caused in the United States. He traces the multifaceted reaction of the U.S. government and evaluates competing explanations for the initial U.S. defeat in space.

Hyten, John E. "A Sea of Peace or a Theater of War? Dealing with the Inevitable Conflict in Space." In *Space Weapons: Are They Needed?*, edited by John M. Logsdon and Gordon Adams. Washington, D.C.: Space Policy Institute, George Washington University, 2003, pp. 297–366. U.S. policy makers ignored space issues in the 1990s, until the Rumsfeld Commission forced them back on the national agenda.

Jasani, Bhupendra, and Gotthard Stein. *Commercial Satellite Imagery and GIS.* New York: Springer, 2002. Satellites are vital tools to monitor compliance with nuclear nonproliferation treaties. Case studies from Canada, France, India, Israel, Russia, and Japan highlight the need for new verification techniques, commercial satellite imagery and processing, and cost-benefit analyses.

Johnson-Freese, Joan. *Space as a Strategic Asset.* New York: Columbia University Press, 2007. How can the United States best protect its vital space-based military assets, especially its numerous satellites? Arguments for and against deploying weapons in space to protect these assets are explored.

Kee, Changdon. "A South Korean Perspective on Strengthening Space Security in East Asia." In *Collective Security in Space: Asian Perspectives*, edited by John M. Logsdon and James Clay Moltz. Washington, D.C.: George Washington University Space Policy Institute, 2008, pp. 14–23. South Korea is emerging as a contender in the global satellite market, but redundant programs such as GPS, Glonass, and Galileo are inhibiting growth and wasting resources.

Nye, Joseph S., and James A. Schear, eds. *Seeking Stability in Space.* Lanham, Md.: University Press of America, 1987. Military experts address the multifaceted potential military uses of space, especially the security implications of ASAT systems.

Print Articles

Amodeo, Christian. "Eyes in the Skies: In the Past Ten Years, the Number of People Making Use of Satellite Technology Has Soared." *Geographical* (February 2003). Satellites are for a variety of remote-sensing applications, but they increasingly face danger from orbiting space debris.

MILITARIZATION OF SPACE

Andrews, Edmund L. "Tiny Tonga Seeks Satellite Empire in Space." *New York Times* (28 August 1990). Tonga exploited a loophole in international law to claim key orbital positions for satellite communications.

Broad, William J. "Debris Spews into Space after Satellites Collide." *New York Times* (12 February 2009). A Russian and a U.S. satellite collided over Siberia, creating fragments that could damage the *International Space Station.*

_____. "NASA Forced to Steer Clear of Junk in Cluttered Space." *New York Times* (31 July 2007). Controllers had to move the *International Space Station* about five miles to avoid damage from an ammonia tank accidentally dropped out of the space shuttle.

_____. "Orbiting Junk, Once a Nuisance, Is Now a Threat." *New York Times* (6 February 2007). China's January 2007 ASAT test may have created enough debris to set the Kessler Syndrome in motion.

Cukurtepe, Haydar, and Ilker Akgun. "Towards Space Traffic Management System." *Acta Astronautica* (September/October 2009), pp. 870–878. As outer space increasingly fills with objects, limiting the availability of prime orbital locations, it is time to consider adopting a space traffic management system.

Jolis, Anne. "Woes Mount at Europe's Bid for GPS Rival." *Wall Street Journal* (17 January 2007). Europe's version of the U.S. global positioning system, Galileo, is behind schedule and considerably over budget. While the U.S. version is owned by the U.S. Air Force, Galileo is funded by a range of investors, including China and European taxpayers.

Lemonick, Michael D. "A Flap over Reactors in Orbit." *Time* (February 20, 1989), p. 80. Soviets have deployed three dozen nuclear-powered satellites; what would happen if one fell to Earth?

_____. "Surging Ahead: The Soviets Overtake the U.S. as the No. 1 Spacefaring Nation." *Time* (October 5, 1987), pp. 64–71. In the final years of the USSR, Soviet space science outpaced the United States in terms of astronomy, astrophysics, space medicine, and space transportation. Yet there are limits; one reason Moscow launched so many satellites is that Soviet-made satellites tend to rapidly disintegrate.

Maran, Stephen P. "Roadside Service: The Right Stuff for Space Clunkers." *Smithsonian* (August 1985), p. 96–140. NASA is developing new technologies and new techniques to capture and repair malfunctioning satellites.

"Noisy Neighbors." *Economist* (December 17, 1988). Nuclear-powered satellites emit gamma rays and anti-matter that would be very dangerous should one fall to Earth.

Reichhardt, Tony. "Satellite Smashers: Space-faring Nations: Clean up Low Earth Orbit or You're Grounded." *Air and Space* (March 1, 2008), pp. 50–53. Scientists use computer-simulation models to predict future orbital debris patterns, as the frequency of collisions increases.

Web Documents

Adams, Cecil. "Is Anybody in Charge of Keeping Satellites from Colliding?" Available online. URL: http://www.straightdope.com/columns/read/2833/is-anybody-in-charge-of-keeping-satellites-from-colliding. Accessed June 10, 2010. This offbeat Web site tackles the issue of satellite collisions with hard data and clear thinking.

Bein, Michael. "Star Wars and Reactors in Space: A Canadian View." Available online. URL: http://www.animatedsoftware.com/spacedeb/canadapl.htm. Accessed April 12, 2010. In 1978 the Soviet Cosmos 954 satellite disintegrated, de-orbited, and spread radioactive debris across northern Canada. The incident prompted discussion of a ban on nuclear-powered objects in space.

David, Leonard. "Space Traffic Control: Steering Clear of Collisions." Available online. URL: http://www.space.com/businesstechnology/technology/space_traffic_040505.html. Despite the increasing amount of orbital debris threatening space assets, the international community should not rush into a space traffic management system.

Krepon, Michael, and Ashley Tellis. "Another Wake-Up Call." Available online. URL: http://www.stimson.org/pub.cfm?ID=764. Accessed June 1, 2010. Space debris is a growing problem and threat to space-based assets, as demonstrated by the February 10, 2009, collision between a dead Cosmos satellite and a revenue-producing Iridium satellite.

Malik, Tariq. "What Happens When Satellites Fall." Available online. URL: http://www.space.com/businesstechnology/090123-falling-satellites.html. Accessed June 1, 2010. Ground controllers have a variety of options to protect Earth and other objects in space when a satellite goes out of control.

"The Sky Is Falling: 10 of the Most Memorable Pieces of Space Junk That Fell to Earth." Available online. URL: http://www.space.com/missionlaunches/080225-top10-debris.html. Accessed June 16, 2010. Satellite debris, rocket parts, and other items once in space have fallen to Earth.

ANTIBALLISTIC MISSILES AND MISSILE DEFENSE SYSTEMS

Books and Book Chapters

Bulkeley, Rip, and Graham Spinardi. *Space Weapons: Deterrence or Delusion?* New York: Polity Press, 1986. While SDI was introduced in 1983, space weapons technology actually dates to at least the 1920s. The authors fit SDI into this historical time line and consider the technical, political, and strategic ramifications of the new program.

FitzGerald, Frances. *Way Out There in the Blue: Reagan, Star Wars, and the End of the Cold War.* New York: Simon and Schuster, 2000. Ronald Reagan proposed a defense program that would create a protective bubble over the United States, shielding it from a Soviet nuclear attack. FitzGerald constructs a psychological portrait of the president through the prism of SDI.

Hey, Nigel. *The Star Wars Enigma: Behind the Scenes of the Cold War Race for Missile Defense.* Dulles, Va.: Potomac Books, 2007. Ronald Reagan's effort to create a shield to protect the United States from a Soviet missile attack has been analyzed from many different perspectives. But Reagan's own national security adviser, Robert McFarlane, praises Hey's work as "the most rigorous scholarship yet to appear on this climactic period of history."

Jasani, Bhupendra, ed. *Space Weapons and International Security: An Overview.* New York: Oxford University Press, 1987. Experts from the United States, USSR, Japan, Europe, and the third world offer perspectives on the political, technical, and legal implications of developing ballistic missile defense systems.

Krepon, Michael. *Cooperative Threat Reduction, Missile Defense, and the Nuclear Future.* New York: Palgrave, 2003. During the cold war the United States and Soviet Union achieved a balance of power through the theory of mutually assured destruction, based on nuclear weapons. But now the United States is the only remaining superpower and its continued reliance on nuclear weapons means it can "outkill" any opponent, giving potential foes incentive to strike first. Krepon argues for Washington to embrace cooperative threat reduction.

Krepon, Michael, and Christopher Clary. *Space Assurance or Space Dominance: The Case against Weaponizing Space.* Washington, D.C.: Stimson Center, 2003. During the cold war the Soviet Union and, especially, the United States began to make increasing use of space-based military assets. Most assets were satellites, but each side explored the possibility of space-based weapons. Krepon and Clary trace the evolution of the international political order from bipolar during the cold war to asymmetric today. They acknowledge that efforts at treaties banning the weaponization of outer space have been only moderately helpful and offers alternative cooperative strategies, such as a space code of conduct.

Lakoff, Sanford A., and Herbert F. York. *A Shield in Space? Technology, Politics, and the Strategic Defense Initiative.* Berkeley: University of California Press, 1989. Reagan's SDI lost support under President George H. W. Bush, in part because the required technology has not yet developed.

Reiss, Edward. *The Strategic Defense Initiative.* New York: Cambridge University Press, 2008. Written 25 years after Ronald Reagan's Star Wars speech, Reiss goes beyond what the program offered to analyze the outcomes and the political and technological reasons so little of SDI was ever deployed.

Print Articles

Baucom, Donald R. "Eisenhower and Ballistic Missile Defense: The Formative Years, 1944–1961." *Air Power History* (2004): 4–17. As president at the dawn of the missile age, Eisenhower set crucial policy maps for development and deployment of bombers, missiles, and missile defense programs.

_____."Space and Missile Defense." *Joint Force Quarterly* (2002): 50–55. U.S. approaches to missile defense date to the Eisenhower administration and have considered space-based missiles, hit-to-kill weapons, and interceptors.

Coleman, Joseph. "U.S., Japan Expand Missile-Defense Plan." *Washington Post* (23 June 2006). As Japan grows increasingly wary of North Korea's ballistic missile program, the United States agrees to transfer an early warning radar to a Japanese military base and to jointly develop a ballistic missile defense shield. The agreement challenges Japan's longstanding ban on arms exports.

Danielson, Dennis L. "Theater Missile Defense and the Anti-Ballistic Missile (ABM) Treaty: Either—Or?" Research Program in Arms Control. University of Illinois at

Urbana–Champaign, February 1995. The 1972 ABM Treaty succeeded in containing the nuclear arms race, but did not reduce the number of U.S. and Soviet strategic warheads. Now the United States should push to create a binding international treaty eliminating nuclear ballistic missiles.

Daniszewski, John. "New Era Dawns in Missile Defense." *Los Angeles Times* (19 December 2002), p. A4. Moscow's initial response to the Bush administration's withdrawal from the ABM Treaty and plans for a new missile defense system is surprisingly subdued.

Dubey, Muchkund. "SDI from the Viewpoint of the Non-Aligned Nations." *Medicine, Conflict, and Survival* (October 1987): 243–248. The Strategic Defense Initiative was usually viewed as a bilateral issue between the United States and USSR, but the nonaligned states would also be affected.

Hitchens, Theresa, and Victoria Samson. "Space-Based Interceptors: Still Not a Good Idea." *Georgetown Journal of International Affairs* (summer/fall 2004). The United States is ready to abandon its longstanding restraint from weaponizing outer space, using several programs that date to the 1983 SDI.

Lemonick, Michael D. "A Flap over Reactors in Orbit." *Time* (February 20, 1989), p. 80. Soviets have deployed three dozen nuclear-powered satellites; what would happen if one fell to Earth?

Warrick, Joby, and R. Jeffrey Smith. "U.S.-Russian Team Deems Missile Shield in Europe Ineffective." *Washington Post* (19 May 2009). New report from East-West Institute concludes that Iran will need another decade to produce a nuclear missile that could reach the United States. Experts also think that proposed U.S. missile defense program would not be able to intercept this type of missile.

Web Documents

Baucom, Donald R. "Missile Defense Milestones, 1944–1997." Available online. URL: http://www.fas.org/spp/starwars/program/milestone.htm. Accessed June 16, 2010. Detailed chronology of the missile defense debate.

Bein, Michael. "Star Wars and Reactors in Space: A Canadian View." Available online. URL: http://www.animatedsoftware.com/spacedeb/canadapl.htm. Accessed April 12, 2010. In 1978, the Soviet Cosmos 954 satellite disintegrated, de-orbited, and spread radioactive debris across northern Canada. The incident prompted discussion of a ban on nuclear-powered objects in space.

Bruno, Greg. "National Missile Defense: A Status Report." Available online. URL: http://www.cfr.org/publication/18792/national_missile_defense.html. Accessed June 16, 2010. Obama's interest in missile defense programs is tempered by a concern for pragmatic, cost-effective technology.

Dinerman, Taylor. "The Bush Administration and Space Weapons." Available online. URL: http://www.thespacereview.com/article/368/1. Accessed June 1, 2010. The George W. Bush administration revived components of SDI and seemed determined to do whatever was necessary to achieve space supremacy.

Grego, Laura, and David Wright. "Bush Administration National Space Policy." Available online. URL: http://www.ucsusa.org/nuclear_weapons_and_global_

security/space_weapons/policy_issues/bush-administration-national.html. Accessed June 1, 2010. The Bush 2006 national space policy has raised alarms about its implied warning that the United States is now willing to weaponize space.

Kotani, Rui. "Japan's Recent Step-up in Missile Defense." Available online. URL: http://www.cdi.org/program/relateditems.cfm?typeID=%288%29&programID=6. Accessed June 10, 2010. After North Korea announced it would test nuclear weapons, Japan announced its intention to acquire a U.S.-made missile defense system. However, a missile defense system would violate Japan's constitution and be technically difficult to deploy.

Lambakis, Steven. "Space Weapons: Refuting the Critics." Available online. URL: http://www.hoover.org/publications/policyreview/3479337.html. Accessed June 15, 2010. Space and national security policy must stick to the core foundations of space law and the longstanding values of the international community, while resisting temptation to follow new trends or concepts.

O'Connor, Sean. "Russian/Soviet Anti-Ballistic Systems." Available online. URL: http://www.ausairpower.net/APA-Rus-ABM-Systems.html. Accessed June 10, 2010. The origins of Russia's contemporary missile defense system can be traced to the Nazi V-2 rocket. Each system is described in detail, accompanied by extensive photographic coverage.

Squeo, Anne Marie. "U.S. Considers European Role for Missile-Defense Programs." *Wall Street Journal* (7 June 2002), p. A4. The George W. Bush administration approached European military contractors about participating in the proposed new U.S. missile defense system. Cooperation would raise several difficult issues, especially as many West European governments opposed Bush's foreign policy. With contracts estimated at $8.2 billion, critics suggested Washington was trying to buy European support.

Toki, Masako. "Missile Defense in Japan." Available online. URL: http://www.thebulletin.org/web-edition/features/missile-defense-japan. Accessed June 10, 2010. Since the 1980s, the United States and Japan have cooperated on missile defense. When North Korea tested a ballistic missile, Japanese leaders decided to strengthen their defenses. After relevant changes in its National Defense Program, Japan began deploying a missile defense system in 2009.

Weeden, Brian. "The Fallacy of Space-Based Interceptors for Boost-Phase Missile Defense." Available online. URL: http://www.thespacereview.com/article/1212/1. Accessed June 1, 2010. While boost-phase missile intercept by space-based assets may seem promising from a political perspective, policy makers tend to overlook the many technical issues involved.

Wright, David, and Lisbeth Gronlund. "Technical Flaws in the Obama Missile Defense Plan." Available online. URL: http://www.thebulletin.org/web-edition/op-eds/technical-flaws-the-obama-missile-defense-plan. Accessed June 1, 2010. President Obama's plan to use the ship-based Aegis defense system over a land-based one in Poland and the Czech Republic may be a bad idea; the technology is unproven and it may upset hawks in Russia and China.

ANTISATELLITE WEAPONS

Books and Book Chapters

Krepon, Michael, with Christopher Clary. *Space Assurance or Space Dominance: The Case against Weaponizing Space.* Washington, D.C.: Stimson Center, 2003. During the cold war the Soviet Union and, especially, the United States began to make increasing use of space-based military assets. Most assets were satellites, but each side explored the possibility of space-based weapons. Krepon and Clary trace the evolution of the international political order from bipolar during the cold war to asymmetric today. They acknowledge that efforts at treaties banning the weaponization of outer space have been only moderately helpful and offer alternative cooperative strategies, such as a space code of conduct.

Nye, Joseph S., and James A. Schear, eds. *Seeking Stability in Space.* Lanham, Md.: University Press of America, 1987. Military experts address the multifaceted potential military uses of space, especially the security implications of ASAT systems.

Stares, Paul. *Space and U.S. National Security.* Washington, D.C.: Brookings Institution, 1987. As the U.S. military becomes increasingly reliant on space-based assets, those assets become highly vulnerable targets. The relative costs and benefits of deploying antisatellite systems are examined and compared for both the United States and the USSR. But in the long term, the United States might be safer if it avoids deploying such weapons.

Yost, David Scott. *Soviet Ballistic Missile Defense and the Western Alliance.* Cambridge, Mass.: Harvard University Press, 1988. Yost challenges the conventional notion that Moscow had to scramble to catch up with U.S. missile defense plans in the wake of SDI. Instead, he demonstrates that modernization and upgrading had begun long before Reagan's Star Wars speech. He also notes a number of potential loopholes in current arms control agreements that need to be modified accordingly.

Print Articles

Frey, Adam E. "Defense of U.S. Space Assets: A Legal Perspective." *Air and Space Power Journal* (2008): 75–84. Responding to China's 2007 decision to destroy one of its own satellites, the author takes a broad look at how ASAT theory and technology follow or clash with existing international space law. The United States has two possible responses: it could move to weaponize space or it could harden its own potential targets to reduce their vulnerability. The latter response would be the more responsible and not trigger a potential arms race in space.

Gill, Bates, and Martin Kleiber. "China's Space Odyssey: What the Antisatellite Test Reveals about Decision-Making in Beijing." *Foreign Affairs* (May/June, 2007). China's 2007 ASAT may tell less about the Chinese space program than the lack of coordination among its many government agencies.

Grier, Peter, and Peter Lubold. "U.S. Missile Shoots Down Satellite: But Why?" *Christian Science Monitor* (22 February 2008), p. 2. Experts are skeptical about official explanation of U.S. Navy's decision to shoot down wayward spy satellite.

Krepon, Michael. "Lost in Space: The Misguided Drive Towards Antisatellite Weapons." *Foreign Affairs* (2001). Donald Rumsfeld's pursuit of antisatellite weapons systems will destabilize the international order.

Morring, Frank, Jr. "Worst Ever: Chinese Anti-Satellite Test Boosted Space-Debris Population by 10% in an Instant." *Aviation Week and Space Technology* (2007): 20. China's 2007 ASAT test not only created one of the largest debris clouds to date, it may set a precedent for other countries to begin potentially hostile applications of space assets.

Web Documents

Baucom, Donald R. "Missile Defense Milestones, 1944–1997." Available online. URL: http://www.fas.org/spp/starwars/program/milestone.htm. Accessed June 16, 2010. Detailed chronology of the missile defense debate.

Easton, Ian. "The Great Game in Space: China's Evolving ASAT Weapons Programs and Their Implications for Future U.S. Strategy." Available online. URL: http://www. project2049.net/documents/china_asat_weapons_the_great_game_in_space.pdf. Accessed May 10, 2010. China's growing ASAT weapons program threatens to shift the balance of power in the Pacific region and possibly in space as well. Easton outlines in detail the technical specifications of China's direct-ascent, kinetic-kill ASAT and their significance for the United States. After an overview of the components of the U.S. satellite system, he concludes that the United States is highly vulnerable to an ASAT attack. In response, the air force has begun an anti-ASAT project.

Krepon, Michael. "After the ASAT Tests." Available online. URL: http://www.stimson. org/pub.cfm?ID=586. Accessed June 1, 2010. These Chinese and U.S. ASAT tests raise similar questions about intent, but few observers are fooled by U.S. attempts to pass a military test off as an environmental project.

———. "Bush's ASAT Test." Available online. URL: http://www.stimson.org/pub. cfm?id=570. Accessed June 1, 2010. U.S. plans to test an ASAT in February 2008 will have long-reaching implications for global security.

———. "Russia and China Propose a Treaty Banning Space Weapons, While the Pentagon Plans an ASAT Test." Available online. URL: http://www.stimson.org/ pub.cfm?ID=568. Accessed June 10, 2010. China offered a draft treaty banning space weapons at the UN Conference on Disarmament, but the main obstacle to its acceptance is the United States.

Krepon, Michael, and Ashley Tellis. "Another Wake-Up Call." Available online. URL: http://www.stimson.org/pub.cfm?ID=764. Accessed June 1, 2010. Space debris is a growing problem and threat to space-based assets, as demonstrated by the February 10, 2009, collision between a dead Cosmos satellite and a revenue-producing Iridium satellite.

Lambakis, Steven. "Space Weapons: Refuting the Critics." Available online. URL: http://www.hoover.org/publications/policyreview/3479337.html. Accessed June 15, 2010. Space and national security policy must stick to the core foundations of space law and the long-standing values of the international community, while resisting temptation to follow new trends or concepts.

DEBRIS AND ENVIRONMENTAL MANAGEMENT
Books and Book Chapters

Pasco, Xavier. "Toward a Future European Space Surveillance System: Developing a Collaborative Model for the World." In *Collective Security in Space: European Perspectives*, edited by John M. Logsdon, James Clay Moltz, and Emma S. Hinds. Washington, D.C.: George Washington University Space Policy Institute, 2007, pp. 51–62. A European space surveillance system would underline the region's identity as a legitimate international actor, build on existing integration efforts, and support the European Security and Defense Policy.

Yang, Junhua. "Developing Space Peacefully for the Benefit of Humanity." In *Collective Security in Space: Asian Perspectives*, edited by John M. Logsdon and James Clay Moltz. Washington, D.C.: George Washington University Space Policy Institute, 2008, pp. 24–32. China is developing its own strategy for mitigating the increasing problem of outer space debris and environmental damage. Yang calls for an improved space object registry and international management of this common problem.

Printed Works

Amodeo, Christian. "Eyes in the Skies: In the Past Ten Years, the Number of People Making Use of Satellite Technology Has Soared." *Geographical* (February 2003). Satellites are for a variety of remote-sensing applications, but they increasingly face danger from orbiting space debris.

Broad, William J. "NASA Forced to Steer Clear of Junk in Cluttered Space." *New York Times* (31 July 2007). Controllers had to move the *International Space Station* about five miles to avoid damage from an ammonia tank accidentally dropped out of the space shuttle.

Cukurtepe, Haydar, and Ilker Akgun. "Towards Space Traffic Management System." *Acta Astronautica* (September/October 2009): 870–878. As outer space increasingly fills with objects, limiting the availability of prime orbital locations, it is time to consider adopting a space traffic management system.

European Space Agency. "Space Junk Could Be a Hazard; As More and More Man-Made Objects Fill Space around the Earth, Space Experts become More Concerned about Potential Collisions." *Satellite Communications* (1989): 72–73. Man-made objects have littered outer space since *Sputnik* entered low earth orbit. While debris is tracked by a variety of agencies, less work is being done to mitigate future debris.

Lemonick, Michael D. "A Flap over Reactors in Orbit." *Time* (February 20, 1989), p. 80. Soviets have deployed three dozen nuclear-powered satellites; what would happen if one fell to Earth?

Maran, Stephen P. "Roadside Service: The Right Stuff for Space Clunkers." *Smithsonian* (August 1985), pp. 96–140. NASA is developing new technologies and new techniques to capture and repair malfunctioning satellites.

Morring, Frank, Jr. "Worst Ever: Chinese Anti-Satellite Test Boosted Space-Debris Population by 10% in an Instant." *Aviation Week and Space Technology* (2007): 20. China's 2007 ASAT test not only created one of the largest debris clouds to date, it may set a precedent for other countries to begin potentially hostile applications of space assets.

Murphy, Jamie, and Jerry Hannifin. "Dodging Celestial Garbage: Right Now There Are 3,800 Pieces of Junk Circling the Earth." *Time* (May 21, 1984), p. 91. In April 1984, astronauts were able to successfully repair the damaged Solar Max satellite. Outer space is littered with thousands of pieces of space junk, and many shards have survived reentry to damage people and property on Earth.

"Noisy Neighbors." *Economist* (December 17, 1988). Nuclear-powered satellites emit gamma rays and anti-matter that would be very dangerous should one fall to Earth.

INTERNATIONAL SPACE LAW
Books and Book Chapters

Aoki, Setsuko. "Japanese Perspectives on Space Security." In *Collective Security in Space: Asian Perspectives*, edited by John M. Logsdon and James Clay Moltz. Washington, D.C.: George Washington University Space Policy Institute, 2008, pp. 47–66. The existing international legal framework provides little security for Asia's needs in outer space. Asia could take the lead in setting regional guidelines that could be a model for a new international regime.

Jasani, Bhupendra, and Gotthard Stein. *Commercial Satellite Imagery and GIS.* New York: Springer, 2002. Satellites are a vital tool to monitor compliance with nuclear nonproliferation treaties. Case studies from Canada, France, India, Israel, Russia, and Japan highlight the need for new verification techniques, commercial satellite imagery and processing, and cost-benefit analyses.

Moltz, James Clay. "Next Steps toward Space Security." In *Collective Security in Space: European Perspectives*, edited by John M. Logsdon, James Clay Moltz, and Emma S. Hinds. Washington, D.C.: George Washington University Space Policy Institute, 2007, pp. 109–130. Varying definitions of "space security" hamper the effort to create binding international agreements to preserve free access to outer space.

Nair, Kiran K. "Space Security: Reassessing the Situation and Exploring Options." In *Collective Security in Space: Asian Perspectives*, edited by John M. Logsdon and James Clay Moltz. Washington, D.C.: George Washington University Space Policy Institute, 2008, pp. 84–93. "Common security," not "collective security," is the more appropriate approach to space security, given the varying levels of development in space programs around the world. Nair concludes that Asia has incentive to cooperate in space, due to being "the most disaster-prone continent in the world."

Annotated Bibliography

Papp, Daniel, and John McIntyre, eds. *International Space Policy: Legal, Economic, and Strategic Options for the Twentieth Century and Beyond.* New York: Quorum Books, 1987. Experts from governments, private corporations, academia, law, and science weigh in on how continued space explorations affect their distinct fields. The papers grew out of a 1985 conference at the Georgia Institute of Technology and highlight the differing military and civilian views.

Pop, Virgiliu. *Who Owns the Moon? Extraterrestrial Aspects of Land and Mineral Resources Ownership.* New York: Springer, 2008. International treaties insist that no single person or country can "own" a celestial body, but Pop explores alternative views and potential claims.

Wadegaonkar, Damodar. *The Orbit of Space Law.* London: Stevens and Stevens, 1984. The often-cited demilitarization clause in the Outer Space Treaty is not only flawed but illogical as well.

White, Irwin. *Decision-Making for Space: Law and Politics in Air, Sea, and Outer Space.* West LaFayette, Ind.: Purdue University Press, 1971. This classic of space law laid the foundation for basing outer space law on the principles governing use of the high seas and national airspace.

Wolter, Detlev. "Common Security in Outer Space and International Law: A European Perspective." In *Collective Security in Space: European Perspectives,* edited by John M. Logsdon, James Clay Moltz, and Emma S. Hinds. Washington, D.C.: George Washington University Space Policy Institute, 2007, pp. 17–28. Europe has a long history of collaborative approaches to common problems and a shared commitment to the peaceful use of space. The European Union, Organization for Security and Cooperation in Europe, and NATO offer frameworks for developing codes of conduct, rules of the road, and other cooperative approaches to managing outer space.

Print Articles

Cukurtepe, Haydar, and Ilker Akgun. "Towards Space Traffic Management System." *Acta Astronautica* (September/October 2009): 870–878. As outer space increasingly fills with objects, limiting the availability of prime orbital locations, it is time to consider adopting a space traffic management system.

Danielson, Dennis L. "Theater Missile Defense and the Anti-Ballistic Missile (ABM) Treaty: Either—Or?" Research Program in Arms Control. University of Illinois at Urbana–Champaign, February 1995. The 1972 ABM Treaty succeeded in containing the nuclear arms race but did not reduce the number of U.S. and Soviet strategic warheads. Now the United States should push to create a binding international treaty eliminating nuclear ballistic missiles.

Deblois, Bruce M. "The Advent of Space Weapons." *Astropolitics* (2003): 29–53. Analyzes the debate over a standard definition of space weapons in international law.

Frey, Adam E. "Defense of U.S. Space Assets: A Legal Perspective." *Air and Space Power Journal* (2008): 75–84. Responding to China's 2007 decision to destroy one of its own satellites, the author takes a broad look at how ASAT theory and technology follow or clash with existing international space law. The United States has two possible responses: it could move to weaponize space or it could harden its own potential targets to reduce their vulnerability. The latter response would be the more responsible and not trigger a potential arms race in space.

Hall, R. Cargill. "The Evolution of U.S. National Security Space Policy and Its Legal Foundations in the 20th Century." *Journal of Space Law* (2007): 1–104. Analysis of every U.S. national security directive on space from Eisenhower to George W. Bush.

Smith, Marcia S. "China's Space Program: A Brief Overview Including Commercial Launches of U.S.-Built Satellites." In *CRS Issue Brief for Congress* (September 3, 1998). China has developed a lucrative business launching satellites manufactured elsewhere. However, the country has also suffered a series of failed launches. U.S. leaders were concerned that China may have received vital data for improving its launch mechanisms from one of its commercial satellite customers. Washington was concerned about this potential transfer of technology, as Beijing could also apply it toward improving its military space program.

Web Documents

"China's Attitude Toward Outer Space Weapons." Available online. URL: http://www.nti.org/db/china/spacepos.htm. Accessed June 1, 2010. China believes that current agreements and treaties banning outer space weapons are inadequate. Therefore it has consistently proposed various draft agreements at the UN Conference on Disarmament's Committee on the Prevention of an Arms Race in Outer Space. Beijing is very concerned about a possible U.S. missile defense system, as it would negate China's nuclear arsenal. China might then be forced into an arms race in order to prevent the United States from achieving global dominance.

Krepon, Michael. "Russia and China Propose a Treaty Banning Space Weapons, While the Pentagon Plans an ASAT Test." Available online. URL: http://www.stimson.org/pub.cfm?ID=568. Accessed June 10, 2010. China offered a draft treaty banning space weapons at the UN Conference on Disarmament, but the main obstacle to its acceptance is the United States.

Krepon, Michael, Theresa Hitchens, and Michael Katz-Hyman. "Preserving Freedom of Action in Space: Realizing the Potential and Limits of U.S. Spacepower." Available online. URL: www.stimson.org/space/pdf/SpacePower-051007.pdf. Accessed June 1, 2010. Given the international community's common vulnerability to space-deployed weapons, the authors propose an international code of conduct to be monitored by satellite systems.

United Nations Office for Outer Space Affairs. "United Nations Treaties and Principles on Space Law." Available online. URL: http://www.oosa.unvienna.org/oosa/en/SpaceLaw/treaties.html. Accessed June 28, 2010. A centralized repository of documents related to space law. Provides links to full text of major international agreements and outlines five sets of legal principles governing activities in outer space.

Annotated Bibliography

COMMERCIAL USE OF OUTER SPACE
Books and Book Chapters

DalBello, Richard. "Commercial Communications Satellites: Assessing Vulnerability in a Changing World." In *Space Weapons: Are They Needed?*, edited by John M. Logsdon and Gordon Adams. Washington, D.C.: Space Policy Institute, George Washington University, 2003, pp. 271–296. The United States should delay the development of space weapons indefinitely in order to preserve its current military dominance.

Hyten, John E. "A Sea of Peace or a Theater of War? Dealing with the Inevitable Conflict in Space." In *Space Weapons: Are They Needed?*, edited by John M. Logsdon and Gordon Adams. Washington, D.C.: Space Policy Institute, George Washington University, 2003, pp. 297–366. U.S. policy makers ignored space issues in the 1990s, until the Rumsfeld Commission forced it back on the national agenda.

Jasani, Bhupendra, and Gotthard Stein. *Commercial Satellite Imagery and GIS.* New York: Springer, 2002. Satellites are a vital tool to monitor compliance with nuclear nonproliferation treaties. Case studies from Canada, France, India, Israel, Russia, and Japan highlight the need for new verification techniques, commercial satellite imagery and processing, and cost-benefit analyses.

Johnson-Freese, Joan. *Space as a Strategic Asset.* New York: Columbia University Press, 2007. How can the United States best protect its vital space-based military assets, especially its numerous satellites? Arguments for and against deploying weapons in space to protect these assets are explored.

Kee, Changdon. "A South Korean Perspective on Strengthening Space Security in East Asia." In *Collective Security in Space: Asian Perspectives*, edited by John M. Logsdon and James Clay Moltz. Washington, D.C.: George Washington University Space Policy Institute, 2008, pp. 14–23. South Korea is emerging as a contender in the global satellite market, but redundant programs such as GPS, Glonass, and Galileo are inhibiting growth and wasting resources.

Nye, Joseph S., and James A. Schear, eds. *Seeking Stability in Space.* Lanham, Md.: University Press of America, 1987. Military experts address the multifaceted potential military uses of space, especially the security implications of ASAT systems.

Print Articles

Amodeo, Christian. "Eyes in the Skies: In the Past Ten Years, the Number of People Making Use of Satellite Technology Has Soared." *Geographical* (February 2003). Satellites are for a variety of remote-sensing applications, but they increasingly face danger from orbiting space debris.

Andrews, Edmund L. "Tiny Tonga Seeks Satellite Empire in Space." *New York Times* (28 August 1990). Tonga exploited a loophole in international law to claim key orbital positions for satellite communications.

Baird, Stephen. "Space: The New Frontier!" *The Technology Teacher* (April 2008): 13–19. Commercial spaceflight offers exciting prospects for space tourism and supporting industries.

Broad, William J. "Debris Spews into Space after Satellites Collide." *New York Times* (12 February 2009). A Russian and a U.S. satellite collided over Siberia, creating fragments that could damage the *International Space Station.*

_____. "NASA Forced to Steer Clear of Junk in Cluttered Space." *New York Times* (31 July 2007). Controllers had to move the *International Space Station* about five miles to avoid damage from an ammonia tank accidentally dropped out of the space shuttle.

_____. "Orbiting Junk, Once a Nuisance, Is Now a Threat." *New York Times* (6 February 2007). China's January 2007 ASAT test may have created enough debris to set the Kessler Syndrome in motion.

Jolis, Anne. "Woes Mount at Europe's Bid for GPS Rival." *Wall Street Journal* (17 January 2007). Europe's version of the U.S. global positioning system, Galileo, is behind schedule and considerably over budget. While the U.S. version is owned by the U.S. Air Force, Galileo is funded by a range of investors, including China and European taxpayers.

Lemonick, Michael D. "A Flap over Reactors in Orbit." *Time* (February 20, 1989), p. 80. Soviets have deployed three dozen nuclear-powered satellites; what would happen if one fell to Earth?

_____. "Surging Ahead: The Soviets Overtake the U.S. as the No. 1 Spacefaring Nation." *Time* (October 5, 1987), pp. 64–71. In the final years of the USSR, Soviet space science outpaced the United States in terms of astronomy, astrophysics, space medicine, and space transportation. Yet there are limits; one reason Moscow launched so many satellites is that Soviet-made satellites tend to rapidly disintegrate.

Maran, Stephen P. "Roadside Service: The Right Stuff for Space Clunkers." *Smithsonian* (August 1985), pp. 96–140. NASA is developing new technologies and new techniques to capture and repair malfunctioning satellites.

Reichhardt, Tony. "Satellite Smashers: Space-faring Nations: Clean up Low Earth Orbit or You're Grounded." *Air and Space* (March 1, 2008), pp. 50–53. Scientists use computer-simulation models to predict future orbital debris patterns, as the frequency of collisions increases.

Smith, Marcia S. "U.S. Space Programs: Civilian, Military, and Commercial." *CRS Issue Brief for Congress* (October 21, 2004). China has developed a lucrative business launching satellites manufactured elsewhere. However, the country has also suffered a series of failed launches. U.S. leaders were concerned that China may have received data vital for improving its launch mechanisms from one of its commercial satellite customers. Washington was concerned about this potential transfer of technology, as Beijing could also apply it toward improving its military space program.

Specter, Michael. "Where Sputnik Once Soared into History, Hard Times Take Hold." *New York Times* (March 21, 1995), p. C1. Russia considers options to save its financially struggling space program, including commercial ventures, privatization, and international partnerships.

Annotated Bibliography

Web Documents

Bein, Michael. "Star Wars and Reactors in Space: A Canadian View." Available online. URL: http://www.animatedsoftware.com/spacedeb/canadapl.htm. Accessed April 12, 2010. In 1978 the Soviet Cosmos 954 satellite disintegrated, de-orbited, and spread radioactive debris across northern Canada. The incident prompted discussion of a ban on nuclear-powered objects in space.

David, Leonard. "Space Traffic Control: Steering Clear of Collisions." Available online. URL: http://www.space.com/businesstechnology/technology/space_traffic_040505.html. Despite the increasing amount of orbital debris threatening space assets, the international community should not rush into a space traffic management system.

Futron Corporation. "Space Transportation Costs: Trends in Price Per Pound to Orbit 1990–2000." Available online. URL: http://www.futron.com/pdf/resource_center/white_papers/FutronLaunchCostWP.pdf. Accessed June 1, 2010. Prices for space tourism are dropping, but not enough for an active, competitive market.

Krepon, Michael, and Ashley Tellis. "Another Wake-Up Call." Available online. URL: http://www.stimson.org/pub.cfm?ID=764. Accessed June 1, 2010. Space debris is a growing problem and threat to space-based assets, as demonstrated by the February 10, 2009, collision between a dead Cosmos satellite and a revenue-producing Iridium satellite.

Malik, Tariq. "What Happens When Satellites Fall." Available online. URL: http://www.space.com/businesstechnology/090123-falling-satellites.html. Accessed June 1, 2010. Ground controllers have a variety of options to protect Earth and other objects in space when a satellite goes out of control.

"SpaceShipOne Soars into Space History." Available online. URL: http://www.spaceto-day.org/Rockets/X_prize.html. Private aircraft company Scaled Composites wins $10 million prize for building aircraft that can reach space.

Spellman, Jim. "Center Encourages Aerospace Entrepreneurs." Available online. URL: http://www.cnn.com/2009/TECH/space/02/23/space.entrepreneurs/index.html. Accessed May 20, 2010. Describes "eSpace: the Center for Space Entrepreneurship," a new facility where small firms can test high-tech products in anticipation of meeting NASA specifications.

CASE STUDIES

United States

BOOKS AND BOOK CHAPTERS

Bulkeley, Rip, and Graham Spinardi. *Space Weapons: Deterrence or Delusion?* New York: Polity Press, 1986. While the Strategic Defense Initiative was introduced in 1983, space weapons technology actually dates to at least the 1920s. The authors fit SDI into this historical time line and consider the technical, political, and strategic ramifications of the new program.

Butrica, Andrew J. *Single Stage to Orbit: Politics, Space Technology, and the Quest for Reusable Rocketry.* Baltimore, Md.: Johns Hopkins University Press, 2003. The race to space has had two main paths: one-time-use rockets and reusable craft. The U.S. decision to use one path in the 1950s and 1960s and then change in the 1970s is the result of changing technologies and politics.

Cadbury, Deborah. *Space Race: The Epic Battle between America and the Soviet Union for Dominion of Space.* New York: HarperCollins, 2006. The U.S.-USSR space race is traced through the perspective of two leading figures: Wernher von Braun and Sergei Korolev.

D'Antonio, Michael. *A Ball, a Dog, and a Monkey: 1957—The Space Race Begins.* New York: Simon and Schuster, 2008. While the early days of the space age drew on hard science, a surprising number of innovations were improvisation.

Dickson, Paul. *Sputnik: The Shock of the Century.* New York: Walker, 2001. Fifty years after the Soviet Union launched the first artificial Earth satellite, Paul Dickson examines the panic it caused in the United States. He traces the multifaceted reaction of the U.S. government and evaluates competing explanations for the initial U.S. defeat in space.

Dolman, Everett C. "Space Power and U.S. Hegemony: Maintaining a Liberal World Order in the 21st Century." In *Space Weapons: Are They Needed?*, edited by John M. Logsdon and Gordon Adams. Washington, D.C.: Space Policy Institute, George Washington University, 2003, pp. 51–114. Dolman suggests that the United States is in fact the "morally superior" choice in terms of which country should seize and control space, because its political values and position as a hegemonic power will allow it to ensure that space is used in ways that are "good for all humankind."

FitzGerald, Frances. *Way Out There in the Blue: Reagan, Star Wars, and the End of the Cold War.* New York: Simon and Schuster, 2000. Ronald Reagan proposed a defense program that would create a protective bubble over the United States, shielding it from a Soviet nuclear attack. FitzGerald constructs a psychological portrait of the president through the prism of his Strategic Defense Initiative.

Hall, R. Cargill. "The Evolution of U.S. National Security Space Policy and Its Legal Foundations in the 20th Century." In *Journal of Space Law* (2007), pp. 1–104. Analysis of every U.S. National Security directive on space from Eisenhower to George W. Bush.

Hey, Nigel. *The Star Wars Enigma: Behind the Scenes of the Cold War Race for Missile Defense.* Dulles, Va.: Potomac Books, 2007. Ronald Reagan's effort to create a shield to protect the United States from a Soviet missile attack has been analyzed from many different perspectives. But Reagan's own national security adviser, Robert McFarlane, praises Hey's work as "the most rigorous scholarship yet to appear on this climactic period of history."

Hitchens, Theresa. "Weapons in Space: Silver Bullet or Russian Roulette? The Policy Implications of U.S. Pursuit of Space-Based Weapons." In *Space Weapons: Are They Needed?*, edited by John M. Logsdon and Gordon Adams. Washington, D.C.: Space Policy Institute, George Washington University, 2003, pp. 115–156. The

Annotated Bibliography

United States has little to gain and much to lose in George W. Bush's campaign to make the country the first to deploy space-based weapons. Such a narrowminded pursuit could ultimately endanger, not protect, national security. Includes data regarding the market for satellites and BMD systems.

Johnson-Freese, Joan. *Heavenly Ambitions: America's Quest to Dominate Space.* Philadelphia: University of Pennsylvania Press, 2009. The United States, like many other countries, has explored outer space, sent probes and spacecraft to study the environment, and deployed satellites. But, so far, no country has crossed the threshold and deployed weapons in outer space. Johnson-Freese warns the United States against seeking military domination of space.

_____. "U.S. Plans for Space Security." In *Collective Security in Space: Asian Perspectives,* edited by John M. Logsdon and James Clay Moltz. Washington, D.C.: George Washington University Space Policy Institute, 2008, pp. 104–119. Reviews the three documents forming the foundation of U.S. space policy and analyzes their suitability in light of China's 2007 ASAT test.

Lambeth, Benjamin S. *Mastering the Ultimate High Ground: Next Steps in the Military Uses of Space.* Santa Monica, Calif.: RAND, 2003. Analyzes conclusions of the Commission to Assess the Ballistic Missile Threat to the United States, chaired by Donald R. Rumsfeld.

Macdonald, Bruce W. *China, Space Weapons, and U.S. Security.* New York: Council on Foreign Relations, 2008. Highlights the dangers and opportunities the United States faces after successful tests of ASAT weapons by China and the United States in 2007 and 2008.

Moore, Mike. *Twilight War: The Folly of U.S. Space Dominance.* Oakland, Calif.: Independent Institute, 2008. The United States may actually make itself more vulnerable to attack if it chooses to deploy weapons in outer space. Given the enormous financial and technological hurdles to cross before that move, the United States should try not to provoke other countries into launching a space-based arms race. Instead, Washington should become a leading advocate of crafting a new, comprehensive space treaty. Not only would the treaty increase U.S. security, it would maintain the traditional U.S. appreciation of the rule of law and potentially stave off tremendous environmental damage from space-based tests or actual battles.

Mueller, Karl P. "Totem and Taboo: Depolarizing the Space Weaponization Debate." In *Space Weapons: Are They Needed?,* edited by John M. Logsdon and Gordon Adams. Washington, D.C.: Space Policy Institute, George Washington University, 2003, pp. 1–49. Should the United States place weapons in space? Mueller examines the advantages and disadvantages of space weaponization, while describing and categorizing the options available. He concludes that the cost and technological advances needed to develop and deploy space weapons are so vast that policy makers should not be frightened by any imminent threat from space.

O'Hanlon, Michael E. "Preserving U.S. Dominance While Slowing the Weaponization of Space." In *Space Weapons: Are They Needed?,* edited by John M. Logsdon and Gordon Adams. Washington, D.C.: Space Policy Institute, George

321

Washington University, 2003, pp. 239–270. The United States should delay the development of space weapons indefinitely in order to preserve its current military dominance.

Reiss, Edward. *The Strategic Defense Initiative.* New York: Cambridge University Press, 2008. Written 25 years after Ronald Reagan's "Star Wars" speech, Reiss goes beyond what the program offered to analyze the outcomes and the political and technological reasons so little of SDI was ever deployed. While new technologies make it almost impossible for a modern military to not rely on space assets, control over access to outer space is weak. The United States should take the lead to establish a more stable and secure environment.

Shayler, David J. *Skylab: America's Space Station.* New York: Springer, 2008. Documents, interviews, and other primary materials remind that Skylab was a triumph of the U.S. space program before it returned to Earth.

Spacy, William, II. "Assessing the Military Utility of Space-Based Weapons." In *Space Weapons: Are They Needed?,* edited by John M. Logsdon and Gordon Adams. Washington, D.C.: Space Policy Institute, George Washington University, 2003, pp. 157–238. Spacy discusses the technical and policy aspects of how well space weapons, especially ballistic missile defense systems, would solve a range of national security objectives. He concludes that existing land, sea, or airborne weapons can provide satisfactory and much cheaper solutions.

Stares, Paul. *The Militarization of Space: U.S. Policy, 1954–1984.* Ithaca, N.Y.: Cornell University Press, 1985. In great detail, Stares chronicles the increasing use of space in military programs since the end of World War II. While both the United States and USSR had so far refrained from deploying weapons in space, Ronald Reagan's Strategic Defense Initiative could have dramatically changed that practice.

_____. *Space and U.S. National Security.* Washington, D.C.: Brookings Institution, 1987. As the U.S. military becomes increasingly reliant on space-based assets, those assets become highly vulnerable targets. The relative costs and benefits of deploying antisatellite systems are examined and compared for both the United States and the USSR. But in the long term, the United States might be safer if it avoids deploying such weapons.

Valasek, Tomas. "The Future of U.S.-European Space Security Cooperation." In *Collective Security in Space: European Perspectives,* edited by John M. Logsdon, James Clay Moltz, and Emma S. Hinds. Washington, D.C.: George Washington University Space Policy Institute, 2007, pp. 63–74. U.S. think tanks have offered a variety of proposals regarding space security cooperation between the United States and Europe, but they have not been received policy. The issues, relative interests, and overlapping interests are outlined.

PRINT DOCUMENTS

Achenbach, Joel. "Mars and Moon Are Out of NASA's Reach for Now, Review Panel Says." *Washington Post* (9 September 2009), p. 1. The Augustine panel recommends that NASA curb its ambitions until better funding becomes available.

Annotated Bibliography

_____. "Where Will NASA's Next Giant Step Take Us?" *Washington Post* (27 October 2009), p. 1. The United States is behind schedule in replacing the space shuttle fleet. Until new systems come on line, Washington may need to send astronauts into space using commercial launch services.

Baucom, Donald R. "Space and Missile Defense." *Joint Force Quarterly* (2002): 50–55. U.S. approaches to missile defense date to the Eisenhower administration and have considered space-based missiles, hit-to-kill weapons, and interceptors.

_____. "Eisenhower and Ballistic Missile Defense: The Formative Years, 1944–1961." *Air Power History* (2004): 4–17. As president at the dawn of the missile age, Eisenhower set crucial policy maps for development and deployment of bombers, missiles, and missile defense programs.

Coleman, Joseph. "U.S., Japan Expand Missile-Defense Plan." *Washington Post* (23 June 2006). As Japan grows increasingly wary of North Korea's ballistic missile program, the United States agrees to transfer an early warning radar to a Japanese military base and to jointly develop a ballistic missile defense shield. The agreement challenges Japan's longstanding ban on arms exports.

Cook, William J., Gareth G. Cook, and Jim Impoco. "When America Went to the Moon." *U.S. News and World Report* (July 3, 1994). Twenty-five years after the lunar landing, the U.S. space program is a shadow of its own self. The United States went to the moon to beat the Soviets, but after they accomplished that goal, interest in—and funding for—space travel dried up. Also includes "where are they now" biographies of U.S. astronauts.

Daniszewski, John. "New Era Dawns in Missile Defense." *Los Angeles Times* (19 December 2002), p. A4. Moscow's initial response to the Bush administration's withdrawal from the ABM Treaty and plans for a new missile defense system is surprisingly subdued.

Desjarlais, Orville F. "The Ultimate High Ground." *Airman* (April 2003). The U.S. space program did not start as a result of *Sputnik,* but had already begun in the mid-1940s.

Frey, Adam E. "Defense of U.S. Space Assets: A Legal Perspective." *Air and Space Power Journal* (2008): 75–84. Responding to China's 2007 decision to destroy one of its own satellites, the author takes a broad look at how ASAT theory and technology follow or clash with existing international space law. The United States has two possible responses: it could move to weaponize space or it could harden its own potential targets to reduce their vulnerability. The latter response would be the more responsible and not trigger a potential arms race in space.

Gildea, Kerry. "Space Commissioner Gives DoD Poor Marks on Space Implementation Plans." *Defense Daily* (March 13, 2002). Pentagon should work more closely with space industry, consider space as a national security priority.

Grier, Peter, and Peter Lubold. "U.S. Missile Shoots Down Satellite: But Why?" *Christian Science Monitor* (22 February 2008), p. 2. Experts are skeptical about official explanation of U.S. Navy's decision to shoot down wayward spy satellite.

Hitchens, Theresa, and Victoria Samson. "Space-Based Interceptors: Still Not a Good Idea." *Georgetown Journal of International Affairs* (Summer/Fall 2004). The

United States is ready to abandon its longstanding restraint from weaponizing outer space, using several programs that date to the 1983 Strategic Defense Initiative.

Logsdon, John M., and Ray A. Williamson. "U.S.-Russian Cooperation in Space: A Good Bet." *Foreign Affairs* (Summer 1995): 39–46. The U.S. and Russian space programs must harness their combined experiences to take human spaceflight beyond Earth orbit.

Moltz, James Clay. "Reining in the Space Cowboys." *Bulletin of the Atomic Scientists* (January/February 2003). George W. Bush's plans for missile defense were examined and evaluated by domestic and international observers, but the Bush White House pushed the limits of allowable military uses of space in other programs as well. Specifically, Bush refused to consider a new outer space treaty or any other efforts at arms control.

Pollpeter, Kevin. "Competing Perceptions of the U.S. and Chinese Space Programs." *China Brief* (January 10, 2007). Compares two major national statements on the weaponization of space in 2006. While China's Defense White Paper stresses the need for peaceful uses of space, the George W. Bush administration has alarmed the international community with a new policy statement saying that the United States will pursue arms deployment in space and oppose efforts at arms control in outer space.

Sawyer, Kathy. "U.S. Proposes Space Merger with Russia." *Washington Post* (5 November 1993), p. A1. As the Russian space program struggles to survive the collapse of the USSR, President Bill Clinton proposes merging the two countries' space programs, allowing Russia to become a partner in the *International Space Station*.

Shreve, Bradley G. "The U.S., the USSR, and Space Exploration, 1957–1963." *International Journal on World Peace* (2003): 67–84. Examines the fledgling U.S. and Soviet space programs as an outgrowth of the cold war, suggesting that the shared concern over the potential damage wrought by nuclear weapons spurred both countries toward a peaceful exploration of space.

Smith, Marcia S. "U.S. Space Programs: Civilian, Military, and Commercial." *CRS Issue Brief for Congress* (October 21, 2004). China has developed a lucrative business launching satellites manufactured elsewhere. However, the country has also suffered a series of failed launches. U.S. leaders were concerned that China may have received data vital for improving its launch mechanisms from one of its commercial satellite customers. Washington was concerned about this potential transfer of technology, as Beijing could also apply it toward improving its military space program.

WEB DOCUMENTS

Air Force Historical Studies Office. "Evolution of the Department of the Air Force." Available online. URL: http://www.airforcehistory.hq.af.mil/PopTopics/Evolution.htm. Accessed June 10, 2010. Traces the origins of the U.S. Air Force from the Wright Brothers through the late 1990s. Also provides history of aircraft in warfare, including balloons.

Annotated Bibliography

Black, Samuel. "U.S. Space Diplomacy." Available online. URL: http://www.stimson.org/pub.cfm?id=692. Accessed June 1, 2010. The United States has a history of preferring space diplomacy as the primary method of resolving issues on outer space. While this line of approach has not always been successful, it should not be abandoned

Bruno, Greg. "National Missile Defense: A Status Report." Available online. URL: http://www.cfr.org/publication/18792/national_missile_defense.html. Accessed June 16, 2010. Obama's interest in missile defense programs is tempered by a concern for pragmatic, cost-effective technology.

David, Leonard. "E-Weapons: Directed Energy Warfare in the 21st Century." Available online. URL: http://www.space.com/businesstechnology/060111_e-weapons.html. Accessed June 15, 2010. New high-tech weapons, such as lasers, particle beams, and high-powered microwaves, may have more impact on warfare than the atomic bomb.

Dinerman, Taylor. "The Bush Administration and Space Weapons." Available online. URL: http://www.thespacereview.com/article/368/1. Accessed June 1, 2010. The George W. Bush administration has revived components of the SDI missile defense system and seems determined to do whatever necessary to achieve space supremacy.

Easton, Ian. "The Great Game in Space: China's Evolving ASAT Weapons Programs and Their Implications for Future U.S. Strategy." Available online. URL: http://www.project2049.net/documents/china_asat_weapons_the_great_game_in_space.pdf. Accessed May 10, 2010. China's growing ASAT weapons program threatens to shift the balance of power in the Pacific region, and possibly in space as well. Easton outlines in detail the technical specifications of China's direct-ascent, kinetic-kill ASATs and their significance for the United States. After an overview of the components of the U.S. satellite system, he concludes that the United States is highly vulnerable to an ASAT attack. In response the air force has begun an anti-ASAT project.

Grego, Laura, and David Wright. "Bush Administration National Space Policy." Available online. URL: http://www.ucsusa.org/nuclear_weapons_and_global_security/space_weapons/policy_issues/bush-administration-national.html. Accessed June 1, 2010. The Bush 2006 national space policy has raised alarms about its implied warning that the United States is now willing to weaponize space.

Krepon, Michael. "After the ASAT Tests." Available online. URL: http://www.stimson.org/pub.cfm?ID=586. Accessed June 1, 2010. These Chinese and U.S. ASAT tests raise similar questions about intent, but few observers are fooled by U.S. attempts to pass a military test off as an environmental project.

———. "Bush's ASAT Test." Available online. URL: http://www.stimson.org/pub.cfm?id=570. Accessed June 1, 2010. U.S. plans to test an ASAT in February 2008 will have long-reaching implications for global security.

Krepon, Michael, Theresa Hitchens, and Michael Katz-Hyman. "Preserving Freedom of Action in Space: Realizing the Potential and Limits of U.S. Spacepower." Available online. URL: www.stimson.org/space/pdf/SpacePower-051007.pdf. Accessed

June 1, 2010. Given the international community's common vulnerability to space-deployed weapons, the authors propose an international code of conduct to be monitored by satellite systems.

Spellman, Jim. "Center Encourages Aerospace Entrepreneurs." Available online. URL: http://www.cnn.com/2009/TECH/space/02/23/space.entrepreneurs/index.html. Accessed May 20, 2010. Describes "eSpace: the Center for Space Entrepreneurship," a new facility where small firms can test high-tech products in anticipation of meeting NASA specifications.

Union of Soviet Socialist Republics/Russian Federation
BOOKS AND BOOK CHAPTERS

Arbatov, Alexei G., and Vladimir Dvorkin. *Outer Space, Weapons, Diplomacy, Security.* Moscow: Carnegie Moscow Center, 2009. Russian scholars address the fundamental question of space security. "Will space become an arena for armed conflict, or will it remain a sphere of international cooperation?"

Brzezinski, Matthew. *Red Moon Rising: Sputnik and the Hidden Rivalries That Ignited the Space Age.* New York: Times Books, 2007. Traces the early years of the space age from both Soviet and U.S. perspectives, highlighting the contributions of key individuals.

Cadbury, Deborah. *Space Race: The Epic Battle between America and the Soviet Union for Dominion of Space.* New York: HarperCollins, 2006. The U.S.-USSR space race is traced through the perspective of two leading figures: Wernher von Braun and Sergei Korolev.

D'Antonio, Michael. *A Ball, a Dog, and a Monkey: 1957—The Space Race Begins.* New York: Simon and Schuster, 2008. While the early days of the space age drew on hard science, a surprising number of innovations were improvisation.

Dickson, Paul. *Sputnik: The Shock of the Century.* New York: Walker, 2001. Fifty years after the Soviet Union launched the first artificial Earth satellite, Paul Dickson examines the panic it caused in the United States. He traces the multifaceted reaction of the U.S. government and evaluates competing explanations for the initial U.S. defeat in space.

Harford, James. *Korolev: How One Man Masterminded the Soviet Drive to Beat America to the Moon.* New York: John Wiley, 1997. Although his identity was long a USSR state secret, Sergei Korolev was the leading force behind the Soviet space program in the 1950s and 1960s.

Ivanovich, Grujica S. *Salyut—The First Space Station: Triumph and Tragedy.* New York: Springer, 2008. Traces the development of the first Soviet space station, launched in 1971, the tragedy causing its redesign, the mysterious Almaz military version, and the replacement station, *Mir.*

Jasani, Bhupendra, and Gotthard Stein. *Commercial Satellite Imagery and GIS.* New York: Springer, 2002. Satellites are a vital tool to monitor compliance with nuclear nonproliferation treaties. Case studies from Canada, France, India, Israel, Russia, and Japan highlight the need for new verification techniques, commercial satellite imagery and processing, and cost-benefit analyses.

Annotated Bibliography

Lo, Bobo. *Axis of Convenience: Moscow, Beijing, and the New Geopolitics.* Washington, D.C.: Brookings Institution, 2008. China and Russia are reluctant allies in matters of space security. While neither wants the United States to weaponize space, none wants to lag behind either. The two countries are more rivals than strategic partners.

Mizin, Victor. "Russian Perspectives on Space Security." In *Collective Security in Space: European Perspectives,* edited by John M. Logsdon, James Clay Moltz, and Emma S. Hinds. Washington, D.C.: George Washington University Space Policy Institute, 2007, pp. 75–108. Russia is proud of its historic contributions to space exploration. Despite political and funding hardships, Moscow is committed to preserving open access to space for security and commercial purposes.

Stares, Paul. *Space and U.S. National Security.* Washington, D.C.: Brookings Institution, 1987. As the U.S. military becomes increasingly reliant on space-based assets, those assets become highly vulnerable targets. The relative costs and benefits of deploying antisatellite systems are examined and compared for both the United States and the USSR. But in the long term, the United States might be safer if it avoids deploying such weapons.

Yost, David Scott. *Soviet Ballistic Missile Defense and the Western Alliance.* Cambridge, Mass.: Harvard University Press, 1988. Yost challenges the conventional notion that Moscow had to scramble to catch up with U.S. missile defense plans in the wake of the Strategic Defense Initiative. Instead, he demonstrates that modernization and upgrading had begun long before Reagan's "Star Wars" speech. He also notes a number of potential loopholes in current arms control agreements that need to be modified accordingly.

PRINT ARTICLES

Daniszewski, John. "New Era Dawns in Missile Defense." *Los Angeles Times* (19 December 2002), p. A4. Moscow's initial response to the Bush administration's withdrawal from the ABM Treaty and plans for a new missile defense system is surprisingly subdued.

Lemonick, Michael D. "Surging Ahead: The Soviets Overtake the U.S. as the No. 1 Spacefaring Nation." *Time* (October 5, 1987), pp. 64–71. In the final years of the USSR, Soviet space science outpaced the United States in terms of astronomy, astrophysics, space medicine, and space transportation. Yet there are limits; one reason Moscow launched so many satellites is that Soviet-made satellites tend to rapidly disintegrate.

_____. "A Flap over Reactors in Orbit." *Time* (February 20, 1989), p. 80. Soviets have deployed three dozen nuclear-powered satellites; what would happen if one fell to Earth?

Logsdon, John M., and Ray A. Williamson. "U.S.-Russian Cooperation in Space: A Good Bet." *Foreign Affairs* (Summer 1995): 39–46. The U.S. and Russian space programs must harness their combined experiences to take human spaceflight beyond Earth orbit.

Oberg, James E. "Disaster at the Cosmodrome." *Air & Space Magazine* (December 1990). Moscow long suppressed any account of the 1960 Nedelin launchpad fire.

Sawyer, Kathy. "U.S. Proposes Space Merger with Russia." *Washington Post* (5 November 1993), p. A1. As the Russian space program struggles to survive the collapse of the USSR, President Bill Clinton proposes merging the two countries' space programs, allowing Russia to become a partner in the *International Space Station.*

Shreve, Bradley G. "The U.S., the USSR, and Space Exploration, 1957–1963." *International Journal on World Peace* (2003): 67–84. Examines the fledgling U.S. and Soviet space programs as an outgrowth of the cold war, suggesting that the shared concern over the potential damage wrought by nuclear weapons spurred both countries toward a peaceful exploration of space.

Specter, Michael. "Where Sputnik Once Soared into History, Hard Times Take Hold." *New York Times* (21 March 1995), p. C1. Russia considers options to save its financially struggling space program, including commercial ventures, privatization, and international partnerships.

Stone, Richard. "Russia's Last Shot at Space." *Science* (1997), pp. 1,780–1,782. Instead of launching a new era of Russian space research, the 1996 probe to Mars fell back to Earth four hours after launch. With lost defense contracts and a plummeting budget, the Institute for Space Research in Moscow is struggling to survive.

WEB ARTICLES

Bein, Michael. "Star Wars and Reactors in Space: A Canadian View." Available online. URL: http://www.animatedsoftware.com/spacedeb/canadapl.htm. Accessed April 12, 2010. In 1978 the Soviet Cosmos 954 satellite disintegrated, de-orbited, and spread radioactive debris across northern Canada. The incident prompted discussion of a ban on nuclear-powered objects in space.

Kramnik, Ilya. "Baikonur Space Center Marks 55th Anniversary." Available online. URL: http://www.spacedaily.com/reports/Baikonur_Space_Center_Marks_55th_Anniversary_999.html. Accessed June 12, 2010. The Soviet space center in Baikonur outlasted the USSR but now faces severe economic shortages.

O'Connor, Sean. "Russian/Soviet Anti-Ballistic Systems." Available online. URL: http://www.ausairpower.net/APA-Rus-ABM-Systems.html. Accessed June 10, 2010. The origins of Russia's contemporary missile defense system can be traced to the Nazi V-2 rocket. Each system is described in detail, accompanied by extensive photographic coverage.

China

BOOKS AND BOOK CHAPTERS

Banerjee, Dipankar. "Indian Perspectives on Regional Space Security." In *Collective Security in Space: Asian Perspectives*, edited by John M. Logsdon and James Clay Moltz. Washington, D.C.: George Washington University Space Policy Institute, 2008, pp. 120–130. The 1991 Gulf War prompted China and India to develop space programs for military use, but it is not too late for the two countries to

Annotated Bibliography

cooperate. The Shanghai Cooperation Organization and ASEAN are cited as possible models.

Chang, Iris. *Thread of the Silkworm.* New York: Basic Books, 1995. Chang profiles Chinese engineer Qian Xuesen (Tsien Hsue-shen), considered to be the father of the Chinese space program. While working at the U.S. government's Jet Propulsion Laboratory after World War II, Qian became caught up in the Red Scare and ultimately was deported to China, where he established the rocketry and space program in his native country.

Harvey, Brian. *China's Space Program: From Conception to Manned Spaceflight.* New York: Springer, 2004. Harvey adds to his surveys of space programs around the world by focusing on China, now the third largest space program in the world. Beijing began a missile program with Soviet help and the space program has always had significant military involvement. The book ends with China's first manned launch, making it somewhat dated.

Johnson-Freese, Joan. "U.S. Plans for Space Security." In *Collective Security in Space: Asian Perspectives,* edited by John M. Logsdon and James Clay Moltz. Washington, D.C.: George Washington University Space Policy Institute, 2008, pp. 104–119. Reviews the three documents forming the foundation of U.S. space policy and analyzes their suitability in light of China's 2007 ASAT test.

Lo, Bobo. *Axis of Convenience: Moscow, Beijing, and the New Geopolitics.* Washington, D.C.: Brookings Institution, 2008. China and Russia are reluctant allies in matters of space security. While neither wants the United States to weaponize space, none wants to lag behind either. The two countries are more rivals than strategic partners.

MacDonald, Bruce W. *China, Space Weapons, and U.S. Security.* New York: Council on Foreign Relations, 2008. While new technologies make it almost impossible for a modern military to not rely on space assets, control over access to outer space is weak. The United States should take the lead to establish a more stable and secure environment.

Nishihara, Masahi. "The Changing Asian Security Environment." In *Collective Security in Space: Asian Perspectives,* edited by John M. Logsdon and James Clay Moltz. Washington, D.C.: George Washington University Space Policy Institute, 2008, pp. 7–13. Asia faces a range of space security issues, from nuclear ambition in North Korea to China's increasing military presence and political instability in many states. Despite the common problems, most countries still put national interests ahead of regional stability.

Pollpeter, Kevin. "The Chinese Vision of Space Military Operations." In *China's Revolution in Doctrinal Affairs: Emerging Trends in the Operational Art of the Chinese People's Liberation Army,* edited by James C. Mulvenon and David Finkelstein. Alexandria, Va.: CNA Corporation, 2005, pp. 329–369. Military leaders in modern China believe that war in space in inevitable, a conviction reinforced by the bellicose statements from the administration of U.S. president George W. Bush. However, Chinese strategists are aware of the U.S. dependence on space-deployed military assets, such as satellites, and think China might actually win a space war

329

by attacking these vulnerable targets. The report draws from numerous Chinese statements, newspaper reports, and military documents on various form of military space options, including the importance of space in fighting wars, the concept of space war, and its characteristics, the platforms and weapons used in space war and their military applications.

Yang, Junhua. "Developing Space Peacefully for the Benefit of Humanity." In *Collective Security in Space: Asian Perspectives,* edited by John M. Logsdon and James Clay Moltz. Washington, D.C.: George Washington University Space Policy Institute, 2008, pp. 24–32. China is developing its own strategy for mitigating the increasing problem of outer space debris and environmental damage. Yang calls for an improved space object registry and international management of this common problem.

Zhong, Jing. "China and Space Security." In *Collective Security in Space: Asian Perspectives,* edited by John M. Logsdon and James Clay Moltz. Washington, D.C.: George Washington University Space Policy Institute, 2008, pp. 75–93. China supports the peaceful use of space, but until a new space arms control treaty is enacted, Beijing will pursue a "limited defense capability" in space.

PRINT ARTICLES

Broad, William J. "Orbiting Junk, Once a Nuisance, Is Now a Threat." *New York Times* (6 February 2007). China's January 2007 ASAT test may have created enough debris to set the Kessler Syndrome in motion.

Gill, Bates, and Martin Kleiber. "China's Space Odyssey: What the Antisatellite Test Reveals about Decision-Making in Beijing." *Foreign Affairs* (May/June 2007). China's 2007 ASAT may tell less about the Chinese space program than the lack of coordination among its many government agencies.

Hennock, Mary, Adam B. Kushner, and Jason Overdorf. "The New Space Race." *Newsweek* (October 20, 2008). China and India are locked in a space race where the prize is national prestige and commercial payoff. Japan is watching the race with high interest.

Pollpeter, Kevin. "Competing Perceptions of the U.S. and Chinese Space Programs." *China Brief* (January 10, 2007). Compares two major national statements on the weaponization of space in 2006. While China's Defense White Paper stresses the need for peaceful uses of space, the George W. Bush administration has alarmed the international community with a new policy statement saying that the United States will pursue arms deployment in space and oppose efforts at arms control in outer space.

_____. "Motives and Implications behind China's ASAT Test." *China Brief* (January 24, 2007). China's ASAT weapon test in January 2007 seems to contradict its long-held opposition to space weapons.

_____. "The Stars of China's Space Program: The Rise of a "Space Gang"?" *China Brief* (September 19, 2007). Profiles key figures in Chinese space program, including Zhang Qingwei, head of the Commission on Science, Technology, and Industry for National Defense; Chi Wanchun, political commissar of the General Arms Department; and others.

Saunders, Phillip C., and Charles D. Lutes. "China's ASAT Test: Motivations and Implications." *Joint Forces Quarterly* (Fall 2009): 39–45. The confusing and contradictory official statements coming from Beijing following China's January 2007 destruction of an aging weather satellite suggests that the country's space program remains highly secret and separate from other governmental bodies.

"Shooting the Moon." *Economist* (September 27, 2008). China has been highly secretive about its manned spaceflight program, but a successful space walk could change international perceptions of the country.

Smith, Marcia S. "China's Space Program: A Brief Overview Including Commercial Launches of U.S.-Built Satellites." *CRS Issue Brief for Congress* (September 3, 1998). China has developed a lucrative business launching satellites manufactured elsewhere. However, the country has also suffered a series of failed launches. U.S. leaders were concerned that China may have received data vital for improving its launch mechanisms from one of its commercial satellite customers. Washington was concerned about this potential transfer of technology, as Beijing could also apply it toward improving its military space program.

Snyder, Scott, and See-won Byun. "Year of China-DPRK Friendship; North's Rocket Fizzles." *Comparative Connections* (2009). Explores the 60-year history of relations between North Korea and China, with extensive information on trade and economic ties.

WEB DOCUMENTS

Baker, John C., and Kevin Pollpeter. "A Future for U.S.-China Space Cooperation?" Available online. URL: http://www.rand.org/commentary/121304SN.html. Accessed April 20, 2010. While opportunities exist for China to cooperate with the United States and Western Europe on a variety of space programs, Washington is worried about possible transfers of military technology.

Brown, Peter J. "China Fears India-Japan Space Alliance." Available online. URL: www.atimes.com/atimes/South_Asia/JK12Df02.html. Accessed June 15, 2010. Beijing is alarmed that Japan and India have agreed to cooperate on a disaster management–imaging program. China fears that the same satellites used to photograph disaster zones could also be used to observe sensitive military sites. India has already reported about Chinese efforts to disguise troop locations with tunnels. Japan is increasing its satellite program since the Space Basic Law now allows deployment of defense assets.

"China's Attitude Toward Outer Space Weapons." Available online. URL: www.nti.org/db/china/spacepos.htm. Accessed April 20, 2010. China believes that current agreements and treaties banning outer space weapons are inadequate. Therefore it has consistently proposed various draft agreements at the UN Conference on Disarmament's Committee on the Prevention of an Arms Race in Outer Space. Beijing is very concerned about a possible U.S. missile defense system, as it would negate China's nuclear arsenal. China might then be forced into an arms race in order to prevent the United States from achieving global dominance.

Easton, Ian. "The Great Game in Space: China's Evolving ASAT Weapons Programs and Their Implications for Future U.S. Strategy." Available online. http://www.project2049.net/documents/china_asat_weapons_the_great_game_in_space.pdf. Accessed May 10, 2010. China's growing ASAT weapons program threatens to shift the balance of power in the Pacific region, and possibly in space as well. Easton outlines in detail the technical specifications of China's direct-ascent, kinetic-kill ASAT and their significance for the United States. After an overview of the components of the U.S. satellite system, he concludes that the United States is highly vulnerable to an ASAT attack. In response the air force has begun an anti-ASAT project.

Futron Corporation. "China and the Second Space Age." Available online. URL: http://www1.futron.com/resource_center/publications/publications_archive.htm. Accessed March 15, 2010. This white paper on the history and future of China's space program was commissioned by NASA following China's first manned launch. In the next 20 years, China will likely increase its industrial, economic, and commercial systems to underpin its varied space aspirations.

Krepon, Michael. "After the ASAT Tests." Available online. URL: http://www.stimson.org/pub.cfm?ID=586. Accessed June 1, 2010. These Chinese and U.S. ASAT tests raise similar questions about intent, but few observers are fooled by U.S. attempts to pass a military test off as an environmental project.

_____. "Bush's ASAT Test." Available online. URL: http://www.stimson.org/pub.cfm?id=570. Accessed June 1, 2010. U.S. plans to test an ASAT in February 2008 will have long-reaching implications for global security.

Krepon, Michael, and Ashley Tellis. "Another Wake-up Call." Available online. URL: http://www.stimson.org/pub.cfm?ID=764. Accessed June 1, 2010. Space debris is a growing problem and threat to space-based assets, as demonstrated by the February 10, 2009, collision between a dead Cosmos satellite and a revenue-producing Iridium satellite.

Manriquez, Manuel. "Japan's Space Law Revision: The Next Step Toward Re-Militarization?" Available online. URL: www.nti.org/e_research/e3_japan_remilitarization0108. htm. Accessed June 10, 2010. Japan could become a major player in the satellite industry if the government revokes a 1969 ban on military-related research. Also includes analysis of North Korea and China weapons and their influence on Tokyo's policies.

Morring, Frank, Jr. "Worst Ever: Chinese Anti-Satellite Test Boosted Space-Debris Population by 10% in an Instant." *Aviation Week and Space Technology* (2007), p. 20. China's 2007 ASAT test not only created one of the largest debris clouds to date, it may set a precedent for other countries to begin potentially hostile applications of space assets.

Pollpeter, Kevin. "Building for the Future: China's Progress in Space Technology during the Tenth 5-Year Plan and the U.S. Response." Available online. URL: http://www.strategicstudiesinstitute.army.mil/pdffiles/PUB852.pdf. Accessed May 10, 2010. China's increasing proficiency in antisatellite technology, as exhibited by its

Annotated Bibliography

January 2007 destruction of its own weather satellite, may erode the lead in space power that the United States has enjoyed for decades. Pollpeter considers China's increasing space power to be a component of its broader emergence on the world scene as a rising military, economic, and political power.

Rincon, Paul. "China 'Could Reach Moon by 2020.'" Available online. URL: http://news.bbc.co.uk/2/hi/science/nature/7506715.stm. Accessed June 1, 2010. After successful ASAT tests and manned flights, China is setting its sights on a lunar landing. Such a feat would be a major challenge to the longstanding U.S. preeminance in space.

———. "What's Driving China's Space Efforts?" Available online. URL: http://news.bbc.co.uk/2/hi/science/nature/7635397.stm. Accessed June 1, 2010. Economic development is driving China's space program. Evidence of this motive is visible in official reporting on China's first space walk, which stresses that Chinese-made technologies were used in the spacecraft and the spacesuit. Beijing will also gain political and military prestige from the program, although there are fears of an emerging space race in Asia, among China, Japan, North Korea, and India.

Toki, Masako. "Missile Defense in Japan." Available online. URL: http://www.thebulletin.org/web-edition/features/missile-defense-japan. Accessed June 10, 2010. Since the 1980s, the United States and Japan have cooperated on missile defense. When North Korea tested a ballistic missile, Japanese leaders decided to strengthen their defenses. After relevant changes in its National Defense Program, Japan began deploying a missile defense system in 2009.

Valasek, Tomas. "The Future of U.S.-European Space Security Cooperation." In *Collective Security in Space: European Perspectives*, edited by John M. Logsdon, James Clay Moltz, and Emma S. Hinds. Washington, D.C.: George Washington University Space Policy Institute, 2007, pp. 63–74. U.S. think tanks have offered a variety of proposals regarding space security cooperation between the United States and Europe but they have not been received policy. The issues, relative interests, and overlapping interests are outlined.

Wolter, Detlev. "Common Security in Outer Space and International Law: A European Perspective." In *Collective Security in Space: European Perspectives*, edited by John M. Logsdon, James Clay Moltz, and Emma S. Hinds. Washington, D.C.: George Washington University Space Policy Institute, 2007, pp. 17–28. Europe has a long history of collaborative approaches to common problems and a shared commitment to the peaceful use of space. The European Union, Organization for Security and Cooperation in Europe, and NATO offer frameworks for developing codes of conduct, rules of the road, and other cooperative approaches to managing outer space.

Europe

BOOKS AND BOOK CHAPTERS

Biddle, Wayne. *Dark Side of the Moon: Wernher von Braun, the Third Reich, and the Space Race.* New York: W. W. Norton, 2009. Modern rocketry began in Nazi Germany during World War II.

Bonnet, Roger, and Vittorio Manno. *International Cooperation in Space: The Example of the European Space Agency.* Cambridge, Mass.: Harvard University Press, 1994. The formation of the European Space Agency has been an exemplar of international cooperation, registering multiple successes despite sometimes competing technological, economic, and political agendas. The ESA and NASA formed an equally impressive series of collaborative projects, especially the Ulysses solar probe and the international space station.

Harvey, Brian. *Europe's Space Programme: To Ariane and Beyond.* New York: Springer, 2003. Seven European countries decided to merge their nascent space programs in 1974 to maximize assets and research. Harvey traces the origins not only of the European Space Agency, but also the individual national programs, beginning with the Nazi V-2 rocket in the 1930s, and examines why collaboration proved so successful, especially in the case of the Arianne rocket.

Jasani, Bhupendra, and Gotthard Stein. *Commercial Satellite Imagery and GIS.* New York: Springer, 2002. Satellites are a vital tool to monitor compliance with nuclear nonproliferation treaties. Case studies from Canada, France, India, Israel, Russia, and Japan highlight the need for new verification techniques, commercial satellite imagery and processing, and cost-benefit analyses.

Krige, John, and Arturo Russo. *A History of the European Space Agency, 1958–1987.* 2 vols. Noordwijk, the Netherlands: European Space Agency, 2000. This two-volume compendium is the official history of the European Space Agency. It begins with western Europe's reaction to the 1957 launch of *Sputnik* and continues through the ambitious—and now largely abandoned—agenda put forth at the 1987 ministerial meeting at the Hague. Each volume carefully details the leaders and projects that were able to span cultural differences and national rivalries to create a respected space program. The ESA History Project Web site (http://www.esa.int/esaCP/ESAQRHPZ9NC_index_0.html#subhead9) has the book available in PDF form, along with separate online resources such as 30 years of related photographs and diagrams.

Pasco, Xavier. "Toward a Future European Space Surveillance System: Developing a Collaborative Model for the World." In *Collective Security in Space: European Perspectives*, edited by John M. Logsdon, James Clay Moltz, and Emma S. Hinds. Washington, D.C.: George Washington University Space Policy Institute, 2007, pp. 51–62. A European space surveillance system would underline the region's identity as a legitimate international actor, build on existing integration efforts, and support the European Security and Defense Policy.

PRINT ARTICLES

Jolis, Anne. "Woes Mount at Europe's Bid for GPS Rival." *Wall Street Journal* (17 January 2007). Europe's version of the U.S. global positioning system, Galileo, is behind schedule and considerably over budget. While the U.S. version is owned by the U.S. Air Force, Galileo is funded by a range of investors, including China and European taxpayers.

Annotated Bibliography

Squeo, Anne Marie. "U.S. Considers European Role for Missile-Defense Programs." *Wall Street Journal* (7 June 2002), p. A4. The George W. Bush administration approached European military contractors about participating in the proposed new U.S. missile defense system. Cooperation would raise several difficult issues, especially as many west European governments opposed Bush's foreign policy. With contracts estimated at $8.2 billion, critics suggested Washington was trying to buy European support.

WEB DOCUMENTS

Amos, Jonathan. "Berlin Unveils 'Crewed Spaceship.'" Available online. URL: http://news.bbc.co.uk/2/hi/science/nature/7419793.stm. Accessed June 1, 2010. The European Aeronautic Defense and Space Company has a new spaceship capable of carrying crew to the *International Space Station* and returning with cargo. EADSC bills the craft as an upgraded version of the non-manned *Jules Verne* transport used by the European Space Agency. Such a craft would allow Europeans to reach the station without U.S. or Russian help.

———. "European Spaceport's Sky-High Ambition." Available online. URL: http://news.bbc.co.uk/2/hi/science/nature/7278409.stm. Accessed June 1, 2010. The European Space Agency opens its new space launch facility in French Guiana, paving the way for it to expand its already extensive manifest of satellite launches into cargo and perhaps crew shuttle services for the *International Space Station*.

———. "Huge Space Truck Races into Orbit." Available online. URL: http://news.bbc.co.uk/2/hi/science/nature/7285796.stm. Accessed June 1, 2010. The *Jules Verne*, a European-made spacecraft, made its successful debut launching atop an Ariane rocket from the Kourou spaceport in French Guiana. The spacecraft will allow the European Space Agency to ferry cargo to the *International Space Station* without help from the United States or Russia.

"Arianespace: 30th Anniversary, New Momentum." Available online. URL: http://www.space-travel.com/reports/Arianespace_30th_Anniversary_New_Momentum_999.html. Accessed June 1, 2010. Arianespace continues to register successes with its Ariane launcher and Guiana Space Center.

Emerging Players
BOOKS AND BOOK CHAPTERS

Aoki, Setsuko. "Japanese Perspectives on Space Security." In *Collective Security in Space: Asian Perspectives*, edited by John M. Logsdon and James Clay Moltz. Washington, D.C.: George Washington University Space Policy Institute, 2008, pp. 47–66. The existing international legal framework provides little security for Asia's needs in outer space. Asia could take the lead in setting regional guidelines that could be a model for a new international regime.

Banerjee, Dipankar. "Indian Perspectives on Regional Space Security." In *Collective Security in Space: Asian Perspectives*, edited by John M. Logsdon and James Clay Moltz. Washington, D.C.: George Washington University Space Policy Institute,

2008, pp. 120–130. The 1991 Gulf War prompted China and India to develop space programs for military use, but it is not too late for the two countries to cooperate. The Shanghai Cooperation Organization and ASEAN are cited as possible models.

Berner, Steven. *Japan's Space Program: A Fork in the Road?* Santa Monica, Calif.: RAND, 2003. Berner traces the development of the Japanese satellite launch program. But while once it was a world leader in launch technology, in the late 1990s it experienced a wave of launch failures.

Biddington, Brett. "An Australian Perspective on Space Security." In *Collective Security in Space: Asian Perspectives*, edited by John M. Logsdon and James Clay Moltz. Washington, D.C.: George Washington University Space Policy Institute, 2008, pp. 94–103. Australia has not pursued its own space program, but it hosts systems for the U.S. and UK militaries. But as land-use demands increase, the government will have to readdress its policies on space-related technologies.

Chubin, Shahram. *Iran's Nuclear Ambitions.* Washington, D.C.: The Carnegie Endowment for International Peace, 2006. Clear, evenhanded analysis of pressures leading Iran to seek nuclear weapons, of the choices open to the United States and other countries in confronting Iran's nuclear program.

Cordesman, Anthony H., and Khalid R. Al-Rodhan. *Iran's Weapons of Mass Destruction: The Real and Potential Threat.* Washington, D.C.: The CSIS Press, Center for Strategic and International Studies, 2006. Covers all facets of Iran's weapons of mass destruction, analyzing Iran's motivation for acquiring WMD capabilities; the history of its WMD program; its chemical, biological, and nuclear capabilities; and its delivery options, including its missile program, air force, and Revolutionary Guards.

Harvey, Brian. *The Japanese and Indian Space Programmes: Two Roads into Space.* New York: Springer, 2000. While China's developing space program has received a considerable amount of public attention, Japan and India have built strong reputations for building and launching satellites. Together they could be a formidable balance to Beijing's growing international role.

Jasani, Bhupendra, and Gotthard Stein. *Commercial Satellite Imagery and GIS.* New York: Springer, 2002. Satellites are a vital tool to monitor compliance with nuclear nonproliferation treaties. Case studies from Canada, France, India, Israel, Russia, and Japan highlight the need for new verification techniques, commercial satellite imagery and processing, and cost-benefit analyses.

Kee, Changdon. "A South Korean Perspective on Strengthening Space Security in East Asia." In *Collective Security in Space: Asian Perspectives*, edited by John M. Logsdon and James Clay Moltz. Washington, D.C.: George Washington University Space Policy Institute, 2008, pp. 14–23. South Korea is emerging as a contender in the global satellite market, but redundant programs such as GPS, Glonass, and Galileo are inhibiting growth and wasting resources.

Kim, Kyung-Min. "South Korean Capabilities for Space Security." In *Collective Security in Space: Asian Perspectives*, edited by John M. Logsdon and James Clay Moltz. Washington, D.C.: George Washington University Space Policy Institute, 2008, pp. 67–74. The Korean peninsula is late to enter the space security arena. South Korea now seeks to develop a comprehensive system to monitor North Korea's nuclear and missile programs.

Lochan, Rajeev. "Some Reflections on Collective Security in Space." In *Collective Security in Space: Asian Perspectives*, edited by John M. Logsdon and James Clay Moltz. Washington, D.C.: George Washington University Space Policy Institute, 2008, pp. 33–46. India has used its space program to promote economic development and insisted on home-grown technology for satellites and launchers. Lochan calls on other countries to follow India's example and eschew deploying weapons in outer space.

Menon, Raja, ed. *Weapons of Mass Destruction.* Thousand Oaks, Calif.: Sage, 2004. India has a distinct view and a range of policy options should a state or non-state actor use chemical, nuclear, or biological weapons.

Nishihara, Masahi. "The Changing Asian Security Environment." In *Collective Security in Space: Asian Perspectives*, edited by John M. Logsdon and James Clay Moltz. Washington, D.C.: George Washington University Space Policy Institute, 2008, pp. 7–13. Asia faces a range of space security issues, from nuclear ambition in North Korea to China's increasing military presence and political instability in many states. Despite the common problems, most countries still put national interests ahead of regional stability.

Suzuki, Kazuto. "Japanese Steps toward Regional and Global Confidence Building." In *Collective Security in Space: Asian Perspectives*, edited by John M. Logsdon and James Clay Moltz. Washington, D.C.: George Washington University Space Policy Institute, 2008, pp. 131–145. Japan's space program is increasingly forcing it to decide between its official pacifism and the reality of hostile, nuclear-armed neighbors.

PRINT ARTICLES

Andrews, Edmund L. "Tiny Tonga Seeks Satellite Empire in Space." *New York Times* (28 August 1990). Tonga exploited a loophole in international law to claim key orbital positions for satellite communications.

Cha, Victor D. "Pyongbang! Washington's Korea Conundrum." *Foreign Affairs* (April 10, 2009). A former U.S. delegate to the Six-Party Talks evaluates North Korea's latest missile test and discusses policy options for the Obama administration.

Coleman, Joseph. "U.S., Japan Expand Missile-Defense Plan." *Washington Post* (23 June 2006). As Japan grows increasingly wary of North Korea's ballistic missile program, the United States agrees to transfer an early warning radar to a Japanese military base and to jointly develop a ballistic missile defense shield. The agreement challenges Japan's longstanding ban on arms exports.

Guiney, Jessica. "India's Space Ambitions: Headed Toward Space War?" *CDI Policy Brief* (2008). ASAT tests by the United States and China have opened the door for India to deploy space-based military assets, but so far New Delhi has maintained its stance on peaceful uses of outer space. But with the country's thriving satellite industry, and the possibly changing international norms on weaponization, India could pursue its own ASAT program. The paper includes a roster of India's current satellites.

Hennock, Mary, Adam B. Kushner, and Jason Overdorf. "The New Space Race." *Newsweek* (October 20, 2008). China and India are locked in a space race where the prize is national prestige and commercial payoff. Japan is watching the race with very high interest.

Hildreth, Steven A. "North Korean Ballistic Missile Threat to the United States." *CRS Issue Brief for Congress* (October 18, 2005). North Korea claims to have tested several ballistic missiles and at least one underground nuclear test, but they have not successfully combined the two technologies.

Hwang, Chin Young. "Space Activities in Korea: History, Current Programs, and Future Plans." *Space Policy* (August 2006): 194–199. Despite its short history in space, Korea has been increasing its technological capabilities with the successful experience of several national projects. The Korean government established a long-term space development plan in 1996, which suggests a clear way forward for space development up to 2015.

Katzman, Kenneth. "Iran's Strategic Capabilities and Weapons of Mass Destruction." *CRS Issue Brief for Congress* (September 2007). Despite UN warnings, Iran continues to seek weapons of mass destruction.

Murphy, John, and Yumiko Ono. "Japanese Pacifism Is Put to the Test." *Wall Street Journal* (5 April 2009). North Korea's missile launches toward Japanese territory have prompted Tokyo to reconsider its postwar opposition to military systems.

Natsios, Andrew. "The Politics of Famine in North Korea." *USIP Special Report* (August 2, 1999). The North Korean famine of the 1990s was the result of government mismanagement and deliberate decisions to provide food aid to some regions but not others. Includes famine timeline.

Ramstad, Evan. "Korean Blast Draws Outrage." *Wall Street Journal* (26 May 2009). North Korea successfully detonated a nuclear warhead and tested three short-range missiles in May 2009, prompting an emergency session of the United Nations and international condemnation.

Sims, Calvin. "South Korea Plans to Begin Rocket Program." *New York Times* (15 January 2000), p. A3. South Korea's announcement that it sought domestic satellite launching capacity significantly changes the security calculus on the Korean peninsula. The United States feared that North Korea would react negatively, as the same rockets could also launch missiles at Seoul's northern neighbor.

Smith, R. Jeffrey, and Stella Kim. "S. Korean Launch Raises Questions." *New York Times* (18 August 2009), p. A8. When the United States refused to help South Korea create a domestic satellite launch program, Seoul turned to Russia. The $200 million investment paid off, when the first South Korean launch vehicle lifted off on August 18. North Korea, however, complained about its receiving a much harsher international reaction when it tests its own missiles.

Snyder, Scott, and See-won Byun. "Year of China-DPRK Friendship; North's Rocket Fizzles." *Comparative Connections* (2009). Explores the 60-year history of

Annotated Bibliography

relations between North Korea and China, with extensive information on trade and economic ties.

Warrick, Joby, and R. Jeffrey Smith. "U.S.–Russian Team Deems Missile Shield in Europe Ineffective." *Washington Post* (19 May 2009). New report from East-West Institute concludes that Iran will need another decade to produce a nuclear missile that could reach the United States. Experts also think that proposed US missile defense program would not be able to intercept this type of missile.

WEB DOCUMENTS

Albright, David, and Paul Brannan. "The North Korean Plutonium Stock, February 2007." Available online. URL: www.isis-online.org/publications/dprk/DPRKplutoniumFEB.pdf. Accessed May 10, 2010. North Korea began accumulating plutonium in 1986 and is believed to have enough to manufacture between five and 12 crude nuclear weapons. However, it may not have an appropriate launch vehicle, as any nuclear weapon would be too large for the current Nodong medium-range missile. The article examines the debates around international sanctions and construction of the Yongbyon reactor and includes aerial photographs of the reactor facility.

Bajoria, Jayshree. "North Korea after Kim." Available online. URL: www.cfr.org/publication/17322. Accessed May 10, 2010. North Korea is already one of the most closed political systems today, but it is even harder to know what is happening with the country's mercurial leader, Kim Jong-Il. Reports of Kim's health suggest a succession might happen soon, but the possible candidates seem unlikely to bring the country into the mainstream international society.

Bein, Michael. "Star Wars and Reactors in Space: A Canadian View." Available online. URL: http://www.animatedsoftware.com/spacedeb/canadapl.htm. Accessed April 12, 2010. In 1978 the Soviet Cosmos 954 satellite disintegrated, de-orbited, and spread radioactive debris across northern Canada. The incident prompted discussion of a ban on nuclear-powered objects in space.

Brown, Peter J. "China Fears India-Japan Space Alliance." Available online. URL: www.atimes.com/atimes/South_Asia/JK12Df02.html. Accessed June 15, 2010. Beijing is alarmed that Japan and India have agreed to cooperate on a disaster management-imaging program. China fears that the same satellites used to photograph disaster zones could also be used to observe sensitive military sites. India has already reported about Chinese efforts to disguise troop locations with tunnels. Japan is increasing its satellite program since the Space Basic Law now allows deployment of defense assets.

Clark, Stephen. "Dawn Launch Sends Japanese Orbiter on the Way to Venue." Available online. URL: http://spaceflightnow.com/h2a/akatsuki/100520launch/. Accessed May 20, 2010. Japan launches Akatsuki, the world's first interplanetary weather satellite, on two-year mission to study Venus. Sent aloft from an H-2A rocket at the Tanegashima Space Center, the launch demonstrates Japan's impressive rocketry and satellite building skills.

Hwang, Chin Young. "Space Activities in Korea: History, Current Programs, and Future Plans." *Space Policy* (August 2006): 194–199. Despite its short history in space, Korea has been increasing its technological capabilities with the successful experience of several national projects. The Korean government established a long-term space development plan in 1996, which suggests a clear way forward for space development up to 2015.

"IRIS & Iran's Emerging Space Program." Available online. URL: http://www.fas.org/nuke/guide/iran/missile/iris.htm. Accessed June 1, 2010. Iran's growing technological advances have international policy makers worried about its motives.

Kotani, Rui. "Japan's Recent Step-up in Missile Defense." Available online. URL: http://www.cdi.org/program/relateditems.cfm?typeID=%288%29&programID=6. Accessed June 10, 2010. After North Korea announced it would test nuclear weapons, Japan announced its intention to acquire a U.S.-made missile defense system. However, a missile defense system would violate Japan's constitution and be technically difficult to deploy.

Manriquez, Manuel. "Japan's Space Law Revision: The Next Step Toward Re-Militarization?" Available online. URL: www.nti.org/e_research/e3_japan_remilitarization0108.htm. Accessed June 10, 2010. Japan could become a major player in the satellite industry if the government revokes a 1969 ban on military-related research. Also includes analysis of North Korea and China weapons and their influence on Tokyo's policies.

"North Korea's Missile Program." Available online. URL: http://newsvote.bbc.co.uk/mpapps/pagetools/print/news.bbc.co.uk/2/hi/asia-pacific/2564241.stm?ad=1. Accessed June 28, 2010. North Korea's missile program began in 1976 and has drawn upon foreign technology to develop its current arsenal. Includes specifications for major missile programs.

"Opposing Weapons in Space." Available online. URL: http://www.ploughshares.ca/libraries/monitor/mons02a.html. Accessed June 15, 2010. Canada plays a leading role in the international effort to prevent weapons in outer space, both in terms of formal definitions of the issues and proposed treaties at the UN Conference on Disarmament.

Toki, Masako. "Missile Defense in Japan." Available online. URL: http://www.thebulletin.org/web-edition/features/missile-defense-japan. Accessed June 10, 2010. Since the 1980s, the United States and Japan have cooperated on missile defense. When North Korea tested a ballistic missile, Japanese leaders decided to strengthen their defenses. After relevant changes in its National Defense Program, Japan began deploying a missile defense system in 2009.

Zissis, Carin, and Jayshree Bajoria. "The Six-Party Talks on North Korea's Nuclear Program." Available online. URL: http://www.cfr.org/publication/13593/. Accessed May 5, 2009. The Six-Party Talks are the primary international forum for discussing North Korea's nuclear program, but the forum faces obstacles from the North Korean regime, varying national approaches, U.S. resistance to bilateral talks, and the status of the nuclear reactors in existing international agreements.

Chronology

MILITARIZATION OF SPACE

1904

- *January:* Wright brothers offer to provide aircraft to War Department; their offer is turned down.

1907

- *August 1:* U.S. Signal Corps creates Aeronautical Division.

1908

- U.S. Signal Corps orders Zeppelin-style dirigible balloon.
- U.S. Signal Corps reconsiders Wrights' offer and orders airplane.

1911

- U.S. Congress makes first appropriation for military aviation.

1914

- *July 18:* U.S. Signal Corps upgrades Aeronautical Division into an Aviation Section.

1918

- Great Britain creates Royal Air Force, using air divisions within the army and navy.

1926

- *July 2:* Congress passes Air Corps Act to strengthen military aviation.

1927

- *June:* Society for Space Travel (Verein Für Raumschiffahrt) is founded in Germany.

1942

- *October 3:* Nazi Germany successfully launches a V-2 rocket.

1945

- *October:* U.S. Navy forms committee to research rockets.

1946

- *May:* Von Braun team tests first V-2 in New Mexico.

1947

- *July 25:* Overhaul of U.S. military structure announced, creating a supra-service Department of Defense to provide civilian oversight.

Chronology

- *September 18:* U.S. Air Force is created as a separate service.
- *October:* The Soviet Union launches their own version of a V-2 rocket from Kazakhstan.

1949

- *September 23:* The Soviet Union tests atomic bomb. (The United States orders a hydrogen bomb soon after.)
- *October:* U.S. Army moves Wernher von Braun's team to Redstone Arsenal in Huntsville, Alabama, from New Mexico; launch operations transferred to Florida.

1950

- *June 2:* North Korea invades South Korea.

1952

- *November 1:* The United States explodes first hydrogen device in the Marshall Islands.

1953

- *August:* The Soviet Union successfully tests "Joe-4 bomb," a much lighter hydrogen bomb than previously tested and based on lithium-deuteride.
- *October:* The United States approves "New Look" strategy, vowing to use reconnaissance satellites to monitor—and quickly react to—any large movements of Soviet troops.

1954

- *March 1:* U.S. explodes its own lithium-deuteride bomb at the Bikini Atoll.
- *May 26:* Korolev proposes launching artificial Earth satellite using R-7 rocket.
- *June 21:* U.S. president Dwight D. Eisenhower orders production of intercontinental ballistic missile.

1955

- *May 26:* President Eisenhower authorizes scientific satellite project as part of the International Geophysical Year (NSC 5520).
- *July 21:* President Eisenhower proposes "Open Skies."
- *July 29:* The White House announces plan to launch small, unmanned Earth satellite.
- *August 30:* Soviet government approves the Korolev satellite proposal.

MILITARIZATION OF SPACE

1956

- The United States begins U-2 overflights; Soviet radar immediately detects aircraft.
- *May 16:* President Eisenhower signs NSC 1553, authorizing arms control verification using satellites.
- *September 20:* The United States launches first Jupiter-C rocket from Cape Canaveral.
- *October 8:* China creates Rocket Research Institute.
- *December 8:* First U.S. test launch of Vanguard rocket.

1957

- *January 25:* Kremlin orders satellite launch.
- *May 1:* Second Vanguard test launch of a dummy satellite.
- *July 1:* International Geophysical Year begins.
- *August 3:* First successful launch of Soviet R-7 rocket, carrying an ICBM.
- *August 8:* The United States launches Jupiter-C rocket at Cape Canaveral; the nose cone becomes the first object recovered from space.
- *September 20:* Fifth test launch of Thor rocket is a success.
- *October 4:* USSR launches *Sputnik.*
- U.S. general James Gavin orders von Braun to begin work on an antisatellite weapon capable of destroying *Sputnik.*
- *October 15:* China and the USSR sign a defense-technical agreement.
- *October 16:* Air force launches Aerobee rocket in New Mexico with enough thrust to escape Earth's gravity.
- *October 23:* The United States successfully launches Vanguard test vehicle.
- *October 25:* *Sputnik* goes silent.
- *November 3:* USSR launches *Sputnik-2,* carrying Laika the dog, who dies several hours later.
- *November 8:* The White House reactivates army's Project Orbiter.
- *December 6:* Navy Vanguard rocket bursts into flames during test launch as media watches. The failure is dubbed "kaputnik."

1958

- *January 4:* *Sputnik* is destroyed reentering Earth's atmosphere.
- *January 22:* President Eisenhower issues NSC 1846, detailing priorities in missile and space programs.

Chronology

- *January 31:* Juno rocket launch at Cape Canaveral puts army's *Explorer 1* (formerly Deal) satellite into orbit.
- *February 5:* Second launch of Vanguard rocket fails.
- *February 6:* Senate forms Committee on Space & Aeronautics, chaired by Lyndon Johnson.
- *March 5:* Launch of *Explorer II* satellite fails.
- *March 15:* Moscow proposes transferring U.S.-USSR space talks to United Nations.
- *March 26:* *Explorer III* satellite is launched carrying 18.83-pound payload.
- *March 17:* Vanguard rocket deploys *Vanguard 1,* the second U.S. satellite.
- *April 14:* *Sputnik II* returns to Earth after 162 days.
- *May 1:* *Explorer I* and *III* confirm existence of Van Allen radiation belts.
- *May 15:* *Sputnik III* launched, an "orbiting lab" weighing nearly 3,000 pounds.
- *May 23:* *Explorer I* goes silent and orbits for another dozen years.
- *June 20:* Draft U.S. policy on outer space is released, favoring international cooperation and peaceful uses.
- *July 26:* *Explorer IV* satellite is launched.
- *August 1:* The United States fires Redstone missile in nuclear test in Pacific Ocean; this is the first attempt at missile defense.
- *August 25:* *Explorer V* launch is unsuccessful due to rocket flaws.
- *September 24:* Navy destroys Polaris missile when it goes off course over Florida.
- *October 1:* The United States creates the National Aeronautics and Space Administration (NASA).
- *October 7:* NASA announces manned spaceflight program, later named Project Mercury.
- *October 11:* *Pioneer I* lunar probe is launched but is unable to leave Earth's orbit.
- *November 8:* *Pioneer II* lunar probe is launched.
- *December 5:* *Pioneer III* lunar probe is launched aboard Juno II rocket.
- *December 15:* UN Resolution 1348 creates ad hoc Committee on the Peaceful Uses of Outer Space (COPUS), which is boycotted by Czechoslovakia, Poland, and the USSR for alleged Western bias.
- *December 18:* Atlas B rocket launches Project SCORE satellite.
- *December 19:* Recorded Christmas message from President Eisenhower is broadcasted from SCORE satellite.
- *December 31:* International Geophysical Year ends (after 18 months).

1959

- *January 2:* USSR launches *Luna I*, which fails to reach Moon.
- *February 17: Vanguard II* weather satellite is launched.
- *February 28:* United States launches *Discovery I* spy satellite.
- *March 3: Pioneer 4* satellite escapes Earth's orbit but fails to reach the Moon.
- *April 9:* Seven men are named as Mercury astronauts.
- *May 28:* United States sends to and returns from orbit two monkeys: Ben and Able.
- *August 7: Explorer 6* transmits television images from space.
- *September 12:* Soviet *Luna II* is launched.
- *September 14:* Soviet *Luna II* lands on the Moon.
- *September 18: Vanguard III* satellite is launched to map Earth's magnetic field.
- *October 4:* The Soviet Union launches *Luna III*.
- *October 7: Luna III* records images of far side of Moon.
- *October 21:* Von Braun team reassigned to NASA.
- *December 12:* USSR granted more seats on AHCOPUOS
- *December:* AHCOPUOS is reorganized into a permanent UN Committee on the Peaceful Uses of Outer Space (COPUOS).

1960

- *March 10: Pioneer 5* satellite is launched into interplanetary space.
- *March 15:* Part of Redstone Arsenal is redesignated Marshall Space Flight Center to develop a Saturn rocket for civilian use.
- *April 1: TIROS 1* weather satellite is launched by NASA.
- *May 19:* Unmanned Vostok spacecraft *(Sputnik 4)* is stranded in space.
- *June 11:* Star City cosmonaut facility opens.
- *July 28:* NASA announces Apollo Moon program.
- *August:* Sino-Soviet split; cooperation between Beijing and Moscow ends.
- *August 12:* First CORONA project film canister is recovered from Pacific Ocean.
- *August 12:* NASA launches *Echo 1*, a radio-reflector satellite.
- *August 18:* Air force recovers first CORONA project film canister in midair.
- *August 19: Sputnik 5* launches and returns two dogs to Earth: Strelka and Belka.
- *October 4:* U.S. Army launches *Courier 1B*, which can relay ground signals.

- **October 24:** Soviet R-16 rocket explodes on Baikonur launchpad, resulting in 126 fatalities.
- **November 5:** China launches its first rocket, Dong Feng-1.

1961

- France establishes national space agency.
- **January 18:** President Eisenhower signs NSC 6108, authorizing ICBM, Polaris, and reconnaissance satellite programs; a clause requires presidential approval of any ASAT test.
- **January 31:** U.S. Mercury-Redstone 2 mission puts chimpanzee into suborbital flight.
- **February 4:** *Sputnik 7* mission to Venus fails to escape Earth's atmosphere.
- **March 9:** *Sputnik 9* successfully tests launching and returning a dog and mannequin.
- **April 12:** Soviet cosmonaut Yuri Gagarin makes one complete orbit of the Earth aboard a *Vostok* spaceship for 89 minutes, becoming first man in orbit.
- **May 5:** U.S. astronaut Alan Shepard completes 15-minute suborbital flight aboard *Freedom 7*.
- **May 25:** President John F. Kennedy announces a plan to reach the Moon by 1970.
- **July 21:** Astronaut Gus Grissom enters suborbital flight aboard *Liberty Bell 7*; capsule floods on splashdown.
- **August 6:** Cosmonaut German Titov becomes the second man in orbit, staying aboard *Vostok 2* for 23 hours, 32 minutes.
- **September 25:** President Kennedy suggests expanding UN jurisdiction to entire universe.
- **October 27:** Apollo program begins by testing *Saturn 1* launch vehicle.
- **November 20:** UN General Assembly acts on President Kennedy's formulation, enacting Resolution 1721 and applying international law to outer space.
- **December 20:** UN General Assembly Resolution 1721 adopts treaty on International Cooperation in the Peaceful Use of Outer Space.

1962

- **February 20:** U.S. astronaut John Glenn orbits Earth three times aboard *Friendship 7*.
- **May 26:** President Kennedy instructs Department of State to assess ongoing U.S.-USSR talks, including discussion on peaceful uses of outer space.

- *July 9:* President Kennedy approves NSC 2454, which seeks to defuse potential Soviet ASAT tests by encouraging atmosphere of international cooperation rather than pursuing a test ban.
- *July 10: Telstar,* first television broadcast satellite, is launched.
- *August 11:* Cosmonaut Andrian Nikolayev is sent into orbit on *Vostok 3.*
- *August 12:* Cosmonaut Pavel Popovich is sent into orbit on *Vostok 4.*
- *August 27:* The United States launches *Mariner 2* satellite to Venus.
- *September 29:* The United States launches Canadian satellite, *Alouette 1.*
- *November 1:* Soviets launch *Mars 1* probe.
- *December 14: Mariner 2* passes Venus.

1963

- USSR creates anti-space defense agency (PKO), charged with "destroying the enemy's cosmic means of fighting."
- *May 15:* Gordon Cooper orbits Earth 22 times aboard *Faith 7* and deploys a small satellite to study Earth's outer environment.
- *May 21:* Mercury program ends with White House reception.
- *June 10:* President Kennedy uses American University commencement address to call for a treaty limiting nuclear weapons tests. He declares moratorium on U.S. tests even if international agreement is not reached.
- *June 16:* Soviet cosmonaut Valentina Tereshkova becomes the first woman in space during a three-day mission.
- *July 25:* The United States and the USSR agree to ban nuclear testing in outer space, the atmosphere, and under water—but not underground.
- *August 5:* Limited Test Ban Treaty is signed.
- *October 10:* UN passes Resolution 1884 to prevent deployment of weapons of mass destruction in orbit. USSR stops fight to ban espionage satellites.
- *December 13:* UN passes Resolution 1962, "Declaration of Legal Principles Governing the Activities of States in the Exploration and Use of Outer Space," which asserts that outer space and celestial bodies are common to all humankind and cannot be claimed by governments; also declares that all countries have a right to explore outer space.

1964

- *March 3:* U.S. president Lyndon B. Johnson approves NASA bid to expand cooperation with USSR on weather data; laying groundwork for future Apollo-Soyuz mission.

Chronology

- **March 20:** European Space Research Organization (ESRO) and European Launcher Development Organization (ELDO) are established.
- **June 4:** First successful launch by Europa project.
- **April 8:** U.S. launches unmanned *Gemini* capsule into orbit.
- **July 31:** *Ranger 7* lunar probe reaches the Moon and sends pictures before crashing.
- **August 1:** NASA launches *Syncom 3*, a geostationary communications satellite.
- **August 20:** International Telecommunications Satellite Organization (INTELSAT) is established among 11 countries.
- **October 12:** USSR launches *Voskhod 1* spacecraft with three cosmonauts aboard. It orbits for one day.
- **November 28:** *Mariner 4* Mars probe is launched.

1965

- **February 10:** Kremlin approves development of *N1* Moon rocket.
- **March 18:** USSR launches *Voskhod 2*; cosmonaut Alexei Leonov makes the first space walk.
- **March 23:** Gus Grissom and John Young orbit Earth three times aboard *Gemini 3*.
- **April 23:** USSR launches *Molniya 1-01*, a satellite with a highly elliptical orbit.
- **June 3:** Ed White performs first tethered U.S. space walk from *Gemini 4*.
- **July 14:** *Mariner 4* reaches Mars.
- **August 29:** *Gemini 5* returns from an eight-day mission, confirming feasibility of Moon mission.
- **November 26:** France launches *A-1 Asterix* satellite aboard Diamant A rocket.

1966

- USSR begins to deploy antiballistic missile defense system around Moscow.
- China successfully tests nuclear missile.
- **March 16:** *Gemini 8*, carrying Neil Armstrong and David Scott, successfully docks two spacecraft.
- **June 2:** U.S. *Surveyor 1* lunar probe successfully lands on the Moon.
- **June 16:** The United States and the USSR submit separate draft treaties to the UN regarding the exploration of outer space. The U.S. version is confined to celestial bodies while the USSR version applies to entire realm of outer space.
- **July 18:** *Gemini 10* successfully docks with target vehicle.

- *July 29:* President Johnson agrees to help the new European Launcher Development Organization with training, testing, and licensing, but restricts technology transfers.
- *August 10: Lunar Orbiter 1* transmits detailed photos of the Moon's surface.
- *September 12: Gemini 11* launches and docks with target vehicle.

1967

- *January 27:* Launchpad electrical fire destroys *Apollo 1* spacecraft, killing astronauts Gus Grissom, Ed White, and Roger Chaffee; 64 countries endorse UN Outer Space Treaty.
- *March:* U.S. president Lyndon B. Johnson announces Soviet interest in beginning arms control negotiations.
- *April 24:* Parachute fails to deploy on Soviet *Soyuz 1* spacecraft, killing cosmonaut Vladimir Komarov.
- *July 12:* President Johnson issues NSAM 338, detailing restrictions on U.S. technology transfers to other countries, including allies, and foreign firms.
- *September 18:* The United States announces that it will begin to deploy a "thin" antiballistic missile system.
- *November 9: Apollo 4* test launches Saturn V rocket.

1968

- *January 22: Apollo 5* tests lunar module engines in Earth orbit.
- *April 22:* The United States signs "Agreement on the Rescue of Astronauts, the Return of Astronauts, and the Return of Objects Launched into Outer Space."
- *July 1:* Non-Proliferation Treaty is signed; President Johnson announces agreement with Moscow to begin arms control talks on limiting systems, nuclear weapons, and ABM systems.
- *August 20:* USSR invades Czechoslovakia; negotiations with United States freeze.
- *September 18:* Soviet unmanned *Zond 5* probe flies around the Moon.
- *October 3:* NASA rocket launches first European satellite *(ESRO 1A)* into orbit.
- *October 11:* U.S. launches *Apollo 7,* first manned flight of the Moon program.
- *December 21: Apollo 8* orbits the Moon; astronauts Frank Borman and James A. Lovell make Christmas Eve broadcast and take the first photos of Earth from space.

Chronology

1969

- Japanese parliament imposes ban on military use of space.
- *January 16:* Soyuz 4 and 5 dock in orbit and exchange crews during spacewalk.
- *January 20:* Inauguration of U.S. president Richard Nixon; Moscow indicates interest in arms control talks.
- *June 28:* Test launch of U.K. *Black Arrow* satellite launcher fails.
- *March 3:* Apollo 9 tests lunar module in Earth orbit.
- *May 21:* Apollo 10 orbits the Moon and gets 10 miles above its surface.
- *July 16:* Apollo 11 Moon mission launches aboard Saturn V rocket.
- *July 20:* U.S. astronaut Neil Armstrong and Buzz Aldrin take the first Moon walk.
- *July 31:* U.S. Mars probe, *Mariner 6,* conducts fly-by of the Red Planet.
- *October:* U.S. president Nixon announces that the USSR is interested in resuming talks.
- *November:* Strategic Arms Limitation Talks (SALT) begin between the USSR and the United States.
- *November 14–24:* Apollo 12 lunar mission; Pete Conrad and Alan Bean make second Moon walk.
- *November 17–December 22:* Preliminary U.S.-USSR arms control talks begin in Helsinki.

1970

- *February 11:* Japan launches its first satellite, the *Ohsumi.*
- *March 10:* France launches Diamant B rocket from French Guiana.
- *April 11–17:* Ill-fated *Apollo 13* fails to reach the Moon due to a damaged oxygen tank, but returns to Earth safely.
- *April 24:* China launches the Chang Zheng-1 rocket, carrying the *Dong Fang Hong-1 (Mao 1* or *China 1)* satellite.
- *June 1:* Soyuz 9 begins an 18-day manned mission.
- *July 10:* President Nixon signs funding authorization of Apollo-Soyuz program.
- *October 25–28:* U.S. and Soviet aeronautics experts launch cooperation project.
- *December 15:* Soviet *Venera 7* probe lands on Venus, recording a temperature of 470°C.

1971

- *January 31–February 9:* Apollo 14 mission is a success; Alan Shepard and Ed Mitchell make Moon walks.

- *April 19:* The USSR launches unmanned *Salyut 1* space station into orbit.
- *May 14:* The United States launches *Skylab,* aboard *Saturn 5* rocket.
- *May 30:* The United States launches *Mariner 9* Mars probe.
- *June 6:* Three-man crew aboard *Soyuz 11* docks with *Salyut 1* and begins 23-day stay.
- *June 29: Soyuz 11* crew is killed during reentry.
- *July 26–August 5: Apollo 15* lunar mission; lunar rover is first used.
- *July 29:* The United Kingdom cancels Black Arrow project.
- *October 10: Salyut 1* is destroyed on reentry.
- *October 28:* Last *Black Arrow* launch deploys *Prospero* satellite.
- *November 15:* The USSR launches *Intersputnik* communications satellite.

1972

- *January 5:* President Nixon announces the space shuttle program.
- *March 2:* NASA launches Jupiter probe *Pioneer 10.*
- *March 29:* Liability Convention opens for international signatures.
- *April 6:* The United States and the USSR form the Apollo-Soyuz Test Project.
- *April 16–27: Apollo 16* explores lunar highlands.
- *May 26:* The United States and the USSR sign the Anti-Ballistic Missile Treaty.
- *May 25:* Last *CORONA* launch occurs.
- *May 26:* U.S. president Richard Nixon and USSR general secretary Leonid Brezhnev sign SALT and ABM Treaties in Moscow.
- *July 23:* NASA deploys *ERTS-1,* the first remote-sensing satellite.
- *August 30:* President Nixon offers launch-assistance services to partners other than the European Space Agency.
- *December 7–19:* Apollo 17 lunar mission, the last in the Apollo program.

1973

- *April 3: Salyut 2* is launched, but it quickly loses contact with ground control.
- *May 14:* U.S. launches unmanned *Skylab* space station.
- *May 25: Skylab 2* carries crew to space station; major repairs are needed.
- *July 28: Skylab 3* mission.
- *November 3: Mariner 10* Mercury probe is launched.
- *November 16:* Final mission of *Skylab,* which lasts 84 days.
- *December 3:* NASA probe *Pioneer 10* flies past Jupiter.

Chronology

1974

- USSR cancels N1 Moon project.
- *May 17:* NASA deploys first geostationary weather satellite, *SMS 1*.
- *June 24: Salyut 3* is placed in orbit. Crew arrives one week later.
- *November 12:* The UN passes the Registration Convention requiring states to log all launches.
- *December 26: Salyut 4* space station is launched into orbit.

1975

- *January 11: Soyuz 17* delivers crew to *Salyut 4* station.
- *April 19:* USSR launches Indian satellite, the *Aryabhata*.
- *May 31:* Ten European countries establish European Space Agency (ESA).
- *July 15–24:* Apollo-Soyuz mission; U.S. and Soviet craft dock in space.
- *August 9:* ESA's first satellite, *Cos-B*, is launched by the United States.
- *November 26:* China launches a surveillance satellite.

1976

- *February 12:* Kremlin green-lights *Buran*, a Soviet space shuttle.
- *June 22: Salyut 5* station is launched into orbit.
- *July 20: Viking 1* probe lands on Mars.
- *August 7: Viking 2* probe lands on Mars.

1977

- *January 18:* National Security Memo 345 orders the Pentagon to develop a low-altitude antisatellite weapon.
- *March 28:* U.S. president Jimmy Carter orders the review and expansion of ASAT research.
- *August 20:* NASA launches *Voyager 2* probe to explore outer planets and moons.
- *September 29: Salyut 6* is launched; it has docking stations at both ends.

1978

- *January 22:* The Soviet Union sends an unmanned supply vehicle, *Progress 1*, to *Salyut 6*.
- *March 2:* Czechoslovakia's Vladimir Remek flies aboard *Soyuz 28*.
- *March 10:* President Carter authorizes ASAT tests if they are "essential to achieve an ASAT capability."

- *April 7:* Japanese space agency launches television broadcast satellite.
- *May 11:* President Carter issues a new national space policy (PD/NSC 37) that calls for using space only for peaceful purposes.
- *June 8–16:* U.S.-USSR ASAT weapons talks begin in Helsinki.
- *October 10:* PD/NSC 42 clarifies U.S. space policy, including U.S. support for treaties that ensure peaceful uses of outer space.

1979

- *January 23–February 19:* U.S.-USSR ASAT talks continue in Bern, Switzerland.
- *April 23–June 17:* The final round of U.S.-USSR ASAT talks are held in Vienna.
- *June 18:* United States and USSR sign SALT II Treaty.
- *July 11: Skylab* falls from orbit and burns up upon reentry.
- *September 1: Pioneer 11* flies past Saturn.
- *November 16:* PD/NSC 54 delimits uses of civil operational remote sensing devices for land, weather, and ocean data.
- *December 18:* United Nations opens Moon Treaty for signatures.
- *December 24:* Ariane, an ESA launch vehicle, is successfully tested.

1980

- *July 18:* India launches its own satellite, the *Rohini 1B.*

1981

- *April 12:* The United States launches first reusable space craft, the shuttle *Columbia,* for a two-day mission.
- *August 20:* Moscow introduces draft text of a "Treaty on the Prohibition of the Stationing of Weapons of Any Kind in Outer Space," which bans weapons in Earth's orbit.
- *November 18:* With NSDD-8, U.S. president Ronald Reagan designates space shuttle as primary launch system for U.S. military and civilian programs; he cancels future procurement of expendable launch vehicles.

1982

- *April 19:* The USSR launches *Salyut 7,* the last in series of space stations.
- *June 24:* Jean-Loup Chrétien becomes first West European in space, aboard the *Soyuz T-6.*
- *July 4:* With NSDD 42, President Reagan announces an interest in verifiable arms control regime for space and the U.S. intention to pursue ASAT development.

Chronology

- **November 11:** Shuttle *Columbia* deploys two commercial communications satellites on its fifth mission.

1983

- **March 23:** President Reagan announces the Strategic Defense Initiative.
- **April 4:** Second space shuttle, *Challenger,* launched.
- **May 16:** Reagan approves the private/commercial use of expendable launch vehicles.
- **June 18:** *Challenger* deploys another two commercial communications satellites. The crew includes Sally Ride, the first female astronaut.
- **August 19:** The USSR issues a draft treaty, "Prohibition of the Use of Force in Outer Space and from Space against Earth."
- **August 30:** *Challenger* deploys Indian communications satellite.
- **November 28:** Shuttle *Columbia* carries *Spacelab,* a European venture, in its cargo bay.

1984

- **January 6:** President Reagan issues NSDD 119, authorizing Pentagon to research ground-based missile defense systems.
- **January 25:** President Reagan announces plans for U.S. space station *Freedom.*
- **February 3:** *Challenger* successfully tests manned maneuvering unit, allows untethered space walks.
- **April 8:** *Challenger* retrieves Solar Max satellite for in-space repairs.
- **April 8:** China launches *Dong Fang Hong-2,* geostationary communications satellite.
- **August 4:** Ariane 3 launch vehicle ready for use.
- **November 13–15:** *Discovery* shuttle captures two satellites to be repaired on Earth.

1985

- **February 25:** President Reagan issues NSDD 164, designating the space shuttle (not ELVs) as primary space launch system for national security and civilian space programs. The Pentagon is called on to improve ELV options.

1986

- **January 24:** *Voyager 2* passes Uranus.
- **January 28:** Shuttle *Challenger* explodes on takeoff, killing all seven astronauts.
- **February 4:** President Reagan calls for hypersonic "Orient Express" aerospace plane.

- *February 19:* USSR launches *Mir* core module.
- *March 13:* Two cosmonauts begin missions to *Mir* and *Salyut 7* space stations.
- *March 14:* ESA's *Giotto* probe flies near nucleus of Halley's Comet.

1987

- *February 5:* The Soviet Union sends new *Mir* crew aboard *Soyuz-TM2*.
- *April 9:* Soviet *Kvant-1* astrophysics module docks with *Mir*.
- *July 22:* The Soviet Union launches *Soyuz-TM3*, first crew exchange with *Mir*.

1988

- *June 15:* First *Ariane 4* test flight.
- *September 7:* China launches first weather satellite, *Fengyun-1*.
- *September 19:* Israel launches its first satellite.
- *September 28:* The United States resumes shuttle launches with Discovery mission. New satellite deployed, replacing one lost aboard *Challenger*.
- *November 15:* Soviet *Buran* space shuttle makes first (unmanned) flight.
- *December 21:* Cosmonauts Musa Manarov and Vladimir Titov return home after 366 days in orbit.

1989

- *April 7:* The United States tests laser weapon as part of Zenith Star.
- *May 4:* Shuttle *Atlantis* deploys *Magellan* space probe headed for Venus.
- *August 25:* *Voyager 2* passes Neptune.
- *October 18:* Atlantis deploys *Galileo* probe.
- *December 6:* *Kvant-2* advanced life-support unit docks with *Mir*.

1990

- *April 7:* China launches *Asiasat-1* comsat, its first commercial launch.
- *April 25:* Discovery deploys *Hubble Space Telescope*.
- *June 10:* *Kristall* science module docks with *Mir*.
- *August 10:* *Magellan* begins to orbit *Venus*.
- *October 6:* Discovery deploys *Ulysses* solar probe.
- *December 2:* Japanese reporter Toyohiro Akiyama flies to *Mir* aboard *Soyuz TM-11*.

1991

- *May 18:* First British astronaut, Helen Sharman, boards the *Soyuz* spaceship headed to *Mir* space station.

Chronology

1992

* **September 12:** Japanese astronaut *Mamoru Mohri* flies on U.S. space shuttle.

1993

* **June 21:** Endeavor mission carries new Spacelab module.
* **June 23:** U.S. Congress nearly cancels *Freedom* space station, which is now way over budget.
* **September 2:** The United States and Russia agree to merge space station programs *(Freedom* and *Mir-2)* and jointly build new *International Space Station.*
* **November 1:** NASA and Russia announce a joint effort to build joint space station.
* **December:** Shuttle *Endeavour* repairs *Hubble Space Telescope* during five space walks.

1994

* **July:** Regime changes in North Korea with the death of Kim Il Sung.
* **October 11:** *Magellan* ends mission by dropping onto surface of Venus.
* **October 15:** India launches its *IRS P2,* a remote-sensing polar satellite.
* North Korea agrees to freeze its plutonium program.

1995

* **February 3:** *Discovery* launch headed to *Mir* space station.
* **March 13:** Norman Thagard becomes first U.S. astronaut to ride aboard Russian rocket.
* **March 22:** Cosmonaut Valery Polyakov sets an endurance record for the most consecutive days in space, logging 437 days, 17 hours, 58 minutes aboard *Mir.*
* **June 29:** *Atlantis* shuttle docks with *Mir.*
* **June 30:** Russian president Boris Yeltsin cancels *Buran* project.
* **July 13:** The United States launches *Galileo* probe to Jupiter.
* **December 7:** *Galileo* enters Jupiter's orbit, beginning an eight-year study mission.

1996

* **April:** North Korea agrees to join South Korea, China, and the United States in peace talks.
* **April 26:** Last *Mir* primary module, *Priroda,* is connected.

1997

* **February:** Fire aboard *Mir* fills station with smoke.

- *June 25:* Ferry craft damages *Mir* solar panel, reducing power to station.
- *July 4:* Mars *Pathfinder* and *Sojourner* rover land on Mars.
- *December 7:* *Galileo* ends Jupiter mission.

1998

- At age 77 John Glenn returns to space aboard space shuttle *Discovery.*
- *Cassini* Venus probe is launched.
- *January 6:* NASA Luna prospector probe is launched.
- *June 4–8:* *Atlantis* makes its final mission to *Mir.*
- *August:* North Korea launches a satellite aboard Taepo Dong-1 rocket.
- *October 21:* *Ariane 5* comes into service.
- *November 20:* Russia launches *Zarya,* the first module for *ISS.*
- *December 7:* *Endeavour* shuttle attaches U.S. *Unity* module with Russia's *Zarya* cargo block.

1999

- *February 22:* French astronaut Jean-Pierre Haigneré arrives at *Mir.*
- *August 28:* Last crew leaves *Mir.*
- *October 14:* China launches *Ziyuan-1,* a joint China-Brazil remote-sensing satellite.
- *November 20:* China launches unmanned rocket, *Shenzhou 1.*

2000

- *April 6:* Russian team returns to *Mir* and investigates potential for commercial use.
- *July 26:* With Russian *Zvezda* module attached, the space station is deemed ready for crew.
- *November 2:* First *International Space Station (ISS)* crew arrives on *Soyuz TM-31.*

2001

- *January:* The United States conducts the first space-related war game, Title 10/Schriever 2001.
- *March 10:* First U.S. crew arrives at *ISS.*
- *March 23:* Russia's *Mir* space station burns up upon return to Earth.
- *April 18:* India successfully uses a geosynchronous satellite launch vehicle to deploy test satellite.
- *April 30:* First space tourist, Dennis Tito, reaches *ISS.*
- *December 13:* U.S. announces withdrawal from ABM Treaty, effective in six months.

Chronology

2002

- **December:** North Korea expels International Atomic Energy Agency monitors after they discover a program to create enriched uranium for nuclear weapons.

2003

- **January:** North Korea withdraws from Non-Proliferation Treaty.
- **January 5:** China's *Shenzhou 4* rocket returns after a six-day orbit.
- **January 16:** Shuttle *Columbia* damages its wing during liftoff.
- **February 1:** *Columbia* disintegrates during reentry; all seven astronauts perish. Washington will deny space access by enemies
- **August:** Six-Party Talks over North Korean nuclear program begin.
- **October 15:** China launches first taikonaut, Yang Liwei, on *Shenzhou 5.*
- **November 19:** Japan's National Space Development Agency (NASDA), Institute of Space and Astronautical Science (ISAS), and National Aerospace Laboratory (NAL), merge into Japan Aerospace Exploration Agency (JAXA).
- **December 25:** ESA's *Mars Express* probe arrives at Mars, but *Beagle 2* landing craft fails to transmit data.

2004

- **January 4:** NASA's *Spirit* rover lands on Mars; soon joined by NASA's *Opportunity* rover; both search for signs of water on the planet.
- **January 14:** U.S. president George W. Bush announces "Vision for Space Exploration," ending the space shuttle program, but pledging to return to the Moon by 2020.
- **July 1:** *Cassini-Huygens* probe reaches Saturn.
- **October 4:** Privately built spacecraft *SpaceShipOne* reaches orbit.
- **October 23:** Brazil launches a rocket loaded with suborbital test satellite.

2005

- **January 14:** *Huygens* probe lands on Titan.
- **July 3:** NASA *Deep Impact* spaceship successfully creates a crater on comet Tempel 1 by firing an "impactor."
- **October 12:** China launches a two-man crew into orbit aboard *Shenzhou 6.*
- **November:** North Korea pulls out of Six-Party Talks.

2006

- **July:** North Korea attempts to launch Taepodong-2 rocket, but fails shortly after liftoff.

- *October:* North Korea tests nuclear device inside mountain tunnel.
- *October:* The United States and China issue new statements of national space policy. Washington will deny space access by enemies.
- *December:* North Korea returns to Six-Party Talks.
- UN Security Council Resolution 1718 condemns North Korea.

2007

- *January 11:* China shoots down own weather satellite; China's official silence about the shooting infuriates the world.
- *February 13:* North Korea signs the Initial Actions Agreement to close portions of its nuclear facility.
- *July:* North Korea shuts down three of its nuclear facilities.
- *October 3:* North Korea signs the Second Phase Actions Agreement and agrees to disable closed nuclear facilities.
- *December:* North Korea begins to discharge spent fuel rods.

Bush administration formally announces plans to deploy ballistic missile defense components in Poland and Czech Republic.

2008

- *March 8:* The European Space Agency launches *Jules Verne,* an automated transfer vehicle from French Guiana.
- *May:* Japanese parliament lifts a ban on military use of space.
- *September 25:* China launches three-man crew aboard *Shenzhou 7.*
- *September 27:* Chinese astronaut Zhai Zhigang performs the country's first space walk.
- *November 8:* India's *Chandrayaan 1* satellite enters lunar orbit.

2009

- *February:* Iran launches small satellite.
- *April 5:* North Korea claims to successfully launch satellite.
- *September 17:* President Obama changes the schedule and the parameters of Bush BMD plan for Europe.

2010

- China plans to land robotic rover on lunar surface.
- *June 28:* Obama White House issues new national space policy that encourages use of commercial sources whenever possible.

Chronology

2015

- Next generation of U.S. space shuttle scheduled to come online.

2020

- NASA hopes to resume manned missions to Moon, using new Orion spacecraft.

Glossary

Advanced Research Projects Agency   created in 1958 to catch up with USSR in space research and development; renamed DARPA (Defense + ARPA) in 1972.

Aegis ballistic missile defense   a sea-based component of the U.S. missile defense program begun under President George W. Bush.

airborne laser   potential component of a MISSILE DEFENSE SYSTEM; the chemical oxygen iodine laser would be mounted on a modified 747 aircraft and try to intercept missiles in boost phase. A steady, sustained laser beam would damage the outer layer of the missile, so it would not reach its target. The U.S. Missile Defense Agency successfully conducted an airborne laser test in February 2010.

antiballistic missile system   a system to counter strategic ballistic missiles or their elements while in flight trajectory, consisting of launchers, interceptors, and radars.

Anti-Ballistic Missile Treaty   In 1972, the United States and Soviet Union agreed not to develop weapons that would intercept or destroy incoming missiles. The United States withdrew from the treaty in June 2002, citing need to defend homeland from non-Russian threats.

antisatellite weapons (ASAT)   weapons that disable or demolish SATELLITES

apogee   the point when a SATELLITE is farthest from Earth.

Apollo   created in July 1960, successful NASA program to put manned flights into ORBIT around the Earth and eventually to the Moon and back. *Apollo 12* astronauts Neil Armstrong and Buzz Aldrin became the first humans to walk on the surface of the Moon. *Apollo 13* suffered severe structural damage during launch and had to abandon its planned lunar landing and improvise repairs to their spacecraft; the tense days prior to their return are the subject of the 1995 movie *Apollo 13*. Project Apollo ended in 1972 after 17 missions.

Apollo-Soyuz Test Project   the first join U.S.-Soviet space mission culminated in the 1975 docking of the U.S. *Apollo* and Soviet *Soyuz 19* spacecraft. The project provided vital information about cooperative efforts that would prove helpful in future space station programs. The Apollo mission was not formally numbered but is regarded as *Apollo 18*.

AQUA   NASA SATELLITE deployed in near polar LEO to monitor weather conditions and systems on Earth.

Ariane   European launch vehicle, used for commercial launches since 1979.

Army Ballistic Missile Agency   created in February 1956, ABMA inherited Wernher von Braun's ROCKET team. The team's main project was to adapt German V-2 rockets to carry Juno missiles and Orbiter SATELLITES.

Atlas   first U.S. launch vehicles/rockets; debuted in 1958 and manufactured by Corvair Astronautics in San Diego, based on air force INTERCONTINENTAL BALLISTIC MISSILES.

attitude   the orientation of a spacecraft or space station; orientation coordinates can be adjusted.

Baikonur   USSR cosmonaut and launch facility, located in Kazakhstan and now leased to Russia.

ballistic missiles   weapons that are launched by rocket propulsion but fall to Earth due to GRAVITY.

ballistic missile defense (BMD)   system designed to detect, track, identify, and destroy incoming missile attacks.

Bogotá Declaration   1976 document in which Brazil, Colombia, Congo, Ecuador, Indonesia, Kenya, Uganda, and Zaire claimed sovereign rights to the air and space above their territorial expanse.

boost phase   the ascending portion of a missile's flight, when it is accelerating under its own power. Ideally interceptors would reach their targets in this phase, as the heat signature from the launch is easy to detect and any debris would fall on the launching country. However, boost phase interception is technically demanding, partly because there would be very little lead time to launch a counter attack.

Booster   small, additional ROCKET attached to larger rocket; provides additional thrust at launch.

Brilliant Eyes   ASAT project to replace Space Surveillance and Tracking System SATELLITES with smaller missile-tracking satellites.

Brilliant Pebbles ANTISATELLITE WEAPONS program connected to the STRATEGIC DEFENSE INITIATIVE Organization; would deploy thousands of individual interceptors with surveillance capacity. Brilliant Pebbles was intended to replace the space-based interceptor.

Buran Soviet space shuttle, flew only one time.

C³I communications, command, control, and intelligence.

Challenger U.S. space shuttle first launched on April 4, 1983. The *Challenger* exploded 73 seconds after takeoff on January 28, 1996, killing all seven astronauts aboard. As a result, the shuttle program was suspended for nearly two years.

Chang Zheng Chinese ROCKET, translates to "Long March."

chemical lasers potential long-range weapon that could destroy targets using continuous waves of infrared radiation. Both MIRACL and the airborne laser are chemical lasers.

Clementine U.S. MICROSATELLITE program; orbited Moon for two months in 1994; program later canceled by President Bill Clinton. Called Clementine because it only carried enough fuel to reach the Moon, not return, so it was "lost and gone forever."

close-proximity intercept destroying a SATELLITE by detonating an explosion, such as a nuclear weapon, close by instead of hitting the satellite directly.

cold war (1947–91) prolonged political, economic, and military standoff between United States and Soviet Union. While no actual war occurred between the two superpowers, they did compete for allies and amassed enormous stockpiles of nuclear weapons.

Columbia the first space shuttle, launched on April 12, 1981, and conducted 28 missions. On February 1, 2003, the spacecraft disintegrated on reentry after an insulating panel was damaged on takeoff. All flights halted while the tragedy was investigated, and the only way to get to the *INTERNATIONAL SPACE STATION* in the interim was aboard Russian spacecraft.

combustion chamber ROCKET compartment where fuel components are mixed and ignite to generate THRUST as exhaust vents toward rear.

Commission on Outer Space congressionally appointed panel of experts charged with reviewing U.S. space policy. Chaired by Donald Rumsfeld, it is often referred to as the Rumsfeld Commission. Delivering the panel's report to Congress, Rumsfeld warned of an impending Space Pearl Harbor and argued that the United States must be prepared to fight in outer space.

Committee on the Peaceful Uses of Outer Space a standing subgroup of the UN Conference on Disarmament in Geneva. China and Russia and other countries regularly propose draft treaties to ban the deployment of space weapons, but the United States usually abstains from votes on the matter.

common aero vehicle (CAV) a U.S. Air Force concept for a maneuverable hypersonic reentry vehicle that could carry 1,000 pounds of munitions or supplies to anywhere in the world within two hours. The program was redesignated as a "hypersonic technology vehicle" in 2004.

Conference on Disarmament a United Nations panel based in Geneva, Switzerland. Founded in 1979, the conference is designed to be the primary multinational international forum for negotiating arms control issues.

confidence-building measures reduce tension and distrust, improve communications, add safety provisions to possible military uses.

Conventional Forces in Europe (CFE) Treaty established in 1992. With the dissolution of the USSR in 1991 and corresponding end of the COLD WAR, signatories pledged to reduce and destroy their conventional weapons and equipment stockpiles.

co-orbital ASATs small weapons or MICROSATELLITES that can shadow or latch onto SATELLITES in orbit, fixing them in position to eventually disable or destroy the satellite with microwaves, explosives, radio-jamming frequencies or by crashing into it. Also called SPACE MINES.

CORONA U.S. spy SATELLITE program of the 1950s. Under control of air force and CIA, CORONA replaced the U-2 Soviet over flights. Very few written records in order to maintain highest secrecy.

Cosmos series of Soviet/Russian launch vehicles that are often used to place PAYLOADS into LOW-EARTH ORBIT.

cruise missile a short-distance guided missile that uses a terrain-following radar.

Deal the later name for army's PROJECT ORBITER Satellite, which used JUPITER-C ROCKETS.

Defense Science Board founded in 1956 the federal advisory group is a panel of independent scientists designated to provide unbiased analysis of U.S. Department of Defense research programs and proposals.

Delta series of U.S. launch vehicles.

direct ascent ASATS payloads launched to crash into SATELLITES, disabling or destroying them. Missiles may be launched into low-earth

ORBIT by medium-range ballistic missiles or air-launched kill vehicles. More powerful launch vehicles can target objects in geosynchronous orbit. The United States tested a direct ascent ASAT weapon in 1985, while China tested one in 2007.

directed energy systems use laser beams to disable or destroy targets in LOW-EARTH ORBIT.

drone unmanned aircraft guided by remote control.

electromagnetic pulse (EMP) generated by either a nuclear explosion or a short, high-intensity burst of electromagnetic energy, an EMP creates an electromagnetic field in space that can interfere with SATELLITE functions.

equatorial orbit SATELLITE path directly above equator; easy to achieve due to extra THRUST from Earth's rotation.

evolved expendable launch vehicle (EELV) early 2000s U.S. Air Force project to develop cheaper, more efficient methods to launch spacecraft. The system consists of two families of LAUNCH VEHICLES, the Boeing Delta IV and the Lockheed Martin Atlas V, that share common components and infrastructure, as well as having a standardized PAYLOAD and launchpad requirements.

exoatmospheric kill vehicles interceptor portion of antisatellite and antiballistic missile programs, designed to destroy hostile ballistic missiles in mid-flight on impact and outside of Earth's atmosphere.

expendable launch vehicle (EVL) single-use system to launch a military or private PAYLOAD (often antisatellite) into space. ELVs consist of one or more ROCKET stages. After each stage has burned its compliment of propellant, it is jettisoned from the vehicle and left to crash back to Earth.

Explorer U.S. Army antisatellite program under Wernher von Braun, previously named DEAL and, before that, PROJECT ORBITER. Launched January 31, 1958, aboard JUNO rocket.

fission a process that releases energy by splitting atoms.

fractional orbital bombardment system (FOBS) Soviet ANTISATELLITE WEAPONS program to place nuclear bombs in space, then "de-orbit" them to hit targets on Earth. FOBS was canceled to comply with SALT II Treaty.

fusion a process that releases nuclear energy by combining atoms.

Future Imagery Architecture U.S. Department of Defense research and development program to improve antisatellite imaging capabilities and resolution.

Gaither Report findings from 1957 White House panel on civil defense; experts said U.S. unprepared to survive Soviet nuclear missile attack. Called for more fallout shelters, an expanded missile program, and more research and development. First talk of "missile gap."

Galileo a global positioning system under development by the European Space Agency.

geosynchronous orbit (GEO) a route above the Earth's equator that keeps the SATELLITE in a relatively fixed location and taking 24 hours to complete one rotation. At 23,000 miles above the equator GEO is furthest limit of satellites and is often used for missile warning, communications, and electronic surveillance satellites.

global positioning system (GPS) SATELLITE network that creates real-time maps of the Earth's surface. Heavily used in 1991 Gulf War to ease navigation through featureless desert. Developed by the U.S. Department of Defense, GPS needs a minimum of 24 satellites to form a complete picture of Earth.

global protection against limited strikes (GPALS) Revamped version of the STRATEGIC DEFENSE INITIATIVE program offered under President George H. W. Bush.

GLONASS Russia's entry in the global navigation market. The program began under the Soviet system and a full complement of navigational SATELLITES had been deployed by 1995, but Russia could not afford to maintain the system.

graveyard orbit damaged SATELLITES or those at the end of their service life may be maneuvered into a higher ORBIT to minimize the possibility of damaging spacecraft.

gravity natural force that pulls object toward much larger object.

gravity well the pull of GRAVITY exerted by a large body in space, such as a sun, planet, or moon. Larger objects have more mass, and thus a stronger gravity well. The larger the body (the more mass) the more of a gravity well it has. Objects on the surface of the body are considered to be at the bottom of the gravity well. The larger a planet or moon's gravity well is, the more energy it takes to achieve escape velocity and blast a ship off of it.

hardening strengthening a target, especially a SATELLITE, so that it is more difficult to damage, disrupt, or destroy.

heat shield outer skin that protects spacecraft during REENTRY; often a material that burns away.

Glossary

hit-to-kill   ballistic missiles that intercept incoming missiles as part of a missile defense system.

hot-metal kill   ANTISATELLITE WEAPONS procedure developed by USSR. As a weapon approaches a target satellite it detonates, spreading an expanding cloud of shredded metal.

H-Series rocket   Japanese launch vehicle.

Hubble Space Telescope (HST)   Jointly managed by NASA and the EURO- PEAN SPACE AGENCY, *HST* was deployed in April 1990 by the *Discovery* space shuttle. It is the largest space telescope in use and has recorded remarkable images of the solar system.

inclination   the angle (tilt) between the orbital plane of a SATELLITE and the equatorial plane of its planet.

intercontinental ballistic missile (ICBM)   warheads launched from ROCK- ETS toward predetermined targets and have a range of 3,500 miles or more.

intermediate-range ballistic missile (IRBM)   a weapon with a range between 1,500 and 3,437 miles.

International Geophysical Year (IGY)   an international cooperation pro- gram in the Earth sciences that ran from July 1, 1957, to December 31, 1958. The United States and USSR planned to launch artificial Earth SATELLITES during the celebration, but the USSR beat the United States by launching *Sputnik* on October 4, 1957.

International Space Station   a multinational endeavor that combined several projects as a cost-saving solution. These included the Soviet *Mir 2,* and *Freedom,* from a multinational partnership including the United States, the European Space Agency, Japan, and Canada. Russia launched the first module in November 1998 and the first, multinational crew arrived in 2000. Originally the U.S. Space Shuttle and the Russian *Soyuz* spacecraft were to share the ferry duties, but the grounding of the shuttle fleet after the *Columbia* disintegrated in 2003 and the retire- ment of all U.S. shuttles in 2010 will leave Russia as the only ride to and from the *ISS.*

Jet Propulsion Laboratory (JPL)   a NASA research and development center in California cofounded by Qian Xuesen and managed by the California Institute of Technology.

Johnson Space Center   named for U.S. president Lyndon B. Johnson, it is NASA's mission control headquarters in Houston, Texas.

Joint Direct-Attack Munitions GLOBAL POSITIONING SYSTEM–guided precision-strike weapons used by United States in Kosovo, Afghanistan, Pakistan, and Iraq.

Juno rocket modified JUPITER-C ROCKET used to launch U.S. SATELLITES.

Jupiter-C rocket a four-stage version of the army's REDSTONE MISSILE; used to test reentry vehicles.

Karman primary jurisdiction line the lowest point at which an object can maintain its ORBIT due to gravitational pull overtaking aerodynamic flight. This line offers one possible answer to the question of where air ends and space begins.

Kasputin Yar Soviet/Russian rocket development and testing site between Volgograd and Astrakhan. Now known as Znamensk.

Kennedy Space Center NASA's main launch complex at Cape Canaveral, Florida.

Kinetic energy interceptors disable or destroy incoming missiles on impact, rather than by detonating an explosive near the target.

Lagrange Libration Points theoretical points of gravitational anomaly, where the gravitational effects of two orbiting bodies would cancel each other out. Named for French mathematician Joseph-Louis Lagrange, who found five points where the gravity of the Earth and the Moon would cancel out the other. An object orbiting around any one of these five points would remain permanently stable, without the fuel expenditure usually associated with maintaining such a position, giving them immense military and commercial value.

LANDSAT a series of Earth-observing SATELLITE missions jointly managed by NASA and the U.S. Geological Survey. Begun in 1972 LANDSAT has refined remote sensing technology.

large advanced mirror program (LAMP) a U.S. program set up in the 1990s to refine the aperture of laser-projection telescopes.

large optics demonstration experiment (LODE) U.S. Defense Department project to create a segmented mirror that could be folded inside a launch vehicle, then unfurled in space and used to generate and direct laser weapons.

launch vehicle a set of equipment consisting of multiple ROCKET stages and rocket boosters used to launch PAYLOADS.

long march rockets Chinese ROCKETS used to launch SATELLITES and manned spacecraft.

Glossary

low-Earth orbit (LEO) defined as less than 200 miles above Earth. LEO is the most populated zone and the site of the *International Space Station* and the space shuttle.

Lunik/Luna Soviet project created to reach the Moon. *Lunik-2* landed on the Moon's surface, while *Lunik-3* traveled behind the Moon and sent photos from the Moon's far side.

manned-orbiting laboratory a facility that stays in ORBIT and is staffed by rotating crews who travel by spaceship. Manned-orbiting laboratories include *SALYUT*, *SKYLAB*, *MIR*, and the *International Space Station*.

Manned Spaceflight Center established at Langley, Virginia, in 1962, and later transferred to Houston, Texas. The Manned Spaceflight Center is responsible for PAYLOADS and Moon modules.

Marshall Spaceflight Center a U.S. facility for launch-vehicle research and development, located at the REDSTONE Arsenal in Huntsville, Alabama.

medium earth orbit (MEO) a zone between LEO and GEO, or 1,243 to 22,236 miles above Earth's surface. This ORBIT is used for GPS SATELLITES.

medium-range ballistic missile (MRBM) weapons with a range of 500–1,499 miles.

Mercury NASA program to ORBIT and return a human aboard a spacecraft.

microsatellites weigh only a few dozen kilograms and have some ability to maneuver and operate autonomously in space. China, Germany, and the United States have microsatellites that weigh around 70 kilograms.

mid-infrared advanced chemical laser (MIRACL) began as an Army anti-missile program under the STRATEGIC DEFENSE INITIATIVE (SDI), but later adopted for ANTISATELLITE WEAPONS use.

Mir was the Soviet/Russian space station in operation from 1986 to 2001. During that time rotating international crews visited and resided at the station, as did the earliest space tourists.

missile defense system (MDS) a defensive system of missile or other interceptors intended to disable or destroy an incoming missile attack.

missile gap a cold war buzzword for the supposed U.S. lag behind the USSR in production of INTER-CONTINENTAL BALLISTIC MISSILES.

Missile Technology Control Regime an informal and voluntary association of countries to prevent the proliferation of unmanned delivery systems capable of delivering WEAPONS OF MASS DESTRUCTION. To this end, they seek to coordinate national export licensing efforts aimed at preventing their proliferation. Established in 1987 by Canada, France,

Germany, Italy, Japan, the United Kingdom and the United States, the regime now has 34 member states. The association has no formal secretariat, instead changing information through point of contact system sponsored by the French Ministry of Foreign Affairs.

multiple independently targeted reentry vehicles (MIRV) technology allowing one missile to carry many warheads, all directed at different targets. This technology, like MULTIPLE REENTRY VEHICLES, created a loophole in the STRATEGIC ARMS LIMITATIONS TALKS, which had focused on limiting the number of missiles, not bombs.

multiple kill vehicle payload systems attach to an interceptor missile to attack multiple targets, such as the hostile missile's reentry vehicle or countermeasures.

multiple reentry vehicles (MRV) technology allowing one missile to carry many warheads, all directed at the same targets. This technology, like MULTIPLE INDEPENDENTLY TARGETED REENTRY VEHICLES, created a loophole in the STRATEGIC ARMS LIMITATIONS TALKS, which had focused on limiting the number of missiles, not bombs.

multistage rockets contain two or more independently powered missiles that launch in sequence. As fuel is used up, that stage falls away, reducing the weight of the ROCKET.

mutual assured destruction (MAD) a national security strategy based on the premise that both sides could obliterate the other using nuclear weapons. Given adequate warning, country A could launch its own nuclear weapons toward country B even before B's nuclear weapons destroy A. This strategy eliminates the advantages of striking first.

National Aeronautics and Space Council (NASC) a presidential advisory group on national space policy. While NASA focused more on civilian space policy, the NASC focused on both civilian and military applications. Senator Lyndon B. Johnson lobbied President Dwight D. Eisenhower to create the council alongside NASA in 1958 and proposed that Eisenhower serve as chair. President John F. Kennedy later appointed then–vice president Johnson to succeed Eisenhower. The council was comprised of the Secretaries of State and Defense, the NASA Administrator, the Chairman of the Atomic Energy Commission, and any other presidential appointees. Nixon eliminated the council in 1972 and responsibility passed to other agencies.

national missile defense concept to defend a homeland from attacks by states with BALLISTIC MISSILES. A missile defense system will track and destroy weapons in space.

Glossary

national technical means use of reconnaissance satellites to monitor compliance with international agreements, such as the ANTI-BALLISTIC MISSILE TREATY. Signatories also agreed to not jam or otherwise disrupt national technical means or conceal items.

nozzle a narrow opening at the base of a ROCKET that allows hot, expanding gases to escape and thereby propel the rocket. The nozzle concentrates the gas emission, increasing THRUST.

Open Skies U.S. president Dwight D. Eisenhower's 1955 proposal to prevent surprise nuclear attack. Open Skies called for the United States and the USSR to permit aerial reconnaissance. An Open Skies Agreement was finally signed by 25 countries in 1992.

orbit the loop one object makes when traveling around a larger object with stronger gravity.

payload cargo deployed during a space shuttle mission or an explosive charge aboard a rocket or missile.

Peenemünde the hidden research and development headquarters of the Nazi Germany V-2 program.

perigee the portion of ORBIT when an object is closest to Earth.

period the time required to complete one ORBIT.

perturbation a disturbance in the ORBIT of a planet or SATELLITE caused by the GRAVITY of another planet.

phased-array radar series of antennas that gather and reroute signals to increase coverage up to 360 degrees.

picosatellites weigh only a few dozen kilograms and have some ability to maneuver and operate autonomously in space. The United States has deployed two picosatellites.

Pobeda the Soviet version of Nazi V-2 ROCKETS.

polar orbit a vertical ORBITAL patch that passes over the North and South Poles, which is the preferred route for Earth-observation SATELLITES.

Polaris a U.S.-built solid-fuel SUBMARINE-LAUNCHED BALLISTIC MISSILE used by the U.S. and British navies in the 1960s and 1970s. It was replaced with the POSEIDON and TRIDENT series of missiles.

Poseidon a U.S.-built solid-fuel SUBMARINE-LAUNCHED BALLISTIC MISSILE used by the U.S. Navy in the 1970s. It was considered more accurate than the POLARIS SLBM, later was replaced by the TRIDENT missile.

Predator U.S. drone aircraft, used heavily in Afghanistan and Iraq Wars.

Prevention of the Placement of Weapons in Outer Space Treaty (PPWT) draft treaty at Conference on Disarmament, backed by China and Russia in particular.

Project Aphrodite a U.S. project based in Great Britain; sought to destroy Nazi ROCKET facilities with aerial bombs.

Project Orbiter army satellite program under Wernher von Braun. It was cancelled in 1955, only to be later revived as the DEAL.

Project SCORE (Signal Communications by Orbiting Relay Equipment) Project SCORE ultra-top secret ADVANCED RESEARCH PROJECTS AGENCY program which created a "talking satellite" that broadcast President Eisenhower's 1958 Christmas message.

Project Slug the early name of ORBITER; a joint army-navy program to create a five-pound satellite.

Proton Soviet workhorse ROCKET, used for unmanned PAYLOADS such as space station components.

Purcell Report a U.S. panel under Eisenhower that endorsed limited military uses of space, such as reconnaissance, communications, and weather satellites.

R-7 a Soviet INTERCONTINENTAL BALLISTIC MISSILE, used for most launch vehicles for Sputnik, Vostok, and Soyuz projects.

Redstone U.S. Army arsenal in Huntsville, Alabama; base for Wernher von Braun's ROCKET team. Redstone is also the name of an early U.S. rocket developed at the arsenal.

reentry the phase of a mission when a spacecraft passes through Earth's ORBIT. It is dangerous due to extreme atmospheric temperatures.

remote sensing the study of Earth from space via satellite.

revolution in military affairs concept for "next-generation" weapons using advanced technologies (especially SATELLITES) to gather and relay data.

rocket a propulsion system to thrust vehicle upward.

Salyut series of early Soviet space stations.

Sary Shagan USSR/Russian ANTIBALLISTIC MISSILE test site in Kazakhstan.

satellite a small object circling a larger object due to the latter's greater gravitational pull.

Saturn the massive U.S. rocket used for the Apollo program. The Saturn, which initially used cluster of REDSTONE-type rockets, was 10 times more powerful than the JUPITER-C ROCKETS.

Glossary

Shenzhou (**Sacred vessel**)   China's manned spacecraft. The *Shenzhou 5* carried China's first astronaut into Earth orbit on October 15, 2003.

shutter control   provisions in a government contract with a commercial provider of satellite services. In times of national crisis, the government can insist that some imagery not be made public.

Skylab   a U.S. space station launched in 1973. After three missions to the station, NASA parked *Skylab* in a GRAVEYARD ORBIT. It eventually deorbited in 1979, causing speculation about its crashing back to Earth, but most of it burned up on REENTRY.

Slichter Panel   created in 1975 under U.S. president Gerald Ford, this panel analyzed U.S. military use of space and concluded that the United States was too reliant on poorly defended SATELLITES.

sovereignty   the ultimate political authority in a given territory. A sovereign state would not answer to any higher (international) authority or to another state.

Soyuz   a series of Soviet spacecraft used, with modifications, from 1967 to 2002.

Soyuz rocket   an R-7 variation used to launch SOYUZ spacecraft.

space mines   small weapons or MICROSATELLITES that can shadow or latch onto SATELLITES in ORBIT, fixing them in position to eventually disable or destroy the satellite with microwaves, explosives, radio-jamming frequencies or by crashing into it. Also called CO-ORBITAL ANTISATELLITES SYSTEMS.

space probe   an autonomous, unmanned vehicle to explore far reaches of the solar system.

Space Transport System   formal name for the U.S. space shuttle. Missions are identified with the label "STS."

space-based infrared satellites   next generation of U.S. global positioning system, using higher (SBIRS-high) and lower (SBIRS-low) orbits.

space situational awareness   military doctrine that focuses on identifying any object orbiting Earth, the owner, and whether it is hostile or friendly.

Sputnik   first artificial Earth SATELLITE; launched by the USSR on October 4, 1957.

Strategic Arms Limitation Talks (SALT)   a series of negotiations between the United States and Soviet Union to cap inventories of ballistic missiles. Begun in November 1969, the process ended in May 1972 when U.S. president Richard Nixon and USSR general secretary Leonid Brezhnev signed the SALT Treaty in Moscow. The 1972 ANTI-BALLISTIC MISSILE TREATY was also part of this process.

Strategic Defense Initiative (SDI) was President Reagan's "Star Wars" project to destroy incoming missiles. While hardly any of the programs were ever deployed, SDI launched new wave of research into space-based military assets.

submarine-launched ballistic missile (SLBM) a warhead, often nuclear, launched underwater. U.S. SLBMs included the POLARIS, POSEIDON, and TRIDENT.

telemetry electronic status data sent from spacecraft to control station on Earth.

TERRA a NASA-managed, multinational SATELLITE program to study climate and environmental change on Earth.

thrust the forward motion of object generated by firing ROCKETS.

Titan a family of U.S. air force launch vehicles used to propel early INTER-CONTINENTAL BALLISTIC MISSILES, including underground, silo-based ICBMs. The program dates to the 1950s and the Titan was the first U.S. two-stage rocket.

Trident a three-stage U.S.-built solid-fuel SUBMARINE-LAUNCHED BAL-LISTIC MISSILE. The U.S. Navy deployed the Trident 1 in the 1980s and 1990s, while the British navy acquired the newer Trident 2. It replaced the POLARIS missile.

V-2 rocket a single-stage rocket developed by Nazi Germany and brought to U.S. by Wernher von Braun.

Van Allen radiation belts two rings of nuclear particles trapped by Earth's magnetic field, spacecraft or astronauts transiting the belts risk severe damage or even death.

Vanguard the first U.S. ROCKET and SATELLITE, built by the navy and launched in 1958; used civilian-built launch vehicle, designed for scientific exploration, not warfare.

Viking rocket designed by Naval Research Laboratory and rival to army's REDSTONE rocket.

Vostok Soviet launch vehicle, based on R-7 ROCKET and used for manned missions.

Voyager a U.S. probe that took photos of Jupiter, Saturn, Uranus, Neptune and their SATELLITES.

weapons of mass destruction (WMD) weapons designed for maximum human and infrastructure damage, including nuclear, biological and chemical weapons. The Outer Space Treaty bans the deployment of WMD in outer space.

Index

Page numbers in **boldface** indicate major treatment of a subject. Page numbers followed by *c* indicate chronology entries. Page numbers followed by *f* indicate figures. Page numbers followed by *g* indicate glossary entries. Page numbers followed by *m* indicate maps. An italic *b* denotes a biography.

377

Index

Index

Conventional Forces in Europe (CFE) Treaty (1992) 366*g*
Convention on International Liability for Damage Caused by Space Objects. *See* Liability Convention
Convention on Registration of Objects Launched into Outer Space. *See* Registration Convention (1976)
Convention on the Prohibition of Military or Other Hostile Use of Environmental Modification Techniques. *See* Environmental Modification Convention (1977)
Cooper, Gordon 63, 348*c*
cooperation. *See* international cooperation
co-orbital ASATs 10–11, 47, 49, 366*g*
CORONA project 9, 42, 346*c*, 352*c*, 366*g*
Corvair 38
cosmonauts, women 75, 79
Cosmos launch vehicles 366*g*
Cosmos series satellites 11, 77
costs. *See also* budget
 of Apollo program 44
 of first U.S. satellites 38
 of space-based assets **24–25**
 of Strategic Defense Initiative 52, 53–54, 55
 of U.S. ballistic missile defense 43
cruise missiles 366*g*
Cuba, U.S. intelligence gathering in 9, 76
Cuban missile crisis (1963) 76
Curtiss-Wright 39
Czech Republic, U.S. missile defense in 13, 60, 84

D

Deal satellite 366*g*
debris
 amount of 11, 262*f*–263*f*
 from antisatellite weapons 12, 93
 environmental consequences of 27–28
 threat posed by 11–12
 types of 11
Deepwater Horizon oil spill (2010) 6
Defense Department, U.S. 294, 342*c*
 antiballistic missiles in 42, 56
 antisatellite weapons in 49–50
 budget of
  under Obama (2010) 176–177
  space systems as share of 55
 cost of space programs of 47
 in Missile Defense Act 56
 in Strategic Defense Initiative 51–52
Defense Science Board 366*g*
delimitation of space **15**
Delta launch vehicles 366*g*
Deng Xiaoping 88–89

deterrence
 in ABM Treaty 48
 with ballistic missile defense 12
 definition of 37–38
 with nuclear weapons in cold war 28, 37–38
 in Reagan's space policy 51
 as strategy against weaponization **28**
Dickson, Paul 36
Di Pippo, Simonetta 87
direct ascent ASATs 366*g*–367*g*
directed energy systems 24, 367*g*
Dobryin, Anatoly 79, 80
docking systems, international development of common 49
documents. *See* primary documents
dogs 3, 75, 346*c*
Dolman, Everett C. 16, 26, 48, 71
Dong Fang Hong 1 (DFH1) satellite 88
Dong Feng-1 missile 88, 347*c*
Doohan, James "Scotty" 63
Doolittle, James 268*b*
Dornberger, Walter Robert 35, 36
Douglas Aircraft 38
drones 4, 9, 59, 367*g*, 373*g*
Dryden, Hugh L. 268*b*
Dryden Flight Research Center 286
Dulles, Allen 42

E

Earth observation satellites **6**
EastWest Institute 61
Egypt 98–99
Eisenhower, Dwight D. **39–43,** 343*c*–347*c*
 approach to space policy 39–40, 42
 nuclear weapons under 37
 on open skies policy 9, 17, 40
 reconnaissance satellites under 9, 39–40, 42, 43
 scientific research under 40, 57
 on *Sputnik* 41
electromagnetic pulse (EMP) 367*g*
emergency position-indicating radio beacons (EPIRB) 6
emergency rescues. *See* rescues
Endeavor space shuttle 83
endurance missions, Soviet 75, 79
Enduring Freedom, Operation 4
Energia 82, 83
England. *See* United Kingdom
environmental consequences, of weaponization **27–28**
Environmental Modification Convention (1977) **23,** 215–216
equator, launches on 86

381

Index

Index

"Urgent National Needs" speech by (1961) 44, 119–122
  vision for U.S. space program 44, 45
Kennedy, Joseph, Jr. 36
Kennedy Space Center 287, 370*g*
Kerimov, Kerim 270*b*
Khrunichev Enterprises 82
Khrunichev State Research and Production Center 82
Khrushchev, Nikita 42, **73–76**
Killian, James R., Jr. 270*b*
Kim Chang Woo 103–104
Kim Dae Jung 103
Kim Il Sung 98, 357*c*
Kim Jong Il 97, 98, 99, 101
kinetic bombardment 24
kinetic energy weapons 13, 24, 370*g*
Koizumi, Junichiro 96
Komarov, Vladimir 350*c*
Koptev, Yuri N. 270*b*
Korea, division of 98. *See also* North Korea; South Korea
Korea Aerospace Research Institute (KARI) 102–103, 285
Korea Institute for Defense Analyses 93
Korea Mining Development Trading Corp. 100
Korean Central News Agency (KCNA) 104
Korean War 98, 343*c*
Korea Ryongbong General Corp. 100
Korea Space Launch Vehicle (KSLV) 103
Korolev, Sergei P. 74, 270*b*, 343*c*
Ko San 103, 104, 270*b*
Kosygin, Alexei 46
Krepon, Michael 11, 12, 28–29, 62
Krikalev, Sergei K. 82, 271*b*
Kurchatov, Igor 271*b*
Kwangmyonsong-1 satellite 99
Kwangmyonsong-2 satellite 99

L

Lagrange Libration Points 370*g*
Laika (dog) 3, 75, 271*b*, 344*c*
LANDSAT 370*g*
Langley, Samuel P. 271*b*, 341*c*
Lanius, Roger 91
large advanced mirror program (LAMP) 370*g*
large optics demonstration experiment (LODE) 370*g*
lasers
  airborne 363*g*
  in antisatellite weapons 11, 77
  chemical 24, 365*g*
  Chinese 93

difficulty of using 24
  ground-based 93
  Indian 106
  space-based 59
  in Strategic Defense Initiative 13
launch(es)
  on equator 86
  registries of 17, 19, 20
launch services
  Chinese 63, 89
  commercial 62, 63–64, 82, 83
  European Space Agency 54, 61, 70, 83, 87
  Indian 104, 105
  Russian 60–61, 63, 64, 83
  South Korean 102–103
launch vehicles 370*g. See also specific vehicles*
  evolved expendable 367*g*
  expendable 367*g*
  Indian 105
  South Korean 102–103
law of space **13–23**
  Antarctica and **17–18**
  commercial satellites and 8
  delimitation in **15**
  innocent passage in **17**
  v. laws of the sea and air 13–14
  registration in **17**
  sovereignty in **15–16**
  treaties in **18–23**
  weaponization in 27
Laymance, Carol 15
Lenin, Vladimir **72**
Leningrad, ABM system around 77
Leonov, Alexei 271*b*, 349*c*
Lewis, James 62
Lewis, Jeffrey G. 102
liability, registries in establishment of 17
Liability Convention (1972) **19–20, **201–205
Libya 56
Li Changhe 92
Limited Test Ban Treaty (1963) **21, **193–194
Lockheed 82
Logsdon, John 82
Long March (Chang Zheng) rocket 90, 365*g, *370*g*
Lovell, James A. 350*c*
low earth orbits (LEOs) 5, 6, 93, 371*g*
Lunik/Luna project 371*g*
Lutes, Charles D. 91, 93

M

MacDonald, Bruce W. 9, 10, 28
Malenkov, Georgy 73
Manarov, Musa 356*c*

Index

Outer Space Treaty on 19, 20
U.S. landing on 44, 45, 46, 59
Moon Agreement (1984) **20–21,** 27, 216–224
Morrell, Geoff 107
Moscow, ABM system around 12, 22, 23, 46, 77, 78
Mudflap project 44
Mueller, Karl P. 26, 27
multiple independently targetable reentry vehicles (MIRVs) 48, 372*g*
multiple kill vehicle payload systems 372*g*
multiple reentry vehicles (MRVs) 372*g*
multipolarity 26–27
multistage rockets 372*g*
Musharraf, Pervez 107
Musk, Elon 63–64, 272*b*
mutual assistance 19
mutual assured destruction (MAD) 48, 372*g*
MX-774, Project 38

N

NASA. *See* National Aeronautics and Space Administration (NASA)
National Advisory Committee for Aeronautics 41–42
National Aeronautics and Space Act of 1958 (U.S.) 117–119
National Aeronautics and Space Administration (NASA) 345*c*
budget of 54, 59, 60, 63–64
and debris 11, 12
ESA's work with 85, 86, 87
establishment of 42
manned spaceflight program of 44–45
need for 41–42
space shuttle program of 48–49, 54, 59
space station program of 48–49
National Aeronautics and Space Council (NASC) 372*g*
National Air and Space Museum 288
National Aviation and Space Agency 288
nationalism, techno- 26
national missile defense 372*g*
National Missile Defense Act of 1999 (U.S.) 57–58
National Police Agency (Japan) 95–96
National Reconnaissance Office (NRO) 288–289
National Science and Technology Council 57
National Security Council, U.S. 58
National Security Decision Directive (NSDD) Number 42 51, 139–144
National Security Directive Number 172 (1985) 53

National Security Memorandum 345 (1977) 49
National Security Presidential Directives (NSPD)
Number 27 158–162
Number 39 170–176
National Security Study Directive 4-86 (1986) 54
national space policy, Chinese 89–90
national space policy, U.S.
under Bush (2006) 59–60
under Carter (1978) 50
under Clinton (1996) 57
current and future **62–64**
under Obama (2010) 62, 186–190
under Reagan (1982) 139–144
under Reagan (1988) 54
national technical means 8, 22, 48, 78, 373*g*
navigational satellites **6**
Navy, U.S.
role in rocket research 40
satellites of 43
Navy Special Projects Office, U.S. 40
Nedelin explosion (1960) 76
Nelson, Bill 60–61, 64
Newsweek magazine 106
Newton, Sir Isaac 272*b*, 341*c*
New York Times, The 16
Nike, Project 42
Nike-X program 43
Nike-Zeus missile 42, 43, 44
Nikolayev, Andrian 348*c*
Nitze, Paul 53
Nixon, Richard M. **46–49,** 351*c*–352*c*
reconnaissance satellites under 46–47
resignation of 49, 79
space shuttle program under 47, 48–49
text of announcement of (1972) 136–139
treaties signed by 47–48, 77–78
Nodong missile 99
no-fly zones 17
North American Aerospace Defense Command (NORAD) 43
North Dakota 12, 22, 48
North Korea **97–102**
Chinese relations with 97, 98, 100, 101
current space program of **99–100**
establishment of 98
history of space program of **98–99**
Japan threatened by 96, 97, 98, 99
military of 98
nuclear weapons of 96–102, 242–243
satellites of 7, 99–100, 101

North Korea *(continued)*
   South Korean relations with 100–101,
     103, 104
   Soviet relations with 98, 99
   U.S. missile defense against 56, 60, 96, 97
   on weaponization of space **100–102**
Northrup 39
nozzle 373*g*
Nuclear Threat Initiative (NTI) 289
nuclear weapons 343*c*
   Chinese 21, 22, 46
   deterrence with 28, 37–38
   environmental consequences of 28
   Indian 105
   Iranian 61, 107
   Limited Test Ban Treaty on **21**
   North Korean 96–102, 242–243
   Pakistani 107
   proliferation of, caused by weaponization
     of space 26, 65
   radiation from 47
   South Korean 102
   Soviet
     under Brezhnev 76–77
     development of 38, 73, 74
   U.S. **37–39**
     development of 37–39
     under Kennedy 43
     under Nixon 47–48
     under Truman 37–39
     use of, in World War II 37, 73
Nurek space tracking center 77

O

Obama, Barack **61–62,** 360*c*–361*c*
   defense budget of 176–177
   missile defense under 13, 61, 84, 178–181
   national space policy of 62, 186–190
   Russian relations with 84
   on vision for space exploration 182–186
Obering, Henry 63
Oberth, Hermann 272*b*
O'Hanlon, Michael 5, 26
oil spills 6
Olympic Games 89
open skies policy 9, 15, 17, 40, 343*c*, 373*g*
orbit(s) 373*g*
   asynchronous 6
   equatorial 367*g*
   geosynchronous 6, 368*g*
   graveyard **6–7,** 368*g*
   low earth 5, 6, 93, 371*g*
   medium earth 6, 371*g*
   polar 6, 373*g*
   of satellites **6–7**

Orbiter, Project 39, 40, 344*c*, 374*g*
organizations, list of 277–295
Orion spacecraft 59, 60
Outer Space Treaty (1967) **18–19**
   delimitation of space in 15
   excerpt from text of 194–199
   Kennedy on 18–19, 44
   Moon in 19, 20
   negotiations for 18–19
   registries in 19, 20
   rescue operations in 19
   signatories to 16
   sovereignty in 16
   Soviet compliance with 77
   Soviet support for 83
   weaponization of space in 27, 59
   WMDs in 19, 27

P

pacemakers 46
Pakistan **106–107**
   Chinese work with 90, 107
   Predator drone attacks in 4
   satellites of 7, 107
   U.S. intelligence gathering with satellites
     in 9
Pakistan Space and Upper Atmosphere
   Research Commission (SUPARCO) 106–
   107, 289–290
Paris Convention (1919) 15, 17
PAROS resolutions, UN
   of 1997 224–227
   of 2003 227–230
   Chinese statement on (2006) 230–234
   Soviet support for 83
patterns-of-life analysis 9
payloads 373*g*
   registries of 17
   U.S. military, number of 259*f*
peaceful use of space
   in Chinese space policy 89–90
   definition of 43, 50
   of reconnaissance satellites 40, 43
   in Russian space policy 83
   UN resolution on 19, 191–193
   in U.S. space policy 50
Peenemünde 35, 36, 373*g*
People's Liberation Army (PLA). *See* military,
   Chinese in space program
perigee 373*g*
period 373*g*
perturbation 373*g*
phased-array radar 373*g*
picosatellites 373*g*
Pike, John 53

Index

Pillbox system 78
Pinkston, Daniel 104
Pobeda (T-1) rocket 73, 373g
pods, rescue 6
Poland, U.S. missile defense in 13, 60, 84
Polaris submarine-launched ballistic missile
 40, 345c, 373g
polar orbits 6, 373g
policies. See national space policy, Chinese;
 national space policy, U.S.
Pollpeter, Kevin 60, 90
Polyakov, Valery 357c
Pop, Virgiliu 272b
Popovich, Pavel 348c
Poseidon submarine-launched ballistic missile
 373g
Powers, Francis Gary 42
precision-guided weapons 9. See also specific
 types
Predator drones 4, 9, 59, 373g
prevention of an arms race in outer space. See
 PAROS resolutions, UN
Prevention of the Placement of Weapons in
 Outer Space Treaty (PPWT) (2008) 234–237,
 374g
primary documents
 international 191–243
 U.S. 117–190
private sector
 satellites in. See commercial satellites
 U.S., aviation and rocket research by
 34–35, 39
probes, space 375g
 number of 262f–263f
Project Apollo. See Apollo program
Proton launcher 78, 79, 82, 374g
Purcell Report 42, 374g
Putin, Vladimir 58, 60, 84, 106

Q

Qaeda, al- 4, 7
Qian Xuesen 88, 272b
Quadrennial Defense Review (2001) 58
Quarles, Donald 41

R

radar systems
 phased-array 373g
 Russian 84
 Soviet 76, 77, 78
 U.S. 96
radiation 47
Raj, Gopal 105–106
RAND 38

Reagan, Ronald 50–55, 354c–355c
 on Challenger explosion 146–147
 national space policy of 51, 54, 139–144
 in presidential election of 1980 50–51
 Soviet relations with 80–81
 on Soviet threat 50–51
 space shuttle under 51
 State of the Union address (1984) by
 145–146
 Strategic Defense Initiative of 13, 22, 23,
 34, 51–55
 on verification 29
reconnaissance aircraft, U.S. 9, 16, 42, 74
reconnaissance capabilities, by country
 264f
reconnaissance satellites 9
 ABM Treaty on 22
 in cold war vii, 9
 Indian 106
 Japanese 95, 96, 97
 peaceful use of 40, 43
 SALT I treaty on 78
 South Korean 104
 Soviet 79
 U.S.
 under Eisenhower 9, 39–40, 42, 43
 under Nixon 46–47
 uses for 9
Redstone arsenal 374g
Redstone rocket 39, 40, 41, 73, 345c, 374g
reentry 374g
registration
 in law of space 17
 Outer Space Treaty on 19, 20
 UN role in 17, 19, 20
Registration Convention (1976) 20
 excerpt from text of 208–211
 North Korea in 101
 provisions of 17, 20
regulation of space. See law of space
Remek, Vladimir 353c
remote sensing 374g
 U.S. policy on commercial 158–162
res communis 14, 15
rescues, common docking systems for 49
rescue satellites 6
Rescue Treaty (1968) 19, 199–201
res nullius 14, 16
revolution in military affairs 4–5, 374g
Rice, Susan E. 100
Ride, Sally 273b, 355c
rockets 374g. See also specific types
 Chinese 88
 German 35–37, 73
 multistage 372g

Index

Titov, German 275*b*, 347*c*
Titov, Vladimir 356*c*
Tokaty, Grigori 275*b*
Tonga 16
Tongasat 16, 18
Topol-M missile 84
tourism, space 7, 82
traffic, in space 11
tragedy of the commons 14
treaties **18–23**. *See also specific treaties*
    monitoring provisions in 8, 48
    regulation of space through 13, **18–23**
    satellites protected in viii
Treaty Banning Nuclear Weapons Tests in the Atmosphere, in Outer Space and Under Water (1963) **21**, 193–194
Treaty on the Principles Governing the Activities of States in the Exploration and Use of Outer Space, including the Moon and Other Celestial Bodies. *See* Outer Space Treaty (1967)
Trident submarine-launched ballistic missiles 376*g*
troop movements 9
Truman, Harry S. **37–39**
Tsander, Fridrikh Arturovich 72, 275*b*
Tsiolkovsky, Konstantin 71, 275*b*, 341*c*
Tukhachevsky, Mikhail 72–73

U

Ukraine 82
Union of Concerned Scientists (UCS) 13, 25, 93
United Kingdom
    in European Space Agency 85
    in Limited Test Ban Treaty 21
    in World War II 36
United Nations
    Kennedy's last address before (1963) 122–127
    in Liability Convention 19–20
    on North Korea 99–100
    on number of satellites 5
    in Outer Space Treaty 18–19
    in registry process 17, 19, 20
    weaponization treaty submitted to (2008) 234–237
United Nations Conference on Disarmament 83, 92, 293, 366*g*
United Nations Office for Outer Space Affairs (UNOOSA) 293
United Nations resolutions
    against an Arms Race in Outer Space (1997) 224–227
    on Prevention of an Arms Race in Outer Space (2003) 227–230

Resolution 1718 on North Korea (2006) 99
Resolution 1721 on International Cooperation in the Peaceful Uses of Outer Space (1961) 19, 191–193
United States **34–65**
    in ABM Treaty 21–23, 78
        withdrawal from 22–23, 59, 84, 156–158
    on Chinese space program 90, 93–94
    in cold war. *See* cold war
    defense systems of. *See* antiballistic missile (ABM) systems; ballistic missile defense (BMD) systems
    Indian relations with 105, 106
    in *International Space Station* 7–8, 60–61
    Japanese relations with 96, 97
    in Limited Test Ban Treaty 21
    military of. *See* military, U.S.
    national space policy of. *See* national space policy, U.S.
    nuclear program of 3, **37–39**
    in Outer Space Treaty 18–19
    primary documents on 117–190
    in SALT I 8, 47–48, 77–78
    in SALT II 49, 50, 78, 79–80
    South Korean occupation by 98
    on South Korean space program 103, 104
    on *Sputnik* launch 3–4, 34, 41, 74
    in START 53, 81
    weapons of. *See specific weapons*
    in World War II 36–37
United States Institute of Peace (USIP) 9
United States space program. *See also specific missions, spacecraft, and systems*
    aviation in
        in cold war 37, 38
        development of 34–35
    under Bush (George H. W.) **55–56**
    under Bush (George W.) **58–61**
    capitalism in 34
    under Carter **49–50**
    under Clinton **56–58**
    cost of. *See* budget; costs
    current status of **62–64**
    under Eisenhower **39–43**
    under Ford 49
    future of **62–64**
    under Johnson 46
    under Kennedy **43–46**
    under Nixon **46–49**
    number of launches per year 261*f*
    under Obama **61–62**
    public interest in 63
    under Reagan **50–55**
    rockets in, development of 34–43

393